An Atlas
of Fullerenes

An Atlas
of Fullerenes

P. W. Fowler
Department of Chemistry
University of Sheffield

D. E. Manolopoulos
Department of Chemistry
University of Oxford

Dover Publications, Inc.
Mineola, New York

Bibliographical Note

This Dover edition, first published in 2006, is a revised and corrected edition of the work first published by The Clarendon Press, Oxford, in 1995. A new Preface to the Dover Edition has been specially prepared for this volume.

Library of Congress Cataloging-in-Publication Data

Fowler, P. W.
    An atlas of fullerenes / P.W. Fowler and D.E. Manolopoulos.
      p. cm.
    Originally published: Oxford : Clarendon Press ; New York : Oxford University Press, 1955, in series: The international series of monographs on chemistry ; 30.
    ISBN 0-486-45362-6 (pbk.)
     1. Fullerenes. I. Manolopoulos, D. E. II. Title.

QD181.C1F68 2007
546'.681—dc22

                                   2006050251

Manufactured in the United States of America
Dover Publications, Inc., 31 East 2nd Street, Mineola, N.Y. 11501

PREFACE

This book grew out of a fascination with the early experimental results on C_{60} that suggested that thousands of years of practical chemistry had missed an allotrope of one of the most important elements, and from the realization that simple mathematical and pictorial methods had a lot to say about the properties of a new class of molecules of which C_{60} was just the first example. As the title suggests, the aim of the book is to present a comprehensive set of pictures of fullerene structures and to tabulate their properties. In deciding where to stop, we noted the advice given in another atlas (Atlas of finite groups, J. H. Conway, R. T. Curtis, S. P. Norton, R. A. Parker, and R. A. Wilson, Oxford 1985) to 'think how far the reasonable person would go, and then go a step further'. Experiment will no doubt continue to push forward the frontier of what is reasonable, and so we have included a listing of a computer program that will allow the tables to be extended as necessary. We hope that the seven chapters of descriptive material preceding the tables will serve as an introduction to the current state of theory in this new and lively area. The treatment is intended to be self-contained and to be useful to students and research workers alike. Lastly, we hope that an explicit enumeration of fullerene polyhedra will stimulate mathematicians to study this hitherto neglected class.

P.W.F./D.E.M.
Exeter/Nottingham
January 1994

PREFACE TO THE DOVER EDITION

The decade since the first publication of the Atlas has seen spectacular developments in the chemistry and materials science of fullerenes and nanotubes, and continued progress on all of the topics covered in this book. Specifically, the "unsolved problem"(p. 41) of fullerene isomer enumeration and construction has now been solved. A method that is mathematically complete is now available,[1] and it confirms that our face-spiral lists are complete to well beyond the range of our tables (to at least 200 atoms). Many more fullerene graphs without face spirals have been constructed,[2] but the tetrahedral C_{380} (Fig. 2.12) remains the smallest known counterexample to our Conjecture 2.1, even though the smallest *general* trivalent polyhedron without a face spiral has only 18 vertices.[3] Interestingly, the *vertex* spiral has surfaced in a nomenclature for labelling fullerene derivatives,[4] in spite of the example of I_h C_{80} (Fig. 2.10) as a fullerene without a vertex spiral; in fact, 100 fullerene isomers on 100 or fewer vertices have no vertex spiral, the smallest being isomer 622 of C_{56}.[5] The enumeration approach has been extended to a census of nanotube caps[6] and catalogues of fullerene isomerisation and growth patches.[7] Simple arguments based on mathematical independence number have been shown to be useful in describing fullerene addition chemistry.[8] It would take several extra volumes, at least, to do justice to the advances in fullerene chemistry, but others have already written these.[9] We have limited the changes to the original text to completion of references, correction of misprints, and expansion of one footnote.

P.W.F./D.E.M.
Sheffield/Oxford
July 2006

[1] G. Brinkmann and A.W.M. Dress, J. Algorithms 23 (1997) 345; Adv. Appl. Math. 21 (1998) 473.

[2] M. Yoshida and P.W. Fowler, Chem. Phys. Lett. 278 (1997) 256.

[3] G. Brinkmann, Chem. Phys. Lett. 272 (1997) 193.

[4] IUPAC Recommendations 2005: Numbering Of Fullerenes. See: F. Cozzi, W. H. Powell and C. Thilgen, Pure Appl. Chem. 77 (2005) 843.

[5] P.W. Fowler, D. Horspool and W. Myrvold. Vertex spirals in fullerenes and their implications for nomenclature of fullerene derivatives, (2006) to be published.

[6] G. Brinkmann, P.W. Fowler, D.E. Manolopoulos and A.H.R. Palser, Chem. Phys. Lett. 315 (1999) 335.

[7] G. Brinkmann, P.W. Fowler and C. Justus, J. Chem. Inf. Comput. Sci. 43 (2003) 917; G. Brinkmann and P.W. Fowler, J. Chem. Inf. Comput. Sci. 43 (2003) 1837.

[8] P.W. Fowler, P. Hansen, K.M. Rogers and S. Fajtlowicz, J. Chem. Soc. Perkin Trans. 2 (1998) 1531.

[9] E.g., A. Hirsch, Handbook of Fullerenes (2nd Edition), Wiley-VCH, New York, 2004; R. Taylor, Lecture Notes on Fullerene Chemistry: A Handbook for Chemists, Imperial College Press, London, 1999; R. Taylor, The Chemistry of Fullerenes, World Scientific, Singapore, 1995 (1998 paperback).

ACKNOWLEDGEMENTS

Joint papers with a number of colleagues are cited at various places in the main text, but we would like to take this opportunity to thank all of those who have worked in this field with us: S. J. Austin, J. Baker, R. C. Batten, J. E. Cremona, S. E. Down, J. S. Hulme, H. W. Kroto, P. Lazzeretti, M. Malagoli, J. C. May, V. Morvan, G. Orlandi, D. B. Redmond, R. P. Ryan, J. P. B. Sandall, J. Steer, R. Taylor, D. R. M. Walton, D. R. Woodall, J. Woolrich, R. Zanasi and F. Zerbetto. In a book that relies so heavily on pictures, it is a pleasure to thank Robin Batten for his many hours of expert help with computer graphics.

CONTENTS

1 Introduction   1

  1.1 The fullerene hypothesis   2

  1.2 From hypothesis to experimental fact   5

  1.3 The need for a systematic theory   9

  1.4 What this book contains   10

     References and notes   12

2 Fullerene cages   15

  2.1 Fullerene polyhedra   15

  2.2 Fullerene duals   17

  2.3 The Coxeter construction   18

  2.4 Fullerene graphs   22

  2.5 The spiral conjecture   23

  2.6 The spiral algorithm   27

  2.7 How many fullerenes are there?   31

  2.8 A fullerene without a spiral   35

  2.9 Concluding remarks   39

     References and notes   41

3 Electronic structure   43

  3.1 Qualitative molecular orbital theory   44

  3.2 Open, closed, and pseudo-closed shells   47

  3.3 Icosahedral fullerenes   50

  3.4 The leapfrog transformation   51

  3.5 Carbon cylinders   59

  3.6 Sporadic closed shells   62

  3.7 Conclusion   65

     References and notes   66

4 Steric strain   68

  4.1 Steric strain and rehybridization   69

4.2 The isolated-pentagon rule   73

4.3 Pentagon indices for lower fullerenes   75

4.4 Hexagon indices for higher fullerenes   80

4.5 Steric strain in leapfrogs and carbon cylinders   84

4.6 Selected higher fullerene examples   88

   References and notes   93

5 Symmetry and spectroscopy   95

5.1 The fullerene point groups   95

5.2 Topological coordinates   101

5.3 Symmetry assignment   105

5.4 ^{13}C NMR spectra   111

5.5 IR and Raman spectra   113

   References and notes   118

6 Fullerene isomerization   120

6.1 The Stone–Wales rearrangment   120

6.2 Symmetry aspects   125

6.3 Chirality and the Stone–Wales transformation   129

6.4 Isomerization maps   131

6.5 The C_{60} Stone–Wales map   141

6.6 Isomer distributions   145

   References and notes   147

7 Carbon gain and loss   149

7.1 C_2 insertion and extrusion   150

7.2 Symmetry aspects of C_2 processes   153

7.3 Insertion/extrusion maps   157

   References and notes   163

Appendix
The Spiral computer program   165

Atlas tables   177

General fullerene isomers C_{20} to C_{50} 180

Isolated-pentagon isomers C_{60} to C_{100} 254

Index   389

1

INTRODUCTION

The fullerenes are closed carbon-cage molecules containing only pentagonal and hexagonal rings, of which the celebrated icosahedral C_{60} molecule is the archetype. Hollow-shell graphite molecules were mentioned in the 'Daedalus' scientific diversions column [1] as long ago as 1966, and theoretical speculations on the possibility of a sixty-carbon framework based on the truncated icosahedron were published several times over the next three decades. [2-4] The first appearance of this molecule in connection with an experiment was the proposal of its structure by Kroto, Heath, O'Brien, Curl, and Smalley in 1985, in an attempt to explain the pronounced abundance of the C_{60} cluster in their graphite laser vaporization experiment. [5] This proposal was subsequently confirmed in 1990, when Krätschmer, Lamb, Fostiropoulos, and Huffman reported a method for the bulk production of C_{60} in a carbon-arc along with infra-red (IR) spectroscopic evidence for its structure. [6] These two papers have since generated an epidemic of new research into C_{60} and other fullerenes which has infected chemists, physicists, and materials scientists throughout the world and which would already be difficult to summarize in a single volume. [7]

The aim of this book is therefore not to give a complete account of the fullerene story, but rather to give a detailed account of a general aspect that is of interest to fullerene researchers. This aspect is the systematics of the fullerene family as a whole, rather than the specific properties of the exceptional C_{60} cage. Although this might seem at first to be an expansion of the topic, this is not the case. The vast majority of the existing fullerene literature is devoted to the chemical and physical properties of icosahedral C_{60} and its derivatives, with comparatively little mention of other fullerenes. The reason for this is simply that C_{60} is more abundant than other fullerenes in graphitic soot. [6] Consequently, more experiments have been performed on C_{60} than on other fullerenes, and much of the highly detailed information that has been obtained in these experiments lies beyond the scope of this book.

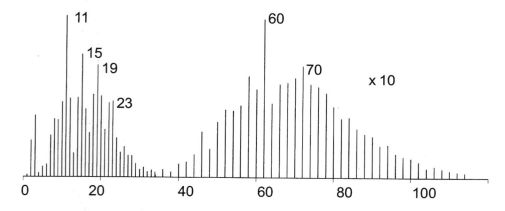

FIG. 1.1. Schematic picture of the early graphite laser vaporization results of Rohlfing, Cox and Kaldor.[8] The abscissa gives the carbon cluster size as the number of atoms n. The ordinate gives the relative time-of-flight mass spectrum ion signal for each C_n^+ cluster. See ref. 8 for more details and for mass spectra obtained under other experimental conditions.

1.1 The fullerene hypothesis

A convenient place to begin a discussion of the fullerenes, which both motivates the fullerene definition and puts the fullerenes in the context of other carbon clusters, is with the early graphite laser vaporization results of Rohlfing *et al.*[8] These results are reproduced here schematically in Fig. 1.1, which shows ion signals from carbon clusters obtained by vaporizing graphite with a high-power laser, cooling and clustering the resulting carbon vapour in a beam of helium, and photoionizing the clusters so obtained for detection by time-of-flight mass spectrometry. The resulting carbon cluster mass distribution is seen to be distinctly bimodal, with both even and odd C_n clusters in the range $n <= 25$ but only even C_n clusters in the range $n >= 40$. Although the details of this mass distribution can vary depending on the experimental conditions, these features are quite general.

Carbon clusters in the range $n = 2$ to 9 are known from *ab initio* and spectroscopic studies to have linear chain structures, with the odd-membered clusters having singlet and the even-membered clusters triplet electronic ground states.[8] However, there is also some *ab initio* evidence that the even-membered clusters in this range may have comparably low-lying electronic states with cyclic equilibrium geometries,[10,11] and carbon clusters in the range $n = 10$ to 25 are believed to have monocyclic ground state structures. This transition from linear chains to monocyclic rings occurs because the extra bonding that comes with ring closure eventually outweighs the strain energy that is incurred on bending a linear polyyne chain to form a ring. The transition point at ten carbon atoms was predicted some time ago on the basis of semi-empirical molecular orbital theory calculations by Pitzer and Clementi[12] and by Hoffmann,[13] and it has since been observed indirectly in photodissociation cross-section measurements on mass-selected carbon cluster ions.[14]

The structures of the carbon clusters in the low-mass region of Fig. 1.1 are thus comparatively simple, with linear chains in the range $n = 2$ to 9 and monocyclic rings in the range $n = 10$ to 25. Even- and odd-membered chains and rings are both possible, which explains why both even and odd mass peaks are seen in the figure. The pronounced 'magic number' C_n^+ ion signals at $n = 11$, 15, 19, and 23 show a periodicity of four which is reminiscent of the Hückel $4n+2$ π-electron rule for monocyclic rings, and the relative abundance distribution in the low-mass region of Fig. 1.1 has been reproduced in detail using a kinetic model with semi-empirical (chain and ring) cluster energies as input.[15] The structures of these carbon clusters can therefore be in little doubt,[16] and all that remains is to find an equally convincing explanation for the high-mass region in Fig. 1.1.

Clearly, any such explanation must account for the fact that only *even* C_n carbon cluster ion signals are seen with any intensity in this region. This immediately rules out a number of possible structures for these clusters such as fragments of the infinite diamond lattice or graphite sheet, which would be expected to show both even and odd mass peaks in common with linear chains and monocyclic rings. However, it does not rule out a number of other possible structures, such as fragments of a hypothetical high-temperature 'carbyne' form of crystalline carbon composed of C_2 units with alternating single and triple carbon–carbon bonds.[8] The key observation which does eliminate such alternatives, and points uniquely in the right direction, is the fact that the C_{60} mass peak (and to a lesser extent also the C_{70} mass peak) is so pronounced. Moreover, these two mass peaks can be made even more prominent under appropriate experimental conditions.[5]

One possible explanation that is compatible with these observations is that the second series of high-mass carbon clusters in Fig. 1.1 are all fullerenes, as defined in the opening sentence of this introduction. This fullerene hypothesis is attractive because closed cages avoid the dangling edge bonds that are expected to destabilize comparably sized fragments of the infinite diamond and graphite lattices,[5] and because trivalent cages satisfy the valence requirements of carbon atoms rather better than linear chains and monocyclic rings. (The infinite graphite sheet can be viewed as a limiting case of a trivalent cage without dangling edge bonds, for example, and graphite is thermodynamically the most stable form of elemental carbon.) It is also natural to assume, by analogy with the transition from linear chains to monocyclic rings at ten carbon atoms, that the strain energy incurred on closing a finite sheet of nominally sp^2 hybridized carbon atoms into a cage will eventually be outweighed by the bonding energy that is released by the closure. This is especially true if the cage contains only pentagonal and hexagonal rings, for which the strain energy is not expected to be particularly severe provided not too many pentagonal rings are directly adjacent. (Heptagonal rings are also possible,[17,18] both on these grounds and on the basis of chemical precedent for the existence of polycyclic cage fragments. However, since each heptagonal ring requires an additional pentagonal ring for cage closure, as a result of Euler's theorem discussed in Chapter 2, and since each additional pentagonal ring implies additional strain, it is likely that the simple fullerene definition captures the most stable class of closed carbon cages.)

These qualitative theoretical arguments in favour of the fullerene hypothesis emerged soon after the initial structural proposal of icosahedral C_{60} in 1985, and they have since been supported by a large number of electronic structure calculations. One particularly important early study was the 1988 survey of elemental carbon cages by Schmalz et al.,[19] in which semi-empirical models were used to compare carbon cages with chains, rings, toroids, and fragments of the infinite diamond and graphite lattices. This study concluded that cage structures would be the most stable elemental carbon clusters with more than about 25 atoms, and moreover that the presence of only five- and six-membered rings and the absence of any abutting five-membered rings were further criteria for cage stability.[19] The preference for fullerenes containing as few adjacent pentagonal rings as possible was also discussed by Kroto in 1987,[20] using heuristic arguments based on chemical precedent. Both of these studies noted that icosahedral C_{60} was the smallest fullerene without adjacent pentagons, and guessed correctly that the next smallest fullerene without adjacent pentagons was a D_{5h}

(a)   (b)

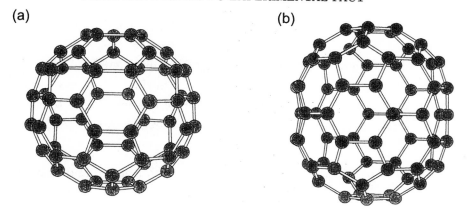

FIG. 1.2. The two most abundant fullerenes produced in experiment: (a) C_{60}:1 (I_h) and (b) C_{70}:1 (D_{5h}).

isomer of C_{70} (Fig. 1.2). The pentagon isolation criterion in fullerene stability will be discussed further in Chapters 3 and 4.

Returning now to the experimental results for carbon clusters in Fig. 1.1, the fullerene hypothesis immediately explains why only even C_n clusters are seen with any intensity in the high-mass range: a closed trivalent C_n cage has a total of $3n/2$ skeletal bonds, and can therefore only exist for even n. The hypothesis also explains why the C_{60} and C_{70} mass peaks are so pronounced,[5] in so far as C_{60} and C_{70} are the first two stable fullerene structures without any pentagon adjacencies. It also gives an appealing overall picture of the stability of carbon clusters, with one-dimensional linear chains at small n giving way to two-dimensional monocyclic rings at intermediate n and then finally three-dimensional fullerene cages at large n.

1.2 From hypothesis to experimental fact

Although the graphite laser vaporization results in Fig. 1.1 are consistent with the fullerene hypothesis, they do not prove its validity. For this proof it is necessary to isolate and characterize representatives of the fullerene family, and this is a challenging experimental problem. Nevertheless, a small number of C_n carbon clusters in the range $n \geq 40$ have been characterized at the time of writing, and these clusters are indeed all fullerenes.

The first fullerene to be characterized was I_h C_{60}, which was originally identified by its four-band IR absorption spectrum by Krätschmer et al.[6] in 1990, and soon afterwards by its one-line ^{13}C nuclear magnetic resonance (NMR) spectrum by Taylor *et al.* [21] in the same year. This second paper also reported the separation of C_{70} from the products of the carbon-arc synthesis by column chromatography on alumina, which was the first time elemental allotropes had been separated using this technique and led to its subsequent use in studies of the higher fullerenes. The ^{13}C NMR spectrum of C_{70} was found to consist of 5 lines with signal integrations 3×10 and 2×20, as expected for the D_{5h} fullerene structure. [21]

A number of other fullerenes have also since been isolated and characterized, including C_{76}, [22] C_{78}, [23,24] and C_{84}. [24,25] The various experimental structures of these higher fullerenes are shown in Fig. 1.3. The feature they share in common with C_{60} and C_{70} is that they all have isolated pentagons, but they do not generally have such high symmetry. Indeed each newly discovered fullerene has come with a new surprise, giving a glimpse of the tremendous diversity that is possible within the fullerene family.

The first higher fullerene to be characterized was chiral D_2 C_{76}, which was identified by its 19 ^{13}C NMR lines of equal intensity by Ettl *et al.* [22] in 1991. Here the surprise was the chirality, which had no precedent in the more symmetric structures of C_{60} and C_{70}; the two enantiomers of D_2 C_{76} have since been resolved in an elegant site-specific osmylation experiment. [26] The second higher fullerene to be characterized was C_{78}, two isomers of which were separated by high-pressure liquid chromatography (HPLC) and identified by ^{13}C NMR spectroscopy by Diederich *et al.* [23] in 1991. An unseparated mixture of three isomers of C_{78} was subsequently identified from its ^{13}C NMR spectrum by Kikuchi *et al.* [24] the following year. Here, there were more surprises: not only was the C_{78} product a mixture of isomers, unlike C_{60}, C_{70}, and C_{76}, but the composition of this mixture was sensitive to the precise experimental conditions (with the dominant isomer in one experiment [24] not being observed at all in the other [23]). These observations suggested that it was possible for several fullerene isomers to have comparable (thermodynamic) stabilities, and moreover that kinetic factors were playing an important role in the formation process. It was therefore again a surprise when several groups found that the ^{13}C NMR spectrum for C_{84}, did *not* vary with the method of preparation, [24,25,27], despite the increased number of structural possibilities for 84 atoms. The most plausible explanation for the observed spectrum is that the C_{84} product consists of an equilibrium mixture of the D_2 and D_{2d} symmetry fullerenes in Fig. 1.3 that is

(a)

(b)

(c)

(d)

(e)

(f)

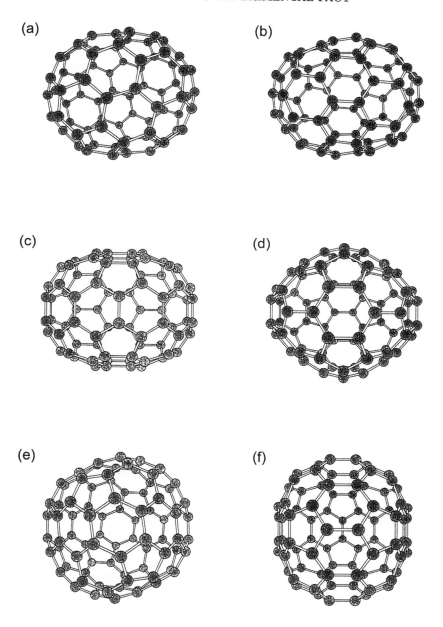

FIG. 1.3. A selection of experimental structures of the higher fullerenes with isolated pentagons: (a) C_{76}:1 (D_2), (b) C_{78}:1 (D_3), (c) C_{78}:2 (C_{2v}), (d) C_{78}:3 (C_{2v}), (e) C_{84}:22 (D_2) and (f) C_{84}:23 (D_{2d}).

insensitive to kinetic factors in the formation process.[25]

These and other[28] higher fullerene experiments have traditionally followed the method of Taylor et al.,[21] with variants of the carbon-arc synthesis[6] being used to generate graphitic soot from which fullerene mixtures are extracted and then separated by HPLC. The most convincing aspect of the structural characterization has usually been ^{13}C NMR spectroscopy. The main differences between the various experiments have been in the details of the method of production of the soot (by carbon-arc or resistive heating) and in the solvents used in the extraction. These differences have been found to affect the relative abundances of the various C_n clusters as well as the fullerene isomer mixtures obtained at each nuclearity. For instance, C_{76}, C_{78}, C_{84}, C_{90}, and C_{94} have been extracted in toluene from one sample of graphitic soot,[27] with little trace of other higher fullerenes, while C_{82} and C_{96} have been extracted in CS_2 from another.[24] Kinetic factors in the formation and/or extraction processes clearly therefore play *some* role in the overall product, even though different conditions always appear to give the same fullerene isomers for C_{60}, C_{70}, C_{76}, and C_{84}.

One important limitation of the production method which has implications for the prospects of isolating and characterizing still higher fullerenes is that there appears to be an upper limit on the size of fullerene cages that will dissolve in organic solvents. This limit is clearly illustrated by Fig. 1 of Yeretzian et al.,[29] which shows laser desorption time-of-flight mass spectra (LDMS) of both raw and toluene-extracted graphitic soot. Whereas the raw soot contains (even) C_n mass peaks from $n = 70$ to well beyond 150, the toluene-extracted soot shows a pronounced drop in cluster abundance beyond $n = 96$. A second limitation is that the graphitic soot itself contains only a comparatively low abundance of fullerenes other than C_{60} and C_{70}, which between them typically account for 95% of the total fullerene yield.[29]

Because of these and other limitations, several groups have recently started looking at alternative methods for synthesis and isolation of fullerenes. Among the more promising of these alternatives are the possibility of producing fullerenes by the high-temperature pyrolysis of naphthalene,[30] and the possibility of separating them by gradient sublimation.[29] The motivation for such studies is that the graphite laser vaporization results and the LDMS analysis of raw graphitic soot both show that there are many other C_n carbon clusters in the range $n \geq 40$ still waiting to be isolated and characterized.

1.3 The need for a systematic theory

The terms 'linear chain' and 'monocyclic ring' both have unambiguous meanings in the sense that a linear C_n chain and a monocyclic C_n ring both have obvious (and unique) bonding connectivities. This is not, however, the case for the term 'fullerene', and in fact the fullerene definition of a closed trivalent C_n cage containing only pentagonal and hexagonal rings can encompass an enormous number of different isomers at each nuclearity.

Despite the fact that there are so many *possible* fullerene isomers, as will become abundantly clear in this book, it is sometimes possible to guess the most stable (and therefore chemically relevant) C_n fullerene cage simply by experimenting with molecular models. This is essentially how the stable I_h C_{60} and D_{5h} C_{70} fullerene structures were found, both in the original theoretical papers on the electronic structure of C_{60} [2-4] and in the subsequent experimental papers of 1985. [5,31] However, the more varied possibilities and generally lower symmetries to be expected for higher fullerenes make the approach less reliable, and in fact the experimental isomers of the higher fullerenes shown in Fig. 1.3 were all first discovered more systematically by computer. [32-34] The computer-generated lists of possible fullerene isomers were then made available to the experimentalists, who were able to use them to 'assign' their observed ^{13}C NMR spectra.

The majority of the theoretical studies of the fullerenes that have been reported to date have actually *not* been systematic, at least in this restricted sense. A typical theoretical approach has been to focus on one or just a few selected fullerene cages and perform calculations with an electronic structure program. Such an approach does have its uses, especially when the considered cages are already known to be experimentally important or interesting for some other reason. However, in order to discover the general properties of the fullerene family as a whole, it is essential to perform electronic structure calculations within a more systematic framework, for example, by comparing *all* the possible isolated-pentagon isomers of a given higher fullerene C_n. Several studies of this type have in fact appeared since the fullerene isomer lists first became available, [35-37] and these studies have been particularly valuable in revealing general patterns in fullerene stability.

A catalogue of the possible isomers is thus the first and most important goal of a systematic theory of the fullerenes. Such a catalogue is useful to experimentalists who want to assign their ^{13}C NMR and other spectra, and to computational chemists who want to predict the most stable C_n isomers with their chosen electronic structure methods, and investigate general trends. The earliest systematic treatment of the fullerene isomer problem was an exploratory 1988 study in which the Coxeter method [38] was used to catalogue fullerenes with high symmetry. [39] This study is discussed along with more recent developments in Chapter 2, which shows how fullerenes of all allowed symmetries may now be generated systematically. The resulting fullerene isomer lists, which contain detailed entries for what are believed to be all possible fullerene isomers in the range $n = 20$ to 50 and all possible isolated-pentagon isomers in the range $n = 60$ to 100, are collected together for reference in the atlas tables.

1.4 What this book contains

Once one has a catalogue of fullerene isomers, the possibilities for further study seem almost endless, and a few of these possibilities are discussed in the remaining chapters of this book. The unifying feature of these discussions is the emphasis on the 'first-order' results that apply to the fullerene class of molecules as a whole. This is actually a unique opportunity. The fullerene field is still comparatively young, and even these first-order results have yet to be collected together. Moreover the very simplicity of the fullerene definition seems to lead to a considerable number of general and useful results.

For example, a brief selection of these results that apply to the fullerenes, and are discussed at various points in this book, is given below.

1. There is at least one fullerene isomer for each even number of atoms $n \geq 20$, with the sole exception of $n = 22$ (Chapter 2).

2. For every fullerene isomer with n atoms, there is another 'leapfrog' fullerene with $3n$ atoms that has the same idealized point group symmetry and a closed-shell electronic structure in Hückel theory (Chapter 3).

3. I_h C_{60} is the only leapfrog fullerene with fewer than 180 atoms that has an optimum distribution of steric strain (Chapter 4).

4. All fullerenes, whether subject to Jahn-Teller distortion or not, belong to one of 28 point groups (Chapter 5).

5. The equilibrium mole fraction of each isomer in a set of isoenergetic and interconverting fullerene isomers is inversely proportional to the order of its molecular point group (Chapter 6).

6. Elimination of a C_2 unit from I_h C_{60} by Stone–Wales rearrangement and C_2 extrusion leads directly to the least-strained isomer of C_{58}; addition of a C_2 unit by Stone–Wales rearrangement and C_2 insertion leads directly to an unstrained isomer of C_{62} (Chapter 7).

In addition to firm results such as these (and this list is not exhaustive), a number of other statements about the fullerenes are still at the stage of conjectures. For example, there is strong empirical evidence that C_{60} has exactly 1812 fullerene isomers (Chapter 2). Many other results are also described here for the first time.

Because of the large number of apparently 'clean' results that can be obtained for the fullerenes, the subject might appear trivial. However, it is not. In the first place, the complete enumeration of fullerene isomers considered in Chapter 2 is a complicated mathematical problem that has still not been solved completely generally, even though it has now been solved well enough for all *practical* chemical purposes. Furthermore, many apparently 'clean' results that apply to small fullerenes seem eventually to break down. For example, the empirical evidence suggested for several years that all fullerenes could be unwound in ring spirals as discussed in Chapter 2, and that the leapfrog and carbon cylinder fullerenes discussed in Chapter 3 were the only fullerenes with properly closed-shell electronic structures in the Hückel approximation. Both of these statements are now known to be false, with the smallest known counterexample to the first statement occuring for a C_n cage with $n = 380$ atoms and the smallest known counter-example to the second occuring at $n = 112$. Finding completely general results for fullerenes such as items 1.–6. above is thus not always easy.

Finally, we should mention some of the many interesting fullerene topics which this book does *not* attempt to cover. These topics include, among others, the addition chemistry of C_{60},[40,41] the structures and properties of endohedral metallofullerene complexes,[42–44] the superconductivity of alkali metal-doped C_{60} phases,[45,46] and the subject of carbon nanotubes.[47,48] (These references are largely to review articles, and are intended to be representative rather than complete. Additional information can also be found in a number of the other books and conference proceedings on fullerenes that are beginning to appear.[49,50]) Some of these topics are mentioned in passing, and others (such as carbon nanotubes) might benefit from an extension of the present approach. However, a full discussion would make this book too long. It would also, for the most part, require attention to second-order effects, whereas the present book is confined by and large to a first-order discussion of the consequences of fullerene topology.

References and notes

[1] D. E. H. Jones, New Scientist 35 (1966) 245.

[2] E. Osawa, Kagaku (Kyoto) 25 (1970) 854.

[3] D. A. Bochvar and E. G. Gal'pern, Proc. Acad. Sci. USSR 209 (1973) 239.

[4] R. A. Davidson, Theor. Chim. Acta 58 (1981) 193.

[5] H. W. Kroto, J. R. Heath, S. C. O'Brien, R. F. Curl, and R. E. Smalley, Nature 318 (1985) 162.

[6] W. Krätschmer, L. D. Lamb, K. Fostiropoulos, and D. R. Huffman, Nature 347 (1990) 354.

[7] T. Braun, Angew. Chem. Int. Ed. 31 (1992) 588.

[8] E. A. Rohlfing, D. M. Cox, and A. Kaldor, J. Chem. Phys. 81 (1984) 3322.

[9] W. Weltner, Jr. and R. J. Van Zee, Chem. Rev. 89 (1989) 1713.

[10] K. Raghavachari and J. S. Binkley, J. Chem. Phys. 87 (1987) 2191.

[11] V. Parasuk and J. Almölf, J. Chem. Phys. 91 (1989) 1137.

[12] K. S. Pitzer and E. Clementi, J. Am. Chem. Soc. 81 (1959) 4477.

[13] R. Hoffmann, Tetrahedron 22 (1966) 521.

[14] M. E. Geusic, T. J. McIlrath, M. F. Jarrold, L. A. Bloomfield, R. R. Freeman, and W. L. Brown, J. Chem. Phys. 84 (1986) 2421.

[15] J. Bernholc and J. C. Phillips, Phys. Rev. B 33 (1986) 7395.

[16] A recent gradient-corrected density functional calculation by K. Raghavachari, D. L. Strout, G. K. Odom, G. E. Scuseria, J. A. Pople, B. G. Johnson, and P. M. W. Gill (Chem. Phys. Lett. 214 (1993) 357) finds C_{20} to be more stable in the form of a monocyclic ring than as the dodecahedral fullerene cage, in accordance with the discussion in this section. Nevertheless, since this book is about the fullerenes, we shall be more concerned with the (higher energy) C_{20} dodecahedron in what follows. The exact transition point from monocyclic rings to fullerene cages is still an open (and interesting) question.

[17] P. W. Fowler and V. Morvan, J. Chem. Soc. Faraday Trans. 88 (1992) 2631.

[18] W. C. Eckhoff and G. E. Scuseria, Chem. Phys. Lett. 216 (1993) 399.

[19] T. G. Schmalz, W. A. Seitz, D. J. Klein, and G. E. Hite, J. Am. Chem. Soc. 110 (1988) 1113.

[20] H.W. Kroto, Nature 329 (1987) 529.

[21] R. Taylor, J.P. Hare, A.K. Abdul-Sada, and H.W. Kroto, J. Chem. Soc. Chem. Comm. (1990) 1423.

[22] R. Ettl, I. Chao, F. Diederich, and R. L. Whetten, Nature 353 (1991) 149.

[23] F. Diederich, R. L. Whetten, C. Thilgen, R. Ettl, I. Chao, and M. M. Alvarez, Science 254 (1991) 1768.

[24] K. Kikuchi, N. Nakahara, T. Wakabayashi, S. Suzuki, H. Shiromaru, Y. Miyake, K. Saito, I. Ikemoto, M. Kainosho, and Y. Achiba, Nature 357 (1992) 142.

[25] D. E. Manolopoulos, P. W. Fowler, R. Taylor, H. W. Kroto, and D. R. M. Walton, J. Chem. Soc. Faraday Trans. 88 (1992) 3117.

[26] J. M. Hawkins and A. Meyer, Science 260 (1993) 1918.

[27] F. Diederich and R. L. Whetten, Acc. Chem. Res. 25 (1992) 119.

[28] R. Taylor, G. J. Langley, A. G. Avent, T. J. S. Dennis, H. W. Kroto, and D. R. M. Walton J. Chem. Soc. Perkin Trans. 2 (1993) 1029.

[29] C. Yeretzian, J. B. Wiley, K. Holczer, T. Su, S. Nguyen, R. B. Kaner, and R. L. Whetten, J. Phys. Chem. 97 (1993) 10 097.

[30] R. Taylor, G. J. Langley, H. W. Kroto, and D. R. M. Walton, Nature 366 (1993) 728.

[31] J. R. Heath, S. C. O'Brien, Q. Zhang, Y. Liu, R. F. Curl, H. W. Kroto and R. E. Smalley, J. Am. Chem. Soc. 107 (1985) 7779.

[32] D. E. Manolopoulos, J. Chem. Soc. Faraday Trans. 87 (1991) 2861.

[33] P. W. Fowler, R. C. Batten, and D. E. Manolopoulos, J. Chem. Soc. Faraday Trans. 87 (1991) 3103.

[34] D. E. Manolopoulos and P. W. Fowler, J. Chem. Phys. 96 (1992) 7603.

[35] J. R. Colt and G. E. Scuseria, J. Phys. Chem. 96 (1992) 10 265.

[36] K. Raghavachari and C. M. Rohlfing, Chem. Phys. Lett. 208 (1993) 436.

[37] B. L. Zhang, C. Z. Wang, and K. M. Ho, J. Chem. Phys. 96 (1992) 7183.

[38] H. S. M. Coxeter, Virus macromolecules and geodesic domes, in: A spectrum of mathematics, J. C. Butcher, ed. (Oxford University Press/Auckland University Press, Oxford/Auckland, 1971), p. 98.

[39] P. W. Fowler, J. E. Cremona, and J. I. Steer, Theor. Chim. Acta 73 (1988) 1.

[40] F. Wudl, Acc. Chem. Res. 25 (1992) 157.

[41] R. Taylor, Phil. Trans. Roy. Soc. Ser. A 343 (1993) 87.

[42] Y. Chai, T. Guo, C. M. Jin, R. E. Haufler, L. P. F. Chibante, J. Fure, L. Wang, J. M. Alford, and R. E. Smalley, J. Phys. Chem. 95 (1991) 7564.

[43] T. Guo, M. D. Diener, Y. Chai, J. M. Alford, R. E. Haufler, S. M. McClure, T. Ohno, J. H. Weaver, G. E. Scuseria, and R. E. Smalley, Science 257 (1992) 1661.

[44] D. S. Bethune, R. D. Johnson, J. R. Salem, M. S. de Vries, and C. S. Yannoni, Nature 366 (1993) 123.

[45] R. C. Haddon, Acc. Chem. Res. 25 (1992) 127.

[46] K. Holczer and R. L. Whetten, Carbon 30 (1992) 1261.

[47] S. Iijima, Nature 354 (1991) 56.

[48] D. Ugarte, Nature 359 (1992) 707.

[49] H. W. Kroto and D. R. M. Walton, eds, The fullerenes: new horizons for the chemistry, physics and astrophysics of carbon (Cambridge University Press, Cambridge, 1993).

[50] H. W. Kroto, J. E. Fischer, and D. E. Cox, eds, The fullerenes (Pergamon Press, Oxford, 1993).

2

FULLERENE CAGES

As emphasised in § 1.3, the first task in a systematic theory of the fullerenes is to identify the possible fullerene isomers of each carbon cage C_n. The reliability of all further discussions of topics such as electronic stability, steric strain, symmetry, spectroscopy, and isomerization depends largely on how well this task is accomplished.

In this chapter we shall discuss two complementary methods for generating fullerene isomers. The first is based on Coxeter's account of icosahedral triangulations of the sphere,[1] which has also been used to study the structures of objects as diverse as virus particles[2] and geodesic domes.[3] This method generates fullerenes of all possible atom counts for a fixed symmetry, and is easiest to apply when the symmetry is high.[4] The second is the ring spiral algorithm,[5] which unwinds the surface of a fullerene in a continuous spiral strip of edge-sharing five- and six-membered rings. This method generates fullerenes of all allowed symmetries for each given atom count, and is easiest to apply when the atom count is low.

Since many of the fullerenes that have been synthesized have moderate atom counts, but only few have especially high symmetry, in practice the ring spiral algorithm is often more useful than the Coxeter method. It also has the advantage of simplicity, in that each generated fullerene isomer is represented as a simple one-dimensional spiral code of five- and six-membered rings from which the more complicated three-dimensional structure is straightforwardly reconstructed. The Coxeter method does, however, have other advantages, as we shall see in § 2.8.

2.1 Fullerene polyhedra

The distinguishing structural property of a molecule is its bonding connectivity, and in order to talk about fullerene structures, we can talk equivalently about their bonding connectivities.

Each carbon atom in a fullerene is bonded to three others to give a closed pseudospherical cage of five- and six-membered rings. This bonding framework forms a polyhedron, with an atom at each vertex, a bond along each edge, and a ring around each face. Fullerene structures are therefore those trivalent spherical polyhedra that contain only pentagonal and hexagonal faces.

One of the best-known properties of polyhedra is Euler's theorem,[6] which states that the numbers of vertices (v), edges (e), and faces (f) in a spherical polyhedron are related by

$$v + f = e + 2. \tag{2.1}$$

For a polyhedron corresponding to a fullerene C_n, the number of vertices is $v = n$ and the number of edges is $e = 3n/2$. The number of faces is therefore $f = n/2 + 2$, as in any spherical polyhedron with trivalent vertices. However, this result can also be taken slightly further by defining the number of pentagons as p and the number of hexagons as h. The total number of vertices is then

$$(5p + 6h)/3 = n, \tag{2.2}$$

and the total number of faces is

$$p + h = n/2 + 2, \tag{2.3}$$

the solution of which is $p = 12$ and $h = n/2 - 10$. Hence all C_n fullerenes contain 12 five- and $n/2 - 10$ six-membered rings.

The trivalent polyhedra that satisfy $p = 12$ and $h = n/2 - 10$ form a subset of the medial polyhedra of Goldberg.[7,8] There is at least one such fullerene polyhedron for each even number of vertices $n \geq 20$, with the sole exception of $n = 22$. Experimentation with models soon reveals that it is impossible to construct a 22-vertex polyhedron with 12 pentagons and a single hexagon, and this result is easily proved.[9] Odd numbers of vertices are also precluded for all trivalent polyhedra, including the fullerenes, because the number of edges $e = 3n/2$ must be an integer. The smallest fullerene polyhedron is the dodeca-hedron, which is the only fullerene with $n = 20$ vertices. However, the number of distinct fullerene polyhedra with a given number of vertices increases rapidly beyond $n = 24$.

Briefly stated, the fullerene isomer problem is to identify and catalogue all fullerene polyhedra with a given vertex count $v = n$. This problem is a math-ematical one, and is independent of whether or not the corresponding carbon cage C_n occurs in nature. It is also independent of theoretical considerations such as the electronic structure of the cage.

2.2 Fullerene duals

Before discussing solutions to the isomer problem it is useful to note a general property of polyhedra which will find application in later chapters, namely that all polyhedra have duals.

The vertices of a polyhedron correspond to the faces of its dual, and vice versa. The edges of the two polyhedra correspond directly to one another, in the sense that an edge of a polyhedron is common to two of its faces and hence to two vertices of its dual. The dual operation is its own inverse, and preserves the point group symmetry of the polyhedron. In effect, this operation corresponds to interchanging v and f in Euler's theorem while leaving e unchanged. The two most often quoted examples of dual pairs are the icosahedron and the dodecahedron, and the octahedron and the cube; the tetrahedron completes the set of Platonic solids and is the only regular polyhedron that is self-dual (Fig. 2.1).

The duals of fullerene polyhedra are deltahedra, which are polyhedra made up exclusively of triangular faces. It is easy to see from the above discussion that the dual of a fullerene polyhedron with n vertices has n triangular faces and 12 five- and $n/2 - 10$ six-valent vertices. These attributes make it a sensible candidate for the skeletal bonding framework of a hypothetical giant *closo*-borane cluster belonging to this restricted coordination class. The simplest example of a fullerene dual is the icosahedron which is the dual of the dodecahedron in Fig. 2.1.

One of the many reasons why fullerene duals are interesting is that it is often easiest to construct a fullerene by first constructing its dual. This trick forms the basis of both of the methods for constructing fullerene isomers discussed below. Once the dual is known, the fullerene can easily be reconstructed, because each set of three mutually adjacent vertices in a fullerene dual encloses a unique triangular face of the dual and hence a unique vertex of the corresponding fullerene polyhedron. (In other words, a fullerene dual cannot have any separating triangles. This is a useful property of fullerene duals which does not extend to the duals of other trivalent polyhedra containing square and/or triangular faces. [10]) It follows that each vertex of the fullerene can be uniquely

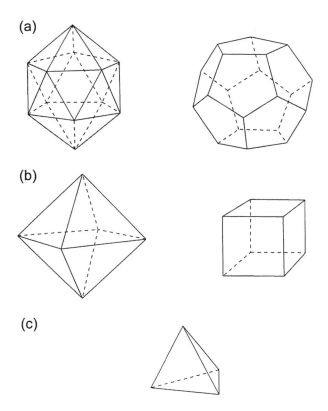

FIG. 2.1. Two dual pairs are (a) the icosahedron and the dodecahedron, (b) the octahedron and the cube. The tetrahedron (c) is the only self-dual Platonic solid.

associated with three mutually adjacent vertices of its dual, and moreover that two fullerene vertices will be adjacent if and only if two of the three dual vertices with which they are associated are the same. These facts allow one to obtain a list of adjacent vertices in the fullerene from a list of adjacent vertices in its dual, which is tantamount to performing the reconstruction. A computer subroutine that reconstructs a fullerene from its dual along these lines is listed in the appendix.

2.3 The Coxeter construction

Since all the vertices of the icosahedron are equivalent, they lie at the same distance from its centre and may be placed on the surface of a sphere. The edges can also be distorted to lie on this sphere, as sections of great circles

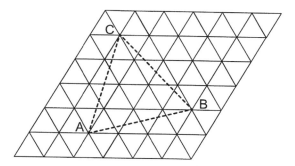

FIG. 2.2. Construction of an icosahedral triangulation of the sphere. The Coxeter parameters in this particular example are $i = 3$ and $j = 1$.

between the vertices. The resulting faces are spherical triangles, and the whole object is an icosahedral triangulation of the sphere.

Coxeter has shown how this triangulation can be generalized to produce larger spherical deltahedra of icosahedral symmetry, and hence larger icosahedral fullerene duals. [1] The construction has two stages which may be summarized as follows. First a pattern of small equilateral triangles is used to decorate a large equilateral triangle, which is taken as a single face of an icosahedron. The same pattern is then repeated on all twenty faces to produce a master icosahedron covered by triangles, which is transferred to the surface of a sphere as above.

The key stage in this procedure is to find a suitable pattern for decorating each face of the master icosahedron. The general solution to this problem is as depicted in Fig. 2.2. Take a plane covered by equilateral triangles, move along i edges from a starting vertex A, change direction by $60°$, and move along j edges to give a second vertex B. Two repetitions of this manoeuvre recover the starting point having marked out a large equilateral triangle. Twenty replicas of the large triangle can be produced by reflection operations and used to assemble the master icosahedron. Vertices of large triangles like A, B, and C become five-valent whereas those interior to a large triangle remain six-valent in the final spherical deltahedron.

Now suppose that each small equilateral triangle in Fig. 2.2 has a side of length one, so that its area is $\sqrt{3}/4$. It is straightforward to show that the area of the large triangle is $\sqrt{3}/4(i^2 + ij + j^2)$, and hence that the spherical deltahedron contains $20(i^2 + ij + j^2)$ triangular faces. Thus its dual, which is an icosahedral fullerene polyhedron, contains $n = 20(i^2 + ij + j^2)$ vertices. Interchanging the integers i and j gives the same deltahedron and fullerene, but these repetitions can be avoided by requiring that $i \geq j$. Hence Coxeter's construction produces an icosahedral fullerene C_n for each n satisfying

$$n = 20(i^2 + ij + j^2), \tag{2.4}$$

with $i > 0$, $j \geq 0$, $i \geq j$, each distinct (i, j) pair giving a distinct isomer.

The fullerenes given by this equation are the only fullerenes with icosahedral symmetry. If the Coxeter parameters i and j are such that the large equilateral triangle in Fig. 2.2 has reflection planes, which happens when $i = j$ or $j = 0$, the fullerene belongs to the full icosahedral point group I_h. Otherwise, it belongs to the rotational subgroup I which contains no reflections. In either event the fullerene is an icosahedral Goldberg polyhedron.[7,8]

Figure 2.3 illustrates the first seven icosahedral fullerenes by showing a single face of the master icosahedron upon which each structure is based.[11] The polyhedra with 20, 60, 80, 180, and 240 vertices in this figure have reflection planes and belong to the point group I_h. The polyhedra with 140 and 260 vertices have no reflection planes and belong to the point group I. The two smallest fullerenes of icosahedral symmetry are the dodecahedron ($i = 1$, $j = 0$) and the truncated icosahedron ($i = 1$, $j = 1$), both of which belong to I_h. However, the relative abundance of structures belonging to the I point group increases with increasing n in Eq. (2.4).

Coxeter's construction can be generalized to give fullerenes of other symmetries, but the effort increases rapidly as the symmetry is lowered. One obvious reason for this is that the generalized nets for lower symmetries contain a number of different equilateral and scalene triangles, so that more than two distinct Coxeter parameters are required. General tetrahedral fullerenes have nets comprising four large equilateral triangles, four small equilateral triangles, and twelve scalene triangles, for example, and are characterized by four distinct Coxeter parameters i, j, k, and l. Fullerenes with fivefold or sixfold axial symmetry can also be characterized by four parameters, but all other symmetries require more.[4]

This proliferation of independent parameters with decreasing symmetry is a fundamental problem, inevitable in any fullerene construction scheme. The reason the Coxeter construction in particular is so difficult to generalize is simply

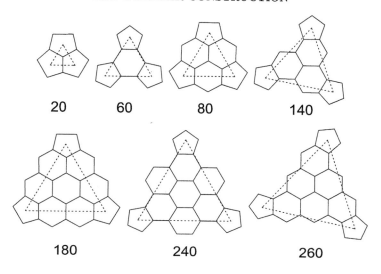

20   60   80   140

180   240   260

FIG. 2.3. The equilateral triangle repeat units that are used within the Coxeter construction to assemble the master icosahedra of the first seven icosahedral fullerenes, shown as dashed lines, with the corresponding fullerene fragments superimposed.

that its parameters are not convenient choices for structures with low symmetry. They are not related to the shape of the fullerene in a transparent way, and are subject to complicated inequality constraints if uniqueness is to be guaranteed. The shape of a low-symmetry fullerene is often more easily visualized from the parameters that arise in the spiral algorithm discussed in § 2.6.

Although it is difficult to extend Coxeter's construction to lower symmetries, it can be done. A detailed account of all tetrahedral and dihedral symmetry cases was given in the early fullerene literature,[4] along with a complete list of fullerenes containing up to 252 atoms belonging to the point groups I_h, I, T_d, T_h, T, D_{6h}, D_{6d}, D_6, D_{5h}, D_{5d}, and D_5. Restrictions on the vertex counts of fullerene polyhedra belonging to these point groups follow naturally from the extended Coxeter method, just as they do for the icosahedral case in Eq. (2.4). For example, each large equilateral triangle in the Coxeter net of a general tetrahedral fullerene contains $i^2 + ij + j^2$ fullerene vertices, each small equilateral triangle $k^2 + kl + l^2$, and each scalene triangle $il - jk$. Therefore, since the net comprises four large and four small equilateral and twelve scalene triangles, the total number n of vertices in a tetrahedral fullerene is restricted to

$$n = 4\left[(i^2 + ij + j^2) + (k^2 + kl + l^2) + 3(il - jk) \right], \tag{2.5}$$

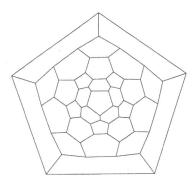

FIG. 2.4. If we imagine the edges to be made of elastic, and the faces to be hollow, this figure shows the result of stretching one of the twelve equivalent pentagonal faces of the familiar icosahedral C_{60} structure to the outside of a planar graph.

where i, j, k, and l are integers satisfying $i > 0$, $j \geq 0$, and $il - jk > 0$. Further rather more complicated restrictions on i, j, k, and l yield a unique parameter set for each polyhedron, and similar results apply to fullerenes with fivefold or sixfold principal axes. [4] However, the Coxeter construction has yet to be applied in a systematic way to fullerenes with lower symmetry.

2.4 Fullerene graphs

Another useful property of polyhedra is that they can be flattened onto a plane in such a way that the edges intersect only at the vertices. The operation proceeds by imagining the edges to be made of elastic, so that any chosen face can be stretched to the outside of a graph. Figure 2.4 shows the result of applying this stretching operation to the familiar truncated icosahedron of C_{60}, one of the 12 equivalent faces of which has been stretched to the outside of a planar graph. The vertices of the graph represent atoms, and the edges represent bonds, just as they do in the polyhedron. Thus the stretching operation transforms a three-dimensional polyhedral representation of C_{60} into a two-dimensional graph (a Schlegel diagram) that contains the same bonding connectivity information.

The correspondence between fullerene polyhedra and fullerene graphs is one to many, because any one of the $n/2 + 2$ faces of a C_n polyhedron can be stretched to the outside of a graph. The resulting graphs are all equivalent or *isomorphic*[12] in the sense that they all correspond to the same polyhedron and hence have the same bonding connectivity. However, they cannot generally be deformed to give the same picture without first leaving the plane on which they

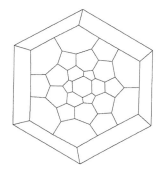

FIG. 2.5. An alternative planar graph of the truncated icosahedron, obtained by stretching a hexagon to the outside. This picture cannot be deformed to give the graph in Fig. 2.4 without crossing edges unless one first leaves the two-dimensional page on which it is drawn. The two graphs are isomorphic nevertheless.

are drawn. An alternative planar graph of truncated icosahedral C_{60} is shown in Fig. 2.5 as an example.

Constructing a fullerene graph is equivalent to constructing its polyhedron, because the bonding connectivities of the two are the same. A list of adjacent vertices in the graph is the same thing as a list of adjacent vertices in the polyhedron, for example, and this is how a computer might represent bonding connectivity. The advantage of the graph is simply that it exhibits this bonding connectivity more clearly on the two-dimensional page.

2.5 The spiral conjecture

The Coxeter construction is based on the fact that a fullerene dual can be represented as a triangulation of the sphere, and is mathematically rigorous. This rigour is what makes it difficult to apply methodically to low-symmetry cases, and is its downfall for practical fullerene applications. The spiral algorithm is based on fullerene graphs, and replaces rigour with a deliberate conjecture that makes low-symmetry cases easier to handle.[5]

The spiral conjecture can be stated equivalently in terms of a fullerene or its dual. The dual version turns out to be more convenient for implementing the spiral algorithm on a computer, for reasons explained below. However, the fullerene version is easier to describe diagrammatically, using pictures of fullerene graphs. This version reads as follows (see, however, § 2.8):

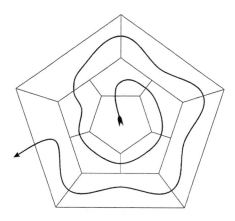

FIG. 2.6. The unique spiral of the dodecahedron, shown superimposed on the corresponding planar graph. Each new face in the spiral after the second is adjacent to both (a) its immediate predecessor and (b) the first face in the preceding spiral that still has an open edge.

Conjecture 2.1: *The surface of a fullerene polyhedron may be unwound in a continuous spiral strip of edge-sharing pentagons and hexagons such that each new face in the spiral after the second shares an edge with both (a) its immediate predecessor in the spiral and (b) the first face in the preceding spiral that still has an open edge.*

As this conjecture is worded, the first face in the spiral can be any one of the $n/2 + 2$ faces of the polyhedron corresponding to a fullerene C_n. The second face can be any one of the (five or six) faces that share an edge with this starting face, and the third face can be either of the two faces that share edges with faces one and two. Thus every C_n fullerene has a total of

$$12 \times 5 \times 2 + (n/2 - 10) \times 6 \times 2 = 6n \qquad (2.6)$$

spiral starts, many of which may be equivalent by symmetry. Once the start has been specified, the remainder of the spiral is determined by constraints (a) and (b) of the spiral conjecture, in accordance with the intuitive notion of a spiral. It may sometimes happen that the entire fullerene surface cannot be unwound from a given start subject to these constraints, in which case the spiral fails. Hence $6n$ is generally an upper bound on the number of successful spirals that can be found in a fullerene C_n.

A few examples should make these comments clearer. Figure 2.6 shows a graph corresponding to the dodecahedron of C_{20}, all twelve pentagonal faces of which are equivalent. In this case all $6 \times 20 = 120$ possible ways of starting the spiral are the same by symmetry, and unwind to give the spiral in the figure. Figure 2.7 shows three graphs corresponding to the truncated icosahedron of C_{60}, each with a spiral superimposed. Here, there are three symmetry-distinct starts, all of which again unwind successfully to give the spirals that are shown. Figure 2.8 shows how the polyhedron corresponding to the experimental D_{5h} C_{70} structure can be unwound in a spiral about its fivefold symmetry axis. In this case, which has lower symmetry, there are also quite a number (20) of other symmetry-distinct spiral starts, all of which again unwind successfully.

The smallest fullerene for which a failing spiral can be found is a D_2 isomer of C_{28}. This has a total of $6 \times 28 = 168$ spiral starts, of which 42 are symmetry-distinct. Of these 42 distinct starts, 41 unwind successfully and one fails. The failing spiral is shown superimposed on a graph of C_{28} in Fig. 2.9. It is constructed from the given start in accordance with constraints (a) and (b) of the spiral conjecture, and comes to a dead end when the last face it enters is surrounded by faces that have already been traversed. The spiral therefore 'shorts' before reaching the final (infinite) face of the graph. An example of a successful spiral is shown in the same figure for comparison.

The numbers of spirals in these examples follow a simple pattern. The total number of ways N_t in which the surface of a given fullerene polyhedron can be successfully unwound in a spiral is equal to the product of the number of symmetry-distinct spirals N_s and the order $|G|$ of the point group of the polyhedron:

$$N_t = N_s |G|. \tag{2.7}$$

For example, the total number of ways in which the dodecahedron can be unwound in a spiral is $N_t = 120$, the number of symmetry-distinct spirals corresponding to the dodecahedron is $N_s = 1$, and the order of the I_h point group is 120. The dodecahedron does not have any failing spirals, but the result does not rely on this. Thus the D_2 C_{28} fullerene in Fig. 2.9 can be unwound in a total of $N_t = 164$ successful spirals, $N_s = 41$ of which are symmetry-distinct, and the order of the D_2 point group is 4.

Equation (2.7) is a symmetry rule, and it can be proved in a straightforward way. The key observation is that a spiral of faces on the surface of a fullerene,

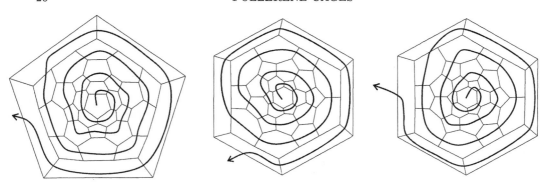

FIG. 2.7. The truncated icosahedron of C_{60} has three symmetry-distinct spirals, all of which unwind successfully.

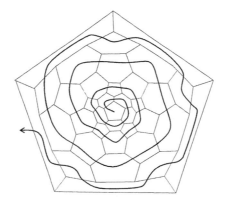

FIG. 2.8. The experimentally observed D_{5h} C_{70} structure with isolated pentagons has 21 symmetry-distinct spirals, all of which unwind successfully. This figure shows the spiral that unwinds around the fivefold symmetry axis as an example.

which is defined with a direction of unwinding in the sense of Figs. 2.6 to 2.9, has site symmetry C_1. (In other words, no symmetry operation of the polyhedron other than the identity operation can take a spiral onto itself.) It follows that each symmetry operation of the polyhedron acts on a given spiral to produce a different symmetry-equivalent copy, and since the number of symmetry operations in a point group is equal to its order, the result in Eq. (2.7) is proved. A corollary is that any one complete set of symmetry-equivalent spiral unwindings of the surface of a fullerene contains sufficient information to characterize its symmetry.

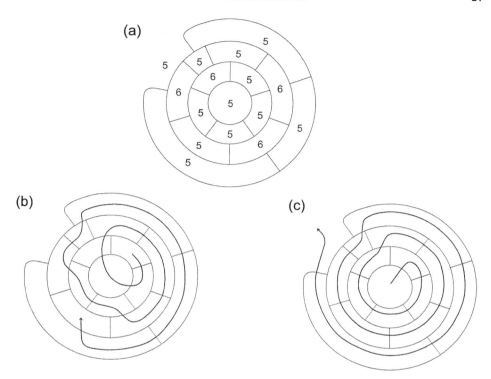

FIG. 2.9. The smallest fullerene for which a failing spiral can be found is a D_2 isomer of C_{28}, a planar graph of which is shown with the pentagonal and hexagonal faces marked for clarity in (a). The offending spiral is shown superimposed on a similar graph in (b). It comes to a dead end when it enters the fifteenth face, which is then surrounded by faces that have already been traversed. The spiral therefore 'shorts' before reaching the final (infinite) face of the graph. An example of a successful spiral is shown superimposed on the same graph in (c).

2.6 The Spiral Algorithm

Within the spiral conjecture, the fullerene isomer problem is straightforwardly solved. One simply generates all possible one-dimensional spiral sequences of pentagons and hexagons consistent with a given C_n atom count, and attempts to wind them up into fullerenes. The spirals that fail are abandoned, and the spiral sequence itself provides a simple way of testing whether each newly generated fullerene is unique.

The first thing to notice is that a fullerene spiral can be represented by a one-dimensional sequence of 5s and 6s which give the positions of the pentagons and hexagons along its path. For example, the dodecahedron spiral in Fig. 2.6

can be represented by the sequence

$$555555555555, \tag{2.8}$$

the three C_{60} spirals in Fig. 2.7 can be represented by the sequences

$$5666665656565656656565656565666665, \tag{2.9a}$$
$$6565656665666566656665665656656566, \tag{2.9b}$$
$$6656565656665665656656656665656566, \tag{2.9c}$$

and the C_{70} spiral in Fig. 2.8 can be represented by the sequence

$$5666666565656565666666666665656565656565. \tag{2.10}$$

The second thing to notice is that, since all C_n fullerene polyhedra have $n/2+2$ faces of which 12 are pentagonal and the remaining $n/2 - 10$ hexagonal, they may be found by searching through a maximum of

$$\frac{(n/2 + 2)!}{12!(n/2 - 10)!} \tag{2.11}$$

such spiral combinations. The spiral sequences corresponding to these combinations are straightforwardly generated on a computer using the positions of the twelve pentagons as nested indices.

Once one has a tentative spiral sequence of 5s and 6s, the next task is to check whether it winds up successfully to give a fullerene. In practice, as mentioned above, it is easiest to do this by first constructing the fullerene dual. The dual can be represented in a computer as a list of its adjacent vertices, which correspond to adjacent faces of the fullerene. This list can then be updated as the tentative spiral is wound up, with each new face in the spiral defining a new vertex of the dual. (It is more difficult to construct a list of adjacent vertices in the fullerene directly because the spiral conjecture does not assume that the fullerene *vertices* can be labelled in spiral order, and indeed sometimes they cannot. For example, the graph of the icosahedral symmetry C_{80} fullerene in Fig. 2.10 does not admit a vertex spiral.)

It is a relatively simple matter to fill in the adjacencies between dual vertices that are implied either directly or indirectly by constraints (a) and (b) of the

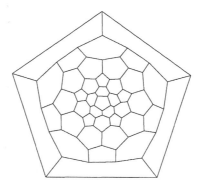

FIG. 2.10. A planar graph of the unique I_h C_{80} fullerene. The vertices in this graph may be divided into two symmetry-independent sets: those which belong to three hexagons, of which there are 20, and those which belong to two hexagons and a pentagon, of which there are 60. Call these vertex types A and B, respectively. Then in order to show that the graph does not admit any *vertex* spirals, that is paths which pass through the vertices of the graph in spiral order, it suffices because of the high symmetry to consider only the four unique vertex spiral starts that begin with ABB, BBA, BAB, and BBB. It is easy to show from the figure that all four of these vertex spirals 'short', in the same sense as in Fig. 2.9 (b), and hence that the vertices of the graph cannot be labelled in spiral order. This explains why the spiral algorithm works with the vertices of the fullerene dual graph and the faces of the fullerene.

spiral conjecture as the spiral winds, and so to test for a possible failure of the spiral along the way. If the spiral fails when fewer than 12 pentagons have been included, the search over spiral combinations in Eq. (2.11) can be streamlined, because all subsequent spiral combinations that are the same as far as the shorting face will obviously fail in the same way. Euler's theorem can also be used to obtain a lower bound on the number of remaining dual vertices required for closure, and hence to detect an open-ended spiral early and shorten the search. In practice, only a very small fraction of the possible spiral combinations in Eq. (2.11) pass these tests and close correctly to give acceptable C_n fullerene duals.

Given an acceptable fullerene dual, one still has to check it is unique. The possible spiral combinations of icosahedral C_{60} in Eq. (2.9) suggest an appealing solution to this problem that is free of additional assumptions. All three spirals wind up to give the same polyhedron, and hence all three spirals are equivalent. However, all three spirals can also be viewed as 32-digit numbers, with numerical values in the order $(2.9a) < (2.9b) < (2.9c)$. Although a given C_n fullerene polyhedron and its dual may be represented by several different spiral combinations, as in this C_{60} example, one always has a smaller value than the rest. This spiral can therefore be defined without ambiguity as the *canonical* spiral of the polyhedron. For example, the canonical spiral of icosahedral C_{60} is given in Eq. $(2.9a)$.

The definition of a canonical spiral solves the uniqueness problem, at least within the spiral conjecture. (Of course this is not an additional restriction, because the only fullerenes we are able to generate using the spiral algorithm are those that obey the conjecture.) Suppose one is given a vertex adjacency list of an acceptable C_n fullerene dual that has been generated from some input test spiral. One then simply unwinds the surface of this fullerene dual in as many of the $6n$ ways in Eq. (2.6) as are possible without shorting, and checks to see whether any of the spirals so obtained is lexicographically smaller. As soon as the first such spiral is found, the test spiral can be discarded as a repetition. If no such spiral has been found after all $6n$ attempts, the test spiral is canonical and corresponds to a new fullerene polyhedron. Finally, if it is required, a list of adjacent vertices in this fullerene can be obtained from the available list of adjacent vertices in its dual using the procedure outlined in § 2.2.

One particularly important subset of fullerenes is formed by those in which all the pentagons are isolated. These isolated-pentagon fullerenes are generally expected to be more stable than their fused-pentagon counterparts, for reasons discussed in Chapters 3 and 4. The smallest isolated-pentagon fullerene is icosahedral C_{60}, for example, which is the most abundant fullerene produced in experiment (see § 1.2). The reason for mentioning this here is that isolated-pentagon fullerenes are easy to construct selectively using a minor modification of the above spiral algorithm. One simply eliminates all spiral combinations from Eq. (2.11) in which two pentagons are directly fused in the spiral sequence, and then checks for and eliminates any extra pentagon adjacencies between successive coils of the winding spiral. This restriction greatly reduces the computational labour, making the algorithm applicable even to quite large fullerenes. It also facilitates an analysis of the results, because only a small fraction of the myriad of possible fullerene isomers belong to the isolated-pentagon class.

These ideas are implemented in the computer program listed in the appendix which searches for fullerene polyhedra with a given vertex count $v = n$ and prints out the canonical spiral for each one. Various symmetry properties of the polyhedron are also output, as these are obtained automatically during the test for a canonical spiral and provide a useful check on the topological coordinate results of Chapter 5. The isolated-pentagon constraint is included as an input option, so that both general and isolated-pentagon polyhedra may be generated with a single program. Appropriately modified versions of this program were used to generate the main fullerene isomer tables and several other tables in this book.

2.7 How many fullerenes are there?

The obvious temptation at this point is to find how many fullerene polyhedra the spiral algorithm generates for each given vertex count $v = n$. This may not be a complete list of all possible fullerenes C_n because the algorithm is based on a conjecture. However, as we shall see below, the existing evidence suggests that the conjecture is likely to be correct throughout the range of n to which the algorithm can reasonably be applied.

The spiral algorithm computer program listed in the appendix has been applied in its unrestricted form to all vertex counts n between 20 and 100. The results are summarized in Table 2.1, which lists the number of C_n fullerene isomers found by the algorithm for each value of n in this range. Enantiomeric isomer pairs are counted in each of two different ways in this table, with the left column corresponding to one entry per pair and the right (bracketed) column to two. Thus the spiral algorithm finds 1812 fullerene isomers of C_{60} when enantiomers are regarded as equivalent, but this number rises to a total of 3532 when they are regarded as distinct. In either event it can be seen from Table 2.1 that the number of fullerene isomers found by the algorithm increases very rapidly with n.

The effect of the isolated-pentagon constraint is shown in Table 2.2, which lists the number of isolated-pentagon fullerene isomers found by the spiral algorithm in the range $n = 60$ to 140. Enantiomeric isomer pairs are again counted in each of the two possible ways in this table, using the same format as in Table 2.1. Comparison of these two tables shows that the isolated-pentagon constraint is highly restrictive, greatly reducing the number of isomers obtained for each fixed vertex count n. However, the total number of isolated-pentagon

Table 2.1 Enumeration of C_n fullerene isomers found by the spiral algorithm in the range $n = 20$ to 100.* The left entry for each value of n is the isomer count obtained when enantiomers are regarded as equivalent, and the right (bracketed) entry is the isomer count obtained when enantiomers are regarded as distinct. $n = 22$ is omitted because C_{22} does not admit a fullerene isomer.

n	Isomers		n	Isomers	
20	1	(1)	62	2 385	(4 670)
24	1	(1)	64	3 465	(6 796)
26	1	(1)	66	4 478	(8 825)
28	2	(3)	68	6 332	(12 501)
30	3	(3)	70	8 149	(16 091)
32	6	(10)	72	11 190	(22 142)
34	6	(9)	74	14 246	(28 232)
36	15	(23)	76	19 151	(38 016)
38	17	(30)	78	24 109	(47 868)
40	40	(66)	80	31 924	(63 416)
42	45	(80)	82	39 718	(79 023)
44	89	(162)	84	51 592	(102 684)
46	116	(209)	86	63 761	(126 973)
48	199	(374)	88	81 738	(162 793)
50	271	(507)	90	99 918	(199 128)
52	437	(835)	92	126 409	(252 082)
54	580	(1 113)	94	153 493	(306 061)
56	924	(1 778)	96	191 839	(382 627)
58	1 205	(2 344)	98	231 017	(461 020)
60	1 812	(3 532)	100	285 913	(570 602)

*In preparing this table, a pentagon start was assumed in the spiral algorithm in order to cut down on computer time. This is reliable for small values of n, and is known to be robust in the present table up to and including C_{76}. However, it is also known that there is at least one (achiral, T_d symmetry) isomer of C_{100} that admits only spirals beginning on hexagons and is therefore missing from the present table. [13]

fullerene isomers also increases rapidly with n beyond a threshold at $n = 70$.

Although Tables 2.1 and 2.2 simply enumerate the fullerene polyhedra found by the spiral algorithm, the shapes and other derived properties of these polyhedra are also readily available from the computer program that was used to find them. Several relevant properties are discussed in the remaining chapters of this book, and illustrated in our main fullerene isomer tables. However, for the moment we are more interested in the number of fullerene polyhedra that

Table 2.2 Enumeration of isolated-pentagon C_n fullerene isomers found by the spiral algorithm in the range $n = 70$ to 140. The left and right (bracketed) entries for each value of n have the same meaning as in Table 2.1. Icosahedral C_{60} is the only isolated-pentagon fullerene with fewer than 70 atoms to be found by the spiral algorithm. Spirals starting on pentagons are assumed beyond C_{100}.

n	Isomers		n	Isomers	
70	1	(1)	106	1 233	(2 401)
72	1	(1)	108	1 799	(3 502)
74	1	(1)	110	2 355	(4 645)
76	2	(3)	112	3 342	(6 658)
78	5	(6)	114	4 468	(8 820)
80	7	(9)	116	6 063	(11 997)
82	9	(12)	118	8 148	(16 132)
84	24	(34)	120	10 774	(21 326)
86	19	(33)	122	13 977	(27 763)
88	35	(56)	124	18 769	(37 313)
90	46	(78)	126	23 589	(46 907)
92	86	(161)	128	30 683	(61 069)
94	134	(252)	130	39 393	(78 476)
96	187	(349)	132	49 878	(99 343)
98	259	(483)	134	62 372	(124 282)
100	450	(862)	136	79 362	(158 258)
102	616	(1 179)	138	98 541	(196 532)
104	823	(1 606)	140	121 354	(242 126)

can be constructed for each given value of n.

One particularly interesting mathematical question concerns the asymptotic dependence of the number of C_n fullerene polyhedra on n. In particular, it is known from general combinatorial methods which do not rely on the spiral conjecture that the number $F(n)$ of C_n fullerene polyhedra will increase with n in such a way that '$F(n)/n^9$ varies between positive constants $k_1 < k_2$ where the variation depends arithmetically on n'.[14]

Figure 2.11 plots the data in the right-hand (bracketed) entries of Table 2.1 on a logarithmic scale for n between 24 and 100. The data appear to be represented by a smooth curve, especially towards the top right-hand corner of the figure. However, a more detailed examination of the numbers in Table 2.1 reveals distinct local 'wiggles' in the curve which are just barely discernible in

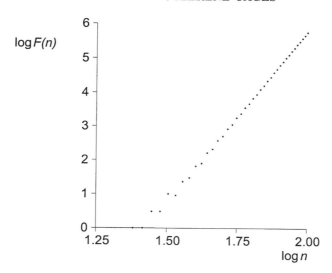

FIG. 2.11. Plot showing the log of the number $F(n)$ of distinct fullerene polyhedra found by the spiral algorithm with a given vertex count n, enantiomers being counted as distinct, as a function of $\log n$ in the range $n = 24$ to 100. See also Table 2.1.

the figure. For example, the isomer counts for $n = 96$ and 98 give a local slope (i.e., exponent of n in $F(n)$) of 9.0, whereas the counts for $n = 98$ and 100 give a local slope of 10.6. This behaviour also seems to be indicative of a more general pattern, in that the local slope in Fig. 2.11 is larger between $n = 4k - 2$ and $4k$ than it is between $n = 4k - 4$ and $4k - 2$.

All of this is clearly consistent with the above combinatorial result, which predicts a gently fluctuating curve in Fig. 2.11 with an average asymptotic slope close to 9.0. In fact, the average slope calculated by linear regression from the data between $n = 60$ and 100 in Table 2.1, which includes an equal number of subintervals of high and low slope, is somewhat higher than the prediction at 10.0. Although $n = 100$ can hardly be regarded as 'asymptotic' this is a clear indication that the spiral conjecture cannot be *drastically* false for the range of vertex counts in Fig. 2.11.

Finally, it can be seen from Table 2.1 that there is at least one fullerene polyhedron for each even number of vertices $n \geq 20$, with the sole exception of $n = 22$, and from Table 2.2, in addition to icosahedral C_{60}, there is at least one isolated-pentagon fullerene polyhedron for each even number of vertices $n \geq 70$. These results can also be obtained using more general combinatorial techniques. [15]

2.8 A fullerene without a spiral

Since the spiral algorithm is based on a conjecture it is important to check its results. The algorithm certainly finds a very large number of fullerene polyhedra as the vertex count n is increased, but how can we be sure it finds them all?

The spiral conjecture was originally motivated by considering the fullerene structures that were available in the early literature,[4,16,17] all of which may indeed be unwound in spirals. Moreover, extensive applications of the spiral algorithm to fullerenes with up to 100 atoms have yet to reveal any inconsistency with regard to the leapfrog construction described in Chapter 3, the Stone–Wales rearrangement described in Chapter 6, or the C_2 intrusion mechanism described in Chapter 7, none of which is enforced by the algorithm *per se*. Each of these transformations takes a given fullerene onto another which is not guaranteed to have a spiral, yet in every case considered in which the starting fullerene has had 100 or fewer atoms, the product fullerene has invariably been found to admit a spiral. It has also recently been *proved* that all fullerenes with fivefold or sixfold axial symmetry, including icosahedral fullerenes, have spirals that wind round this principal axis.[18] However, encouraging as they may be, these results do not rule out the possibility that a fullerene without a spiral may exist in some other symmetry group.

It is certainly not easy to find such a counter-example. The first problem is that, in order to show that there is no successful spiral in the surface of a given fullerene C_n, one has to consider all $6n$ possible spiral starts in Eq. (2.6) and show that all fail. The second problem is that, since most fullerenes do have spirals, one has to search through a large number of tentative candidate structures to find an exception. The two problems together are compounded by the fact that fullerenes without spirals are likely to be rather large. For example, the spiral algorithm counts in Table 2.1 have recently been confirmed all the way up to C_{70} using a different fullerene generation algorithm.[19]

The only apparent way to test the spiral conjecture beyond internal consistency checks is to confine the search for a counter-example to high symmetry. Moreover, the only obvious way to do this is to use the Coxeter tessellation method described in § 2.3, which is known to generate *all* fullerene isomers of a given (high) symmetry and can be applied routinely to the point groups D_{5+}, D_{6+}, and T_+.[4] (Here, and in what follows, G_+ is used to denote a group G and its supergroups within the list of 28 fullerene point groups given in Table 5.1.)

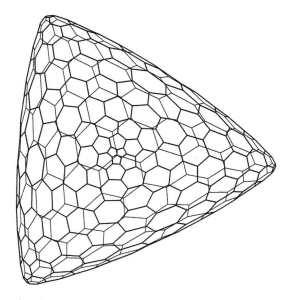

FIG. 2.12. The smallest tetrahedral fullerene without a spiral is a T isomer of C_{380}. This picture of its polyhedron is plotted directly from topological atomic coordinates (see Chapter 5).

The D_{5+} and D_{6+} cases are redundant, because fullerenes belonging to these point groups are known in advance to admit spirals.[18] However, the T_+ cases are more revealing.

While preparing this book and trying to convince ourselves one way or the other about the validity of the spiral conjecture, we used the Coxeter method to generate all T_+ (tetrahedral and icosahedral) symmetry fullerenes with up to 1000 atoms and searched their surfaces for spirals. This search was successful in revealing the first known series of counter-examples to Conjecture 2.1.[13] The smallest tetrahedral fullerene without a spiral was found to have 380 atoms, chiral T symmetry, and four symmetry-equivalent sets of three fused pentagons at the vertices of a twisted master tetrahedron. A picture of its polyhedron is shown in Fig. 2.12, and a graph of its connectivity is shown in Fig. 2.13 with an example of a failing spiral superimposed. The spiral passes through two of the three adjacent pentagons at a vertex of the master tetrahedron in one coil and is trapped by the third in the next. Moreover, it has been verified by computer that all $6n - 1 = 2279$ other spiral starts in the surface of the fullerene lead to spirals that fail in the same way. For want of a better term, this fullerene is *unspirallable*.

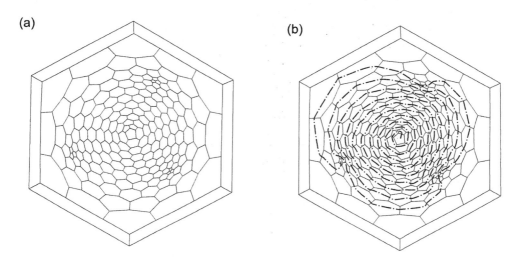

FIG. 2.13. The C_{380} fullerene in Fig. 2.12 has four symmetry-equivalent patches of three fused pentagons at the vertices of a twisted master tetrahedron, which can be seen more clearly here in the planar graph in (a). A spiral starting at any one of these patches gets trapped on its second pass through another, as shown in (b). Moreover all $6n = 2280$ possible starts in the surface of the fullerene lead to spirals that meet the same fate.

This tetrahedral fullerene with 380 atoms is simply the first in a series of larger T symmetry fullerenes without spirals.[13] The next member of the series has 404 atoms, and again corresponds to a twisted master tetrahedron with three fused pentagons at each vertex. The series continues on to higher atom counts, and includes 26 other chiral T symmetry fullerenes with up to 1000 atoms. These unspirallable fullerenes are listed along with their Coxeter parameters in Table 2.3. Since there are no T_d or T_h fullerenes in this table it is conceivable that some modification of the D_{5+}/D_{6+} spiral proof[18] will eventually be found to cover these achiral point groups.

Although the majority of the fullerenes in Table 2.3 have Coxeter parameters $k = -1$ and $l = 1$, and hence four caps of triply fused pentagons,[4] five do not. These five exceptions each have four caps of three isolated pentagons which either border a single hexagon (when $k = -1$ and $l = 2$) or are directly connected to the same atom (when $k = -2$ and $l = 2$). The first of the five, which is therefore the smallest T symmetry isolated-pentagon fullerene without a spiral, is a C_{800} cage with Coxeter parameters $i = 2$, $j = 10$, $k = -2$, and

Table 2.3 Vertex counts (n) and Coxeter parameters (i, j, k, and l) of all unspirallable fullerenes of T symmetry with up to 1000 atoms. [13] The parameters $k, l = -1, 1$ imply a fused-pentagon triple at each vertex of a master tetrahedron. [4] The parameters $k, l = -1, 2$ and $k, l = -2, 2$ imply different arrangements of three isolated pentagons near each vertex, the pentagons being separated by a hexagon in the first case and a fused-hexagon triple in the second. All achiral tetrahedral fullerenes (belonging to groups T_d and T_h) with up to 1000 atoms admit spirals.

n	i	j	k	l	n	i	j	k	l
380	2	7	−1	1	796	4	10	−1	1
404	1	8	−1	1	800	2	10	−2	2
460	2	8	−1	1	808	9	4	−1	2
488	1	9	−1	1	824	3	11	−1	1
524	3	8	−1	1	860	2	12	−1	1
548	2	9	−1	1	884	5	10	−1	1
580	1	10	−1	1	904	1	13	−1	1
616	3	9	−1	1	908	4	11	−1	1
644	2	10	−1	1	916	2	11	−2	2
680	1	11	−1	1	924	10	4	−1	2
692	4	9	−1	1	940	3	12	−1	1
716	3	10	−1	1	964	11	3	−1	2
748	2	11	−1	1	980	2	13	−1	1
788	1	12	−1	1	1 000	5	11	−1	1

$l = 2$. Thus the effect of the isolated-pentagon constraint is to push the first breakdown of the spiral conjecture to a higher atom count for fullerenes with T symmetry.

The most important thing that these examples with T symmetry prove is that the spiral algorithm will not give a complete list of fullerenes containing 380 or more atoms. However, since the algorithm cannot reasonably be applied to count fullerenes this large in any case, this is hardly a practical problem. The only remaining question, on which new work is still being done, is whether it is possible to construct an unspirallable fullerene of any other symmetry containing fewer than 380 atoms. The evidence available at present suggests not, but this evidence (which includes tests of internal consistency with respect to various transformations as discussed above) is still largely circumstantial.

2.9 Concluding remarks

In this chapter we have discussed two systematic methods for generating fullerene isomers: the Coxeter construction in § 2.3 and the spiral algorithm in § 2.6. The Coxeter construction is mathematically rigorous and can be used to generate fullerenes with high symmetry. The spiral algorithm is based on a conjecture that is now known to break down for fullerenes containing 380 or more atoms, but nevertheless seems to be on safe ground throughout its practical range of applicability. This method generates fullerenes of all allowed symmetries for a given atom count, which is often more useful for applications.

These are not the only methods that have been proposed for generating fullerene isomers. However, the more successful of the competing methods may all, by and large, be viewed as generalizations of the spiral algorithm.[20] The basic idea is that, instead of adding each new face in the construction of a fullerene graph directly to its immediate predecessor, so that a spiral of faces ensues, a more general recipe can be chosen for the addition. The safest approach of all, which is guaranteed to be complete, is to try separately each possible location at which each new face can be added to the boundary of the existing graph fragment.[21] However, this brute force approach rapidly becomes very expensive, reducing its practical domain of applicability. A compromise approach, which may be slightly safer than the spiral algorithm and yet is still not too expensive, is to add each new face to the existing boundary in such a way as to minimize the number of open edges after the addition.[20] The location for the addition is not as well defined in this scheme as it is in the spiral algorithm, because several possible locations can usually be found which meet the minimization requirement. A unique location can, however, be specified by labelling the available boundary positions in the current graph fragment in some specified order and choosing the location with the smallest label. This results in an algorithm which retains a one-dimensional code of 5s and 6s for each newly generated fullerene structure, similar to those shown for the spiral algorithm in § 2.6. However, the relationship between the one-dimensional code of 5s and 6s and the three-dimensional fullerene structure is not so transparent in this scheme as it is in § 2.6.

One clear way to see why these alternative schemes might be more robust than the original spiral algorithm is provided by an even simpler modification. Suppose that, rather than requiring each new face in the spiral to be adjacent

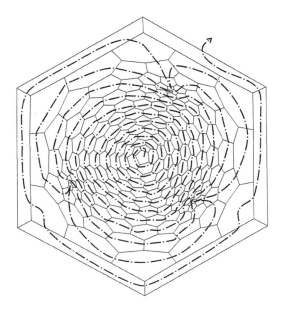

FIG. 2.14. One possible way in which the spiral algorithm might be made more robust. A modified spiral constructed according to Conjecture 2.2 escapes from the trap in the C_{380} counter-example in Fig. 2.13 by retracing its steps at each of the three bottlenecks it encounters.

to its immediate predecessor, as in § 2.5, the spiral conjecture is rephrased as follows:

Conjecture 2.2: *The surface of a fullerene polyhedron may be unwound in a continuous spiral strip of edge-sharing pentagons and hexagons such that each new face in the spiral after the second shares an edge with both (a) the first and (b) the last face in the preceding spiral that still has an open edge.*

It is easy to see that every fullerene which admits a spiral of the type defined in Conjecture 2.1 also admits a spiral of the type defined in Conjecture 2.2. However, this modified conjecture is less restrictive. For example, it allows the spiral to escape from the trap in the (originally) unspirallable C_{380} fullerene as shown in Fig. 2.14. This is not to suggest that Conjecture 2.2 is completely safe, and in fact it too may eventually break down. It is simply to illustrate a general direction in which the spiral algorithm might ultimately be improved.

Of course the point of generating fullerene structures in the first place is to use them to say something interesting. Some authors have expressed the view that, since the fullerene isomer problem has not yet been solved in a way which is both highly efficient *and* mathematically robust, it remains an unsolved problem.[19,20] However, we feel that this view is too pessimistic. The simple unmodified spiral algorithm described in § 2.6 is certainly efficient enough to say something useful, produces an easily and uniquely interpreted code of 5s and 6s for each newly generated fullerene isomer, and is likely to be complete to well beyond its practical limit of applicability. Thus the fullerene isomer problem has already been solved for practical purposes.

References and notes

[1] H. S. M. Coxeter, Virus macromolecules and geodesic domes, in: A spectrum of mathematics, J. C. Butcher, ed. (Oxford University Press/Auckland University Press, Oxford/Auckland, 1971), p. 98.

[2] D. L. D. Caspar and A. Klug, Cold Spring Harbor Symp. Quant. Biol. 27 (1962) 1.

[3] R. B. Fuller, Ideas and integrities (Prentice-Hall, Englewood Cliffs, 1963).

[4] P. W. Fowler, J. E. Cremona and J. I. Steer, Theor. Chim. Acta 73 (1988) 1.

[5] D. E. Manolopoulos, J. C. May, and S. E. Down, Chem. Phys. Lett. 181 (1991) 105.

[6] A detailed discussion of Euler's theorem, containing a number of different proofs, is given in: I. Lakatos, Proofs and refutations (Cambridge University Press, Cambridge, 1976).

[7] M. Goldberg, Tôhoku Math. J. 40 (1934) 226. This paper defines 'medial' polyhedra as trivalent polyhedra composed of at most two kinds of face, b-gons and $(b + 1)$-gons. Medial polyhedra with 12 or more than 13 faces are fullerene cages, and Fig. 1 of this early paper shows, amongst others, both isomers of C_{28} and the icosahedral isomers of C_{60} and C_{80}.

[8] M. Goldberg, Tôhoku Math. J. 43 (1937) 104. This paper discusses 'multi-symmetric' polyhedra made up of two types of face. Icosahedral pentagon + hexagon, octahedral square + hexagon, and tetrahedral triangle + hexagon polyhedra are classified and counted by the tessellation construction.

[9] The proof proceeds by attempting to construct a trivalent planar graph with a single hexagon at the centre and every other face a pentagon. The construction fails because the final (infinite) face of the graph has to be another hexagon, giving a total of two hexagons and 24 vertices – i.e., the unique fullerene structure of C_{24}.

[10] The 'hard part' of the proof that a fullerene dual cannot have any separating triangles consists in showing that a simply connected disk of fused pentagons and hexagons with only three divalent vertices on its boundary must contain at least seven pentagons. We are indebted to D. R. Woodall for communicating an outlined

case-by-case proof of this fact. (The case-by-case proof is necessary because Euler's theorem applied to the disk gives a lower bound of six pentagons, which is not enough to do the job.) It is a simple matter to construct examples of separating triangles in the duals of a trivalent polyhedra containing square and/or triangular faces. For instance, the three four-valent vertices form a separating triangle in the dual of a trigonal prism.

[11] P. W. Fowler, Chem. Phys. Lett. 131 (1986) 444.

[12] The mathematical theory of graphs and their application to chemical problems is discussed in a number of books. See, for example: (a) A. T. Balaban, ed., Chemical applications of graph theory (Academic Press, New York, 1976); (b) D. M. Cvetković, M. Doob and H. Sachs, Spectra of graphs: theory and application (Academic Press, New York, 1979). These two books contain definitions that are more careful than those used in our heuristic approach.

[13] D. E. Manolopoulos and P. W. Fowler, Chem. Phys. Lett. 204 (1993) 1. Added in Dover edition: In fact, the work of Brinkmann and Dress (G. Brinkmann and A.W.M. Dress, J. Algorithms 23 (1997) 345; Adv. Appl. Math. 21 (1998) 473.) shows that canonical spirals starting on a hexagon occur only from C_{100} and beyond, with 1 each at 100, 104, 110, 124, 132, 134, 136, 140, then 2 at 144, ... The first IPR fullerene with a canonical spiral starting on a hexagon is an isomer of C_{206}.

[14] C-H. Sah, A generalised leapfrog for fullerene structures, submitted to Fullerene Science and Technology (1993). This paper discusses the unpublished work of W.P. Thurston (Shapes of polyhedra, Research Report GCG7, Geometry Center, University of Minnesota (1990)), and applies it to the fullerene enumeration problem to obtain the $O(n^9)$ result.

[15] C-H. Sah, Croat. Chem. Acta 66 (1993) 1.

[16] H. W. Kroto, Nature 329 (1987) 529.

[17] T. G. Schmalz, W. A. Seitz, D. J. Klein, and G. E. Hite, J. Am. Chem. Soc. 110 (1988) 1113.

[18] P. W. Fowler, D. E. Manolopoulos, D. B. Redmond, and R. P. Ryan, Chem. Phys. Lett. 202 (1993) 371.

[19] D. Babić, D. J. Klein, and C-H. Sah, Chem. Phys. Lett. 211 (1993) 235.

[20] D. Babić and N. Trinajstić, Comp. & Chem. 17 (1993) 271.

[21] X. Liu, D. J. Klein, T. G. Schmalz, and W. A. Seitz, J. Comp. Chem. 12 (1991) 1252.

3

ELECTRONIC STRUCTURE

The electronic structures of the fullerenes pose an interesting problem in systematics. Brute-force use of expensive *ab initio* methods and powerful computers may produce accurate results for specific cages, but will not be useful for surveying the general features of fullerenes as a class. To deduce the relationship between geometric and electronic structure, a broader, more qualitative approach is needed. The simplest model of fullerene electronic structure that is capable of distinguishing between isomers is based on Hückel molecular orbital (HMO) theory, and that is the model described here. It has the enormous advantage that it is purely topological and free of parameters, so that the pattern of orbital energies is available directly from the bonding connectivity. Magic numbers and electron-counting rules for infinite sequences of fullerene isomers equivalent to Hückel's famous $(4n + 2)$ rule for aromatic hydrocarbons can be obtained exactly within the model. The price to be paid is that in finely balanced cases the model may not be able to cope with the complexities of the real molecule, and these complexities will be discussed in Chapter 4. Nevertheless, for a first general survey of the electronic structures of the fullerenes, HMO theory remains an indispensible tool.

3.1 Qualitative molecular orbital theory

The conventional picture of electronic structure is based on two obvious features of fullerenes: the molecules are more or less spherical, geometrically closed cages, and most of the individual atoms are in graphite-like environments. If the core (1s) electrons are considered to be localized, each carbon atom has four electrons and four valence orbitals (one 2s and three 2p) to contribute to cluster bonding. These are used to produce an electron-precise system of single carbon–carbon σ bonds spanning the edges of the fullerene polyhedron, leaving one 'radial' orbital and one electron per atom to take part in a surface π system. The radial orbital can be a pure 2p orbital if the local environment is perfectly planar, but more generally it is an *exo*-pointing sp^m hybrid (§ 4.1).

In the simplest model the σ and π systems are treated separately, and the Hückel approximations used so successfully for planar unsaturated hydrocarbons[1] are applied directly to the n-orbital, n-electron π system. Each atomic π orbital is assigned a Coulomb integral α and each pair of neighbouring atomic π orbitals a resonance integral β. Only nearest-neighbour interactions are taken into account and the absolute values of the negative energies α and β are not needed, as they serve only to set the origin and the scale, respectively, of the molecular orbital energy-level diagram. The π molecular orbitals and their energies $\epsilon_k = \alpha + x_k\beta$ are found by diagonalizing an $n \times n$ *adjacency matrix* with elements

$$A_{ij} = \begin{cases} 1, & \text{if atoms } i \text{ and } j \text{ are nearest neighbours,} \\ 0, & \text{otherwise,} \end{cases} \qquad (3.1)$$

as explained in more detail in standard textbooks.[1] The key feature of this approach which makes it useful for survey work is that the eigenvalues x_k ($k = 1, \ldots n$) of the adjacency matrix, and hence the π molecular orbital energy levels ϵ_k, are entirely determined by the molecular topology.

As a quantitative model of electronic structure, however, this procedure is undoubtedly oversimplified. It assigns a constant value of α to all atomic π orbitals, regardless of their hybridization, and a constant value of β to all π bonds, irrespective of their length or the mutual orientation of the atomic orbitals. The assumed separability of σ and π manifolds, which is forced by the reflection symmetry in planar conjugated systems, seems less likely *a priori* to be a good approximation in a curved molecule. (In fact, the indications from more sophisticated all-electron calculations[2] are that it does remain a good approximation in C_{60}.) Refinements to take account of overlap, curvature, second-neighbour interactions and electron repulsion can obviously be proposed, and such matters will be considered in Chapter 4. Nevertheless, the HMO model in its crude form gives a valuable qualitative description of the electronic structures of the fullerenes, as we shall now show for C_{60} itself.

All sixty atoms of icosahedral C_{60} are equivalent by symmetry and so 'rigorously' share a single α value, but the bonds fall into two sets (30 hexagon–hexagon and 60 hexagon–pentagon edges, with lengths r_h and r_p, respectively) and should in principle be described by two β parameters. The 60 Hückel

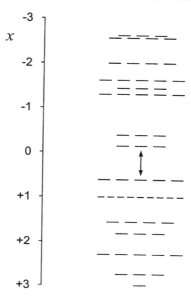

FIG. 3.1. π energy levels of C_{60} in the simple Hückel approximation. Each orbital has energy $\alpha + x_k\beta$ where the values of x_k are listed in the text. β is assumed to be the same for all bonds. $\Delta = 0.7566|\beta|$ is the HOMO–LUMO gap.

molecular orbitals span the reducible representation

$$\Gamma(O_{60}, I_h) = A_g + T_{1g} + T_{2g} + 2G_g + 3H_g + 2T_{1u} + 2T_{2u} + 2G_u + 2H_u, \quad (3.2)$$

and so by taking full account of the symmetry the 60×60 adjacency matrix can be factorized and diagonalized by solving nothing worse than a cubic equation. [3] By this method, or by computer diagonalization, the results are the same: in the single-β approximation, the orbital energies (and symmetries) are $\epsilon_k = \alpha + x_k\beta$ with $x_k = +3$ (a_g), $+2.7566$ (t_{1u}), $+2.3028$ (h_g), $+1.8202$ (t_{2u}), $+1.5616$ (g_u), $+1$ ($g_g + h_g$), $+0.6180$ (h_u), -0.1386 (t_{1u}), -0.3820 (t_{1g}), -1.3028 (h_g), -1.4383 (t_{2u}), -1.6180 (h_u), -2 (g_g), -2.5616 (g_u), -2.6180 (t_{2g}), giving rise to the π molecular orbital energy-level diagram in Fig. 3.1.

Already this crude model is telling us almost everything we would want to know about C_{60}. There are 30 bonding and 30 antibonding π molecular orbitals, so the neutral molecule has exactly the right electron count to give a closed shell with a reasonable HOMO–LUMO gap ($0.7566 |\beta|$, compared with $2|\beta|$ in both benzene and ethylene) and a total resonance energy ($0.5527 |\beta|$ per atom) approaching that of graphite ($0.5761 |\beta|$ per atom in the same model). The

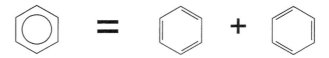

FIG. 3.2. The benzene molecule has a nuclear framework of D_{6h} symmetry, but each of the Kekulé structures has individually only D_{3h} symmetry.

formal bond orders computed from the coefficients of the occupied molecular orbitals are also informative. The total bond order is 1.601 for each r_h bond and 1.476 for each r_p bond, indicating bond alternation within the hexagons with $r_p > r_h$ as observed in the X-ray structures of C_{60} and its complexes.[4] This partial localization of electronic structure is more pronounced in more sophisticated treatments (a bond-length/bond-order correlation gives a ratio of $\sim 1\frac{2}{3}:1\frac{1}{3}$ for the bond orders in the self-consistent-field (SCF) calculation as opposed to 2:1 in a fully localized structure), but already in Hückel theory the description of C_{60} as a three-dimensional aromatic molecule is questionable.[5]

The emerging addition chemistry of C_{60} is fully compatible with a localized electronic structure, with Pt and other metals adding in η^2 fashion across one formal double bond rather than η^5 or η^6 to a face.[6] A symmetry argument also supports the idea of C_{60} as a 'super alkene' rather than a 'super arene': in aromatic benzene it is *necessary* to have resonance between two Kekulé structures to produce a wavefunction with the full sixfold symmetry of the carbon framework (Fig. 3.2); in C_{60} there is already a unique Kekulé structure with the full molecular symmetry, the one with 30 double bonds along r_h edges and 60 single bonds along r_p (Fig. 3.3). There is therefore no absolute need to invoke aromatic stabilization to describe the electronic structure. The indications are that this single Kekulé structure gives a good approximation to the true electron distribution.

The fact that the LUMO of C_{60} is low-lying and triply degenerate (Fig. 3.1) indicates a special stability for the C_{60}^{3-} ion with a high-spin open-shell electronic configuration and an icosahedrally symmetric structure. Because these t_{1u} orbitals are bonding along r_p edges but antibonding along r_h, occupation of the LUMO by three (or six) electrons will reduce bond alternation and produce a fully delocalized anion that is more aromatic in character than neutral C_{60}.[7] The Hückel degeneracy pattern therefore rationalizes the existence of alkali-metal fullerides of stoichiometry A_3C_{60}, and could even be taken to hint at the unusual electric properties of these solids containing large highly charged anions

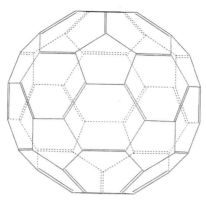

FIG. 3.3. C_{60} has a Kekulé structure with the full icosahedral symmetry of the nuclear framework in which 30 of the bonds (hexagon–hexagon edges) are double and 60 (pentagon–hexagon edges) are single.

with diffuse electron clouds.[8]

As we will see in Chapter 5, the idealized symmetry, approximate atomic coordinates and spectroscopic signature of a fullerene all follow from the eigenvectors of its adjacency matrix and therefore from its Hückel molecular orbitals. Hückel theory can even be claimed to predict the best isomer of C_{60}, since of all 1812 distinct isomers generated by the spiral algorithm, the icosahedral isomer has the largest band gap between HOMO and LUMO and the largest π-bonding resonance energy.[9] It is of course also the only isomer with isolated pentagons.

3.2 Open, closed, and pseudo-closed shells

A basic question to ask about the electronic structure of a general fullerene is whether it has a closed-shell electronic configuration. Qualitative arguments based on the special stabilities of closed shells have been used in many areas of chemistry to rationalize magic numbers and electron counts. The Hückel $(4n + 2)$ rule for cyclic polyenes and Wade's $(n + 1)$ rule for closo, nido, and arachno boron hydrides are examples of the link between specific geometric or topological structures and the number of bonding orbitals that they can sustain.[10,11] All fullerene cages are geometrically closed, but only for a small fraction of them does the number of electrons exactly match the capacity of the bonding orbitals. A closed shell is one indicator of stability, though not the only one, and within the Hückel model an almost complete enumeration of closed shells is given by just two geometrically based magic-number rules.

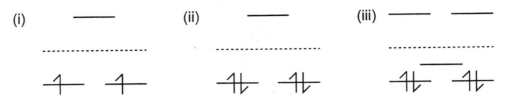

FIG. 3.4. Types of electronic configuration: (i) open, (ii) closed, and (iii) pseudo-closed shells, as defined in the text. The dotted line denotes the π orbital barycentre $\epsilon = \alpha$, $x = 0$, and the arrows show the occupation of the HOMO.

These are the so-called leapfrog and cylinder rules, to be described in detail below, and they guarantee at least one closed shell isomer when the atom count is $n = 60 + 6k$ ($k \neq 1$) or $n = 2p\,(7 + 3m)$ ($p = 5$ or 6). The sporadic 'accidental' closed shells not covered by these rules are rare, as we shall see later, and occur only at large nuclearity.

Three types of electronic configuration can be envisaged for a general fullerene. The examples in Tables A.21 and A.22 suggest that a fullerene cage of n atoms ($n > 20$) always has a surface π system with at least $n/2$ bonding and at most $n/2$ antibonding molecular orbitals in the Hückel model.[12] Non-bonding orbitals may also be present, but it seems that there is almost always room to accommodate all the electrons of the neutral molecule in bonding orbitals. The tendency for the cluster to have 'extra' bonding orbitals presumably arises from the substitution of pentagonal defects for some of the hexagons of the graphite sheet,[8] and leads to molecules with moderate to high electron affinities. In these circumstances, the three important possibilities for the configuration are as listed below (Fig. 3.4):

(i) An open shell in which the bonding HOMO is degenerate and only partly filled.

(ii) A properly closed shell with $n/2$ doubly occupied bonding and $n/2$ empty antibonding (or nonbonding) orbitals.

(iii) A pseudo-closed shell in which $n/2$ bonding orbitals are doubly occupied but some 'extra' bonding orbitals remain empty.

The first two cases need little comment. An open shell is susceptible to Jahn–Teller distortion (see § 5.3). A cage with a pseudo-closed shell may also some-times distort; if possible, it will adopt a geometry in which the binding energy of the occupied orbitals is maximized and the energies of the empty orbitals are pushed up into the antibonding region. The distinction between properly closed and pseudo-closed shells is formally clear in Hückel theory, but it is model-dependent. For example, in a many-electron treatment, occupation of one set of bonding orbitals tends to increase the energy of the next set to be occupied, so pseudo-closed systems have a tendency to become properly closed as the model is made more sophisticated.

These ideas about electronic configurations can be expressed in terms that are more usually applied in inorganic chemistry, namely the chemical potential, hardness, and softness of the charge distribution. The chemical potential μ is defined by Pearson [13] as the rate of change of the total energy E with electron count N

$$\mu = (\partial E / \partial N)_v, \tag{3.3}$$

and the hardness η is the second derivative of the energy or the first derivative of the chemical potential

$$\eta = \frac{1}{2}(\partial \mu / \partial N)_v, \tag{3.4}$$

where the subscript v implies a constant nuclear potential (i.e., fixed nuclei). Softness is just the reciprocal of hardness. In an orbital model, both μ and η can be approximated as functions of the orbital energies,

$$\mu \approx \frac{1}{2}(\epsilon_{LUMO} + \epsilon_{HOMO}), \tag{3.5}$$

$$\eta \approx \frac{1}{2}(\epsilon_{LUMO} - \epsilon_{HOMO}), \tag{3.6}$$

and in Hückel theory these reduce to

$$\mu \approx \alpha + \frac{1}{2}(x_{LUMO} + x_{HOMO})\beta, \tag{3.7}$$

$$\eta \approx \frac{1}{2}(x_{LUMO} - x_{HOMO})\beta. \tag{3.8}$$

Hardness parameters for all isomers can be read off from the main tables. For C_{60} $\mu \approx \alpha + 0.2397\beta$ and $\eta \approx -0.3783\beta$, for C_{70} $\mu \approx \alpha + 0.2647\beta$ and $\eta \approx -0.2647\beta$, and for an open shell $\eta = 0$. In the graphite limit $\mu \rightarrow \alpha$ and $\eta \rightarrow 0$.

In general, large closed-shell fullerenes have decreasing band gaps and hence low values of η. Soft molecules are more reactive and polarizable, and our earlier considerations about pseudo-closed shells can be restated as the principle of maximum hardness: [13] molecules arrange themselves to be as hard as possible

(a) 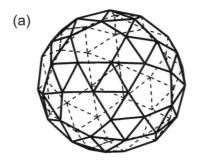   (b)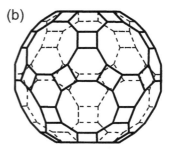

FIG. 3.5. Two icosahedral polyhedra. (a) The snub dodecahedron has 60 vertices, 150 edges and 92 faces of which 12 are pentagonal and 80 triangular. This polyhedron has I symmetry and can therefore be constructed in left- and right-handed forms. (b) The great rhombicosidodecahedron is a polyhedron of I_h symmetry that has 120 vertices, 180 edges and 62 faces of which 12 are decagons, 20 are hexagons and 30 are rectangles. In the equilateral version of the polyhedron the faces are regular decagons, regular hexagons and squares, but the most general version compatible with I_h symmetry has three different edge lengths.

(and so to have maximal HOMO–LUMO gaps). In these terms, fullerenes are all quite soft molecules. The relationships between orbital energies, chemical potential, hardness, and softness are well defined only in a minimal-basis model. In an SCF calculation on a closed-shell system, for example, ϵ_{LUMO} can be made arbitrarily close to zero by addition of sufficiently diffuse basis functions. Nevertheless, the hard/soft classification is a useful way of summarizing qualitative expectations about the chemistry of this class of molecules.

3.3 Icosahedral fullerenes

The highest point group available to a fullerene is I_h, consisting of the 120 symmetry operations of a regular icosahedron or dodecahedron. Next is the I point group, including the 60 proper rotations of the icosahedron but without the inversion or any improper operation. I is appropriate to a twisted (chiral) object such as the Archimedean snub dodecahedron (Fig. 3.5a), whereas I_h describes the achiral Platonic icosahedron, the dodecahedron, and the Archimedean great rhombicosidodecahedron (Fig. 3.5b).

As discussed in § 2.3, the Coxeter tessellation method [14,15] produces an icosahedral fullerene C_n for each $n = 20(i^2 + ij + j^2)$, where i and j are integers satisfying $i \geq j$ and $j \geq 0$. Numbers of this form are either exactly divisible by 60 or leave a remainder of 20 on division by 60, and so icosahedral (I and I_h) fullerenes have either $n = 60k$ or $n = 60k+20$, and these cases give rise between them to eleven I_h isomers and ten I pairs of enantiomers with $n <= 1000$.[16] The fullerenes in these two series are the icosahedral multi-symmetric polyhedra of Goldberg.[17] Those in the second series always have 20 atoms at the vertices of a regular dodecahedron, whereas those in the first series have hexagon centres at these positions.

So far this is an interesting result in geometry but may seem remote from chemistry. The chemistry appears when we consider the π systems of the fullerenes in the two series. Hückel energies are derived from the adjacency matrix (Eq. (3.1)), and the adjacency matrix is completely specified for an icosahedral fullerene by the connections between the small triangles of its Coxeter net. The adjacency matrix can therefore be constructed directly from i and j using an appropriate algorithm to 'fold' the net, and when this is done, the electronic configurations of the icosahedral fullerenes are seen to follow a remarkably simple rule:[16]

(a) for an I or I_h fullerene with $60k$ atoms, a properly closed π shell is found for the neutral molecule;

(b) for an I or I_h fullerene with $60k + 20$ atoms, the neutral molecule has an open shell with two electrons in a fourfold degenerate HOMO.

Thus I_h C_{60} is the first of an infinite magic-number series and C_{20} is the first of a series with a closed shell for the dication but an open shell (implying Jahn–Teller distortion) for the neutral. This in itself is an interesting analogy to the Hückel $4n + 2$ and $4n$ rules for aromatic and antiaromatic hydrocarbons, but it is only part of a bigger pattern. Inspection of the two series shows that there is a general geometrical relationship between them and that any open-shell icosahedral fullerene C_n with $n = 60k + 20$ can be converted into a larger fullerene C_{3n} with $3n = 3(60k + 20) = 60k'$ ($k' = 3k + 1$) and hence a properly closed shell. The conversion is accomplished by the leapfrog transformation[16,18] which derives its significance from the fact that it can be applied to any fullerene (icosahedral or not) and *always* generates a structure with a closed-shell electronic configuration, as we will see in the following section.

3.4 The leapfrog transformation

To see how the leapfrog transformation works, consider the C_{20} dodecahedron (Fig. 3.6). If this structure is capped on every face it becomes a deltahedron with 12 five-valent and 20 six-valent vertices. If the deltahedron is then converted to

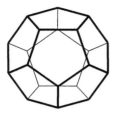  

FIG. 3.6. Leapfrogging a dodecahedron: a dodecahedron is converted to a 32-vertex deltahedron by omnicapping, then to a truncated icosahedron by taking the dual.

its dual, the 32 vertices become face centres (12 pentagons and 20 hexagons) and the 60 triangular faces of the deltahedron become the 60 vertices of a truncated icosahedron (i.e., of icosahedral C_{60}). By omnicapping and dualizing the original fullerene, a new fullerene with three times as many vertices has been created. The procedure is nicknamed the leapfrog transformation because it jumps from one fullerene to another over the intervening deltahedron.[19] The procedure can clearly be used again on C_{60}, or indeed on any fullerene whether icosahedral or not.

Figure 3.7 shows the fate of the various structural components of a fullerene under the transformation. Each p-sided face of the parent appears in the leapfrog but rotated by (π/p) about its normal and surrounded by new hexagons, each edge of the parent survives in the leapfrog but is rotated by $(\pi/2)$ about a line joining its midpoint to the centre of the fullerene, and each vertex of the parent disappears but gives rise to a new hexagonal face in the leapfrog and hence to three new edges and three new vertices (allowing for the fact that hexagons generated from neighbouring vertices of the parent have two vertices in common).

Leapfrogging preserves the symmetry of the fullerene polyhedron and so in the icosahedral series described above it converts $I \to I$ and $I_h \to I_h$. When it acts on any member of series (b) it replaces the 'extra' 20 atoms by hexagon centres and generates an equisymmetric member of series (a) having a multiple of 60 atoms. Every member of series (b) is thus the parent for an infinite sequence of leapfrogs in (a), and every icosahedral fullerene with $60k$ atoms can be reached either directly from (a) or by leapfrogging a smaller fullerene in (b). The chemical corollary is that every closed-shell (but no open-shell) neutral fullerene of icosahedral symmetry is a leapfrog. Whether a neutral fullerene has an open or a closed shell, its leapfrog has a properly closed shell.

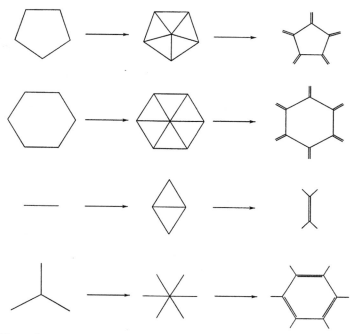

FIG. 3.7. Transformation of faces, vertices, and edges of a fullerene under leapfrogging. The double bonds are placed to indicate a consistent Kekulé structure with the maximum possible number of benzenoid rings, constructed by doubling every edge of the leapfrog that is derived directly from one in the parent.

The remarkable fact is that this last statement remains true regardless of the point-group symmetry of the parent (and the leapfrog). *Every* fullerene isomer leapfrogs to a closed shell. Since fullerene isomers can be constructed for all $n = 20 + 2k$ ($k \neq 1$), and leapfrogging triples the vertex count, the leapfrog rule is that closed shells occur for all $n = 60 + 6k$ ($k \neq 1$). Specifically, for every $n = 60 + 6k$ ($k \neq 1$) there is a closed-shell isomer for each isomer of the fullerene $C_{n/3}$, matching it in symmetry and derived from it by omnicapping and taking the dual. C_{60} is the first leapfrog and the first fullerene with a properly closed shell, and the next few are shown in Fig. 3.8.

A less useful corollary is that there are always at least as many isolated-pentagon isomers of C_n as there are isomers of any kind for $C_{n/3}$ when $n = 60 + 6k$ ($k \neq 1$). In fact, for $k > 1$ there are always many more isolated-pentagon fullerenes than leapfrogs, and isolated-pentagon isomers are not confined to the $60 + 6k$ series.

Why does the leapfrog rule work? It was known to work empirically for many

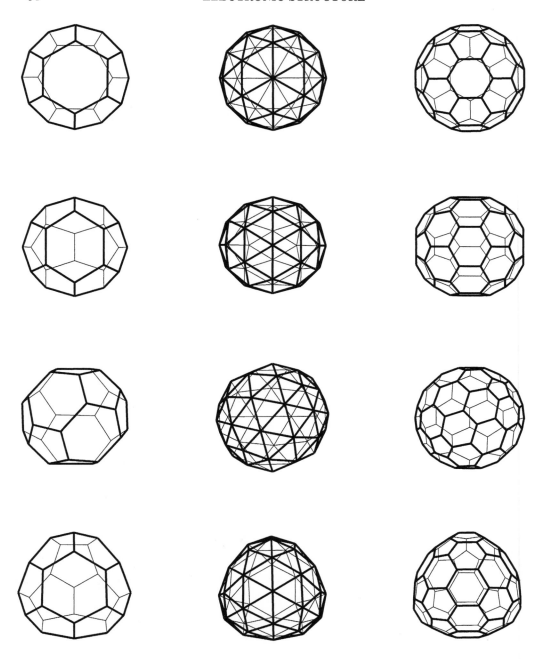

FIG. 3.8. Parents (C_n), intermediate deltahedra ($D_{3n/2+2}$) and leapfrogs (C_{3n}) in the transformation of the smallest fullerenes $C_{24} \rightarrow D_{38} \rightarrow C_{72}$, $C_{26} \rightarrow D_{41} \rightarrow C_{78}$, and $C_{28} \rightarrow D_{44} \rightarrow C_{84}$ (two isomers).

individual cases long before a fully satisfactory mathematical proof was devised, and several heuristic proofs and rationalizations can be offered.[18-21] The full graph-theoretical proof[22] has the advantage that it shows that leapfrogging extends well beyond the fullerene class: in fact the eigenvalue spectrum of an adjacency matrix of any trivalent graph will contain $n/2$ bonding and $n/2$ antibonding energies if the graph has at least one face with $3m + 1$ or $3m + 2$ edges. A cube can be leapfrogged, and fulleroids with heptagonal rings[23] can be leapfrogged to give closed shells. On the other hand, the tetrahedron does not leapfrog to a properly closed shell because all of its faces have three edges, but instead the leapfrog (the truncated tetrahedron) has a nonbonding doubly degenerate orbital which would correspond to the fully occupied HOMO of a neutral C_{12} cluster. Other non-fullerene exceptions to the leapfrog rule are members of the infinite series of polyhedra with four triangular and $(n/2 - 2)$ hexagonal faces that generalize the tetrahedral multisymmetric polyhedra of Goldberg.[17]

More informal rationalizations can give some chemical insight into the leapfrog rule. Two are based on the observation that one third of the edges of the leapfrog are derived directly from edges of the parent. These are the rotated edges mentioned earlier. In one scheme[18] the n atomic π orbitals are paired up into localized bonding and antibonding orbitals along these $n/2$ special edges and it is argued that interaction between the local orbitals to produce the delocalized π MOs of the leapfrog cage will alter the energies but not the numbers and symmetries of the bonding orbitals and hence $n/2$ MOs will be bonding and $n/2$ antibonding (which is broadly true, but breaks down in special cases like the tetrahedron). This scheme has the advantage that it predicts the symmetry of the occupied orbitals of any leapfrog fullerene: the occupied π orbitals of C_n span the same symmetry as the edges of the parent $C_{n/3}$. Thus, for example, the 30 edges of C_{20} span the representation

$$\Gamma(O_{30}, I_h) = A_g + G_g + 2H_g + T_{1u} + T_{2u} + G_u + H_u, \qquad (3.9)$$

and this is precisely the symmetry spanned by the 30 bonding π MOs of C_{60} (see Fig. 3.1). Other general theorems relating the vertex, edge, and face symmetries of parent and leapfrog fullerenes can also be proved.[20,24]

A different rationalization is couched in valence bond (VB) language.[21] If every one of the $n/2$ special edges is assigned a double bond, and every other

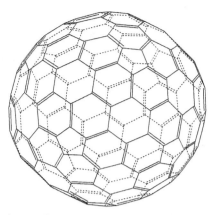

FIG. 3.9. The special ('Fries') Kekulé structure of a leapfrog fullerene has $n/3 = 60$ benzenoid rings in the case shown here of I_h C_{180}, the leapfrog of I_h C_{60}.

edge a single bond, then we will have a Kekulé structure for the leapfrog fullerene that is fully symmetric, that places no double bond inside a pentagon and has the maximum possible number $(n/3)$ of benzenoid hexagonal rings. An example is shown in Fig. 3.9 for C_{180}, the leapfrog of C_{60}. Leapfrog clusters are unique in that they can achieve this maximum. Taylor[25] has proposed the number of benzenoid rings in the Kekulé structure as a qualitative VB criterion for stability of a fullerene isomer; on this criterion, therefore, leapfrogs are the most stable isomers possible. In practice, they may not be, because the (qualitative) VB picture suffers from the same problem as Hückel MO theory in that it does not account for steric strain (Chapter 4).

These arguments also link the theory of fullerenes and benzenoid hydro-carbons. In the early part of this century Fries proposed that those planar benzenoids with maximal numbers of formal benzene rings would be the most stable;[26] the Taylor criterion extends this idea to fullerenes. The special Kekulé structure of a leapfrog fullerene is also known as a Fries structure for this reason. Clar's theory of the aromatic sextet[27,28] is another rationalization of hydrocar-bon stability in terms of valence-bond structures. A leapfrog operation can be defined for benzenoids: cap all the hexagonal faces of the molecular graph and then take its inner dual (Fig. 3.10). Dias[29] shows that this planar operation leads to a benzenoid that can be totally covered by aromatic sextets (i.e., maxi-mally stable on the Clar criterion, and having a properly closed-shell electronic configuration). Leapfrog fullerenes are also totally resonant: the $60 + 6k$ atoms can be divided into $k + 12$ disjoint polygons, 12 of which are pentagons and k

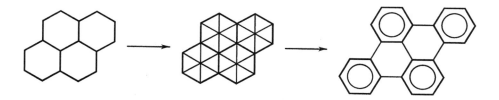

FIG. 3.10. Leapfrogging a planar benzenoid hydrocarbon to give a totally resonant, Clar-sextet, closed-shell product. Each hexagonal face of the carbon framework is capped and then the dual is taken, converting each hexagonal ring of the original to one bearing a Clar sextet and each vertex with three carbon neighbours in the original to an 'empty' hexagon

hexagons. In fact, three entirely equivalent definitions of a leapfrog fullerene are: a fullerene produced by omnicapping and dualizing a smaller fullerene, a fullerene that has a Fries structure, and a fullerene that is a Clar polyhedron.[30]

The tables of isolated-pentagon isomers in this book include leapfrogs at C_{60}, C_{72}, C_{78}, C_{84}, C_{90}, and C_{96}. For C_{60} and C_{72}, there is only a single isolated-pentagon isomer and the correspondence between parent and leapfrog is obvious. For the higher members of the series the relationships are $C_{26}:1 \rightarrow C_{78}:4$, $C_{28}:1 \rightarrow C_{84}:1$, $C_{28}:2 \rightarrow C_{84}:20$, $C_{30}:1 \rightarrow C_{90}:1$, $C_{30}:2 \rightarrow C_{90}:2$, $C_{30}:3 \rightarrow C_{90}:16$, $C_{32}:1 \rightarrow C_{96}:4$, $C_{32}:2 \rightarrow C_{96}:1$, $C_{32}:3 \rightarrow C_{96}:3$, $C_{32}:4 \rightarrow C_{96}:30$, $C_{32}:5 \rightarrow C_{96}:33$, $C_{32}:6 \rightarrow C_{96}:136$.

Leapfrog isomers can be generated without going through the explicit geometric construction, because the adjacency matrix of a leapfrog fullerene follows directly from that of the parent. A suitable computer algorithm takes $A_{ij}(P)$ and analyses it to find the edges and faces of the parent and then constructs an enlarged matrix $A_{ij}(L)$ with the connections appropriate to the leapfrog. In this way we can pick out closed-shell isomers out of the mass of possible isomers for fullerenes beyond the reach of complete enumeration. Table 3.1 lists, for example, the 40 leapfrog isomers of C_{120}. Although there are thousands of isolated-pentagon isomers of C_{120} and hundreds of thousands of general isomers, we can expect a small number of closed-shell isomers: 40 leapfrogs, one carbon cylinder, and one extra sporadic closed shell[31] that will be discussed below. The symmetries, spectroscopic signatures, and Hückel molecular orbital energy levels of the 40 leapfrog isomers are all available once the isomers of C_{40} have been counted.

Table 3.1. Leapfrog isomers of C_{120}. Each isomer is identified by its position in the full list of isolated-pentagon C_{120} fullerenes generated by the spiral algorithm, by its C_{40} parent, by the positions of the 12 pentagons in its canonical ring spiral, and by the point group G common to both parent and leapfrog. The leapfrog HOMO energies are given in units of β. The LUMO energies and band gaps are given in units of $|\beta|$, so that positive values in the HOMO and LUMO columns correspond to bonding and antibonding levels, respectively.

Leapfrog isomer	Parent isomer	Ring spiral	G	HOMO	LUMO	Band gap
120:1	40:1	1 7 9 11 13 15 48 50 52 54 56 62	D_{5d}	0.3097	0.0610	0.3707
120:2	40:3	1 7 9 11 13 18 45 50 52 54 56 62	D_2	0.3590	0.0746	0.4336
120:3	40:2	1 7 9 11 13 24 39 50 52 54 56 62	C_2	0.3867	0.0779	0.4646
120:53	40:5	1 7 9 11 13 37 40 46 52 54 56 62	C_s	0.4269	0.0795	0.5064
120:59	40:4	1 7 9 11 13 37 42 48 53 55 57 60	C_1	0.4260	0.0900	0.5130
120:70	40:6	1 7 9 11 13 40 44 48 51 55 57 59	C_1	0.4228	0.0825	0.5053
120:74	40:7	1 7 9 11 13 40 46 48 51 53 55 57	C_s	0.4096	0.0934	0.5030
120:77	40:8	1 7 9 11 13 42 44 48 51 56 58 60	C_{2v}	0.4119	0.0753	0.4872
120:180	40:9	1 7 9 11 22 25 38 41 52 54 56 62	C_2	0.4621	0.0813	0.5434
120:205	40:12	1 7 9 11 22 25 41 48 51 53 55 57	C_1	0.4378	0.0906	0.5284
120:216	40:13	1 7 9 11 22 25 43 48 51 54 58 60	C_s	0.4592	0.0913	0.5505
120:252	40:10	1 7 9 11 22 35 39 42 50 54 56 59	C_1	0.4461	0.0864	0.5324
120:304	40:11	1 7 9 11 22 39 46 51 53 56 58 61	C_2	0.4403	0.1026	0.5430
120:759	40:14	1 7 9 11 25 36 39 42 46 54 56 62	C_s	0.4585	0.0841	0.5427
120:764	40:15	1 7 9 11 25 36 42 44 48 53 57 60	C_2	0.4295	0.0910	0.5205
120:766	40:18	1 7 9 11 25 36 42 44 50 55 57 60	C_2	0.4285	0.1131	0.5416
120:771	40:16	1 7 9 11 25 36 42 46 48 53 55 60	C_2	0.4572	0.0900	0.5472
120:772	40:17	1 7 9 11 25 36 42 46 50 53 55 57	C_1	0.4495	0.0988	0.5483
120:3347	40:21	1 7 9 13 20 23 41 48 51 55 57 59	C_2	0.4357	0.0833	0.5190
120:3358	40:22	1 7 9 13 20 25 41 46 51 55 57 59	C_1	0.4486	0.0881	0.5367
120:3361	40:23	1 7 9 13 20 25 41 48 53 55 58 62	C_2	0.4583	0.1017	0.5600
120:3383	40:20	1 7 9 13 20 28 46 48 50 54 56 60	C_{3v}	0.4142	0.0796	0.4938
120:3452	40:24	1 7 9 13 20 39 42 47 49 51 55 57	C_s	0.4619	0.0866	0.5485
120:3659	40:33	1 7 9 13 23 25 41 43 53 56 58 61	D_{2h}	0.3937	0.0971	0.4907
120:3661	40:32	1 7 9 13 23 25 41 45 51 53 55 57	D_2	0.3969	0.0937	0.4906
120:3786	40:25	1 7 9 13 23 34 42 44 50 53 58 60	C_2	0.4211	0.0883	0.5095
120:3835	40:26	1 7 9 13 23 39 42 44 46 51 57 59	C_1	0.4557	0.0910	0.5467
120:3841	40:28	1 7 9 13 23 39 42 44 49 55 58 62	C_s	0.4530	0.1038	0.5568
120:3846	40:27	1 7 9 13 23 39 44 46 49 53 56 61	C_2	0.4543	0.0987	0.5530
120:3873	40:34	1 7 9 13 23 42 44 46 49 51 54 60	C_1	0.4456	0.1004	0.5460
120:4269	40:29	1 7 9 13 34 40 43 45 47 51 53 60	C_2	0.4830	0.0896	0.5725
120:4270	40:31	1 7 9 13 34 40 43 45 49 51 53 57	C_s	0.4809	0.0936	0.5745

Table 3.1. *(Continued)*

Leapfrog isomer	Parent isomer	Ring spiral		G	HOMO	LUMO	Band gap
120:4271	40:30	1 7 9 13 34 40 45 47 51 53 55 58	C_3		0.4778	0.1047	0.5825
120:4398	40:19	1 7 9 15 34 38 45 47 51 53 55 58	C_2		0.4714	0.1015	0.5729
120:4814	40:40	1 7 9 23 26 35 41 44 46 50 53 57	T_d		0.4548	0.1089	0.5637
120:10604	40:35	1 7 11 18 21 26 37 43 48 53 57 60	C_2		0.4447	0.0983	0.5430
120:10612	40:37	1 7 11 18 23 26 40 43 48 55 58 62	C_{2v}		0.5031	0.0997	0.6028
120:10613	40:36	1 7 11 18 23 26 40 46 48 53 56 61	C_2		0.4853	0.0972	0.5824
120:10665	40:38	1 7 11 23 26 29 33 43 47 52 58 60	D_2		0.4789	0.1013	0.5803
120:10666	40:39	1 7 11 23 26 29 33 43 49 55 59 62	D_{5d}		0.5207	0.0993	0.6201

3.5 Carbon cylinders

In addition to the leapfrogs, another series of closed-shell fullerene isomers has been identified. Its members have roughly cylindrical shapes in which a central barrel portion is capped by two approximate hemispheres, and they are constructed by a generalization of Coxeter's procedure for icosahedra (§ 2.3) to dihedral polyhedra. [15]

A Coxeter-type net can be generated for any fullerene with a fivefold symmetry axis by stretching the icosahedral net (Fig. 3.11). Two independent lattice vectors (i, j) and (k, l) are now needed to specify the ten equilateral and ten scalene triangles of the net, and the number of atoms in a fivefold symmetric fullerene (I_h, I, D_{5h}, D_{5d}, or D_5) can be shown using methods similar to those described for icosahedral and tetrahedral fullerenes in § 2.3 to be

$$n = 10 \left[k^2 + kl + l^2 + (il - jk) \right], \tag{3.10}$$

with certain restrictions on i, j, k, and l to prevent double counting. [15] Every multiple of ten apart from ten itself occurs in the series. As the net consists of five identical strips, when folded up it gives a cylindrical cluster with pentagons at the north and south poles, and expansion of the net by one more identical strip gives a cluster with a sixfold axis and hexagons at the poles. There is therefore a D_{6+} partner (D_{6h}, D_{6d}, D_6) to every fivefold D_{5+} isomer with

$$n = 12 \left[k^2 + kl + l^2 + (il - jk) \right] \tag{3.11}$$

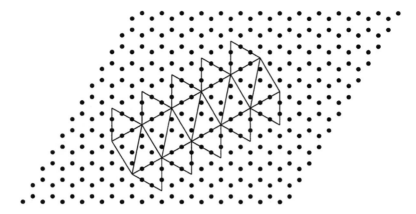

FIG. 3.11. The Coxeter-type net of a cylindrical fullerene with fivefold symmetry can be superimposed on a tessellation of the plane. Each black dot represents the centre of a triangular tile, and therefore the site of a carbon atom in the final fullerene when the net has been folded. Ten of the master triangles are still equilateral as in the icosahedral net, but ten are now scalene.

atoms. The correspondence is $D_{5h} \to D_{6h}$, I_h and $D_{5d} \to D_{6d}$, I and $D_5 \to D_6$. The construction can also be extended to 'fulleroids' with heptagons and even larger rings at the poles.[23]

The smallest examples of cylinder partners are the pair C_{20}:1 (I_h), C_{24}:1 (D_{6d}), and the smallest isolated-pentagon partners are their leapfrogs C_{60}:1 (I_h), C_{72}:1 (D_{6d}). The D_{5h} C_{70} structure with isolated pentagons is a cylinder formed by expansion of C_{60} with a belt of five hexagons (ten extra atoms, 15 extra bonds) at the equator. It has a closed-shell electronic configuration in Hückel theory, with 35 doubly occupied orbitals below an empty nonbonding LUMO (Fig. 3.12). The equivalent sixfold symmetric cluster is C_{84}:24 (D_{6h}) and this has the same general pattern of electronic structure.

Leapfrogs of cylindrical fullerenes are also cylindrical, and so the D_{5+} and D_{6+} series include an infinite number of leapfrog closed shells. Given that a leapfrog fullerene in this series has $60 + 6k$ atoms and also has three times the cylindrical count, leapfrog C_n isomers (with properly closed shells) occur at $n = 60 + 30m$ for D_{5+} and $n = 72 + 36m$ for D_{6+}, with $m = 0, 1, 2, \ldots$

Systematic generation of all the members of the cylinder series for $n <= 50p$ ($p = 5$ for D_{5+}, $p = 6$ for D_{6+}) shows that, apart from the leapfrog members of the series, closed shells appear at the 'magic' numbers $n = 2p(7 + 3m)$, $m = 0, 1, 2, \ldots$[32] Every member of this sequence has a nonbonding LUMO, and all

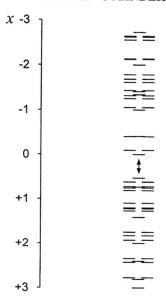

FIG. 3.12. π energy levels of C_{70} in the simple Hückel approximation. Each orbital has energy $\epsilon_k = \alpha + x_k\beta$. Positive x_k values correspond to bonding and negative x_k values to antibonding orbitals. In this model, one π orbital of C_{70} is exactly nonbonding. $\Delta = 0.5293|\beta|$ is the HOMO–LUMO gap.

are expansions of C_{60} $(p = 5)$ or C_{72} $(p = 6)$ (and therefore ultimately all are derived from C_{60} itself). Insertion of hexagons, one layer of p rings at a time, makes the cylindrical portion of the fullerene larger, leaving the caps unchanged but altering the degree of twist between them so that the point-group symmetry alternates $\ldots D_{ph} \rightarrow D_{pd} \rightarrow D_{ph} \rightarrow D_{pd}\ldots$ With every third added layer, and hence alternately with a prism or an antiprism point group, a closed-shell with a nonbonding LUMO is produced.

The presence of a nonbonding orbital (NBO) can be justified by the analogy between the fullerene π system and the model problem of an electron confined to the surface of a cylinder. When the number of layers in the cylindrical stack is just right, one solution is a standing wave with nodal planes on alternate layers and cancellation of bonds and antibonds within layers (Fig. 3.13); this is the NBO. More detailed symmetry arguments explain why there are precisely $n/2$ bonding orbitals below this NBO, which is therefore also the LUMO of the neutral cluster.[32] The same arguments apply to cylinders with fivefold or sixfold symmetry, and so the closed shells are paired (for example, $C_{70}{:}1(D_{5h})$ is

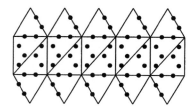 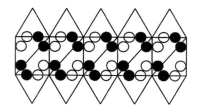

FIG. 3.13. The nonbonding π orbital of C_{70}. The net of C_{70} (left) is decorated with black and white circles (right) to denote positive and negative atomic orbital coefficients, respectively. (The corresponding atomic orbitals are approximately radially directed 2p orbitals.) All non-zero coefficients in this molecular orbital are of equal magnitude $(= 1/\sqrt{40})$.

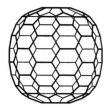

FIG. 3.14. The 'carbon cylinder' isomer of C_{120}. This D_{6d} cage has a closed shell with an empty nonbonding LUMO and is the second member of the closed-shell series that starts with C_{84}:24 (D_{6h}).

paired with C_{84}:24 (D_{6h})). The second member of the sixfold series is the closed-shell D_{6d} isolated-pentagon isomer of C_{120} (Fig. 3.14) for which a symmetrical (palindromic) spiral is

666666656565656565666666666666666666666666666656565656565656666666

with pentagons in positions 8, 10, 12, 14, 16, 18, 45, 47, 49, 51, 53, and 55. This isomer has a HOMO energy of $\alpha + 0.3501\beta$ and a nonbonding LUMO. It completes the set of 41 closed shells predicted by application of the two rules for C_{120}.

3.6 Sporadic closed shells

Both the leapfrog and cylinder rules are prescriptive rather than restrictive. Although they predict closed shells for all carbon cages that belong to certain specified geometric types, they leave open the possibility that other fullerenes

(a)　　　　　　　　　　(b)

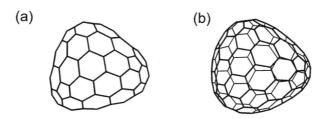

FIG. 3.15. (a) The first of the 'sporadic' closed shells is an isomer of C_{112} with the spiral code listed in Table 3.2. This fullerene has a plane of symmetry. (b) The first 'sporadic' closed shell with the same number of atoms as a leapfrog fullerene. C_{120}:627 has only C_1 symmetry.

might also have closed shells. Direct computer searches using the spiral algorithm show that no closed-shell cage outside the rules is to be found for any isomer in the range $20 <= n <= 100$. However, further searches of isolated-pentagon isomers with n in the range $100 < n <= 140$ show that there are occasional isolated-pentagon cages with closed shells that are neither leapfrogs nor cylinders. The first of these sporadic closed shells is at C_{112} (Fig. 3.15a). It has a HOMO energy of $\alpha + 0.3603\beta$ and a LUMO energy of $\alpha - 0.0041\beta$. Another 48 examples are found in the range $116 <= n <= 140$, and there may be more since the spiral algorithm search that yielded them was restricted to a pentagon start. The spirals, point group symmetries, neighbour indices (§ 4.4), and electronic characteristics of these clusters are listed in Table 3.2. All are characterized by very small LUMO bonding energies of the order of a few thousandths of the Hückel β parameter, and many have only the trivial C_1 symmetry.

C_{120} is the first leapfrog fullerene to have an 'extra' closed-shell isomer, and C_{120}:627 (Fig. 3.15b) completes the tally of 42 properly closed shells found at this nuclearity. A decreasing LUMO energy and a closing of the HOMO–LUMO gap is in general to be expected for the larger fullerenes as the number of hexagonal faces in the polyhedron increases and the graphite-sheet limit is approached. However, the sporadic closed shells are ahead of the trend. C_{120}:627 has a LUMO energy that is less than 0.5% of the average for the 40 leapfrog isomers and a HOMO–LUMO gap that is less than 70% of the average. By $n = 138$ the LUMO energy has fallen slightly, to an average of -0.0832β for the leapfrog isomers, but has risen to an average of -0.0113β for the four sporadic isomers. The sporadic closed shells are thus still outliers from the main trend. Both the carbon cylinders and the sporadic exceptional closed

Table 3.2. Closed shells outside the leapfrog and cylinder rules. Each cage is denoted by its sequence number in the full catalogue of isolated-pentagon fullerenes generated by the spiral algorithm for each C_n. The HOMO and LUMO energies and band gaps are given in β units as in Table 3.1.

Isomer	Ring spiral	G	HOMO	LUMO	Band gap
112:106	1 7 9 11 18 36 38 44 46 49 51 54	C_s	0.3603	0.0041	0.3645
116:2361	1 7 9 13 23 29 42 44 48 50 57 60	C_1	0.3567	0.0059	0.3626
120:627	1 7 9 11 24 37 42 44 51 55 57 59	C_1	0.3639	0.0004	0.3643
122:4037	1 7 9 13 20 36 42 48 52 54 57 59	C_1	0.3594	0.0054	0.3648
124:51	1 7 9 11 13 40 45 47 51 53 59 64	C_1	0.3496	0.0109	0.3604
124:843	1 7 9 11 24 37 43 45 51 54 60 64	C_1	0.3663	0.0113	0.3779
124:6766	1 7 9 15 32 37 40 47 52 58 60 64	C_1	0.3868	0.0042	0.3910
128:360	1 7 9 11 22 36 42 49 53 55 61 66	C_1	0.3490	0.0100	0.3590
128:1677	1 7 9 11 34 37 43 49 51 55 57 64	C_1	0.3673	0.0165	0.3837
128:8670	1 7 9 13 32 39 47 50 52 54 57 59	C_1	0.3630	0.0063	0.3493
130:1148	1 7 9 11 24 37 43 47 51 54 64 66	C_1	0.3672	0.0044	0.3717
130:1150	1 7 9 11 24 37 43 47 54 60 63 65	C_1	0.3742	0.0092	0.3834
130:1775	1 7 9 11 30 37 39 51 56 58 60 67	C_1	0.3591	0.0098	0.3689
132:2140	1 7 9 11 35 38 43 50 55 57 59 63	C_1	0.3649	0.0084	0.3733
132:2251	1 7 9 11 37 39 43 47 52 57 59 68	C_s	0.3559	0.0003	0.3562
132:2269	1 7 9 11 37 43 47 50 52 56 58 64	C_1	0.3701	0.0052	0.3753
132:6433	1 7 9 12 28 35 49 55 57 62 64 67	C_1	0.3467	0.0037	0.3504
134:411	1 7 9 11 18 38 46 50 55 57 60 68	C_s	0.3352	0.0004	0.3356
134:1323	1 7 9 11 24 37 51 55 57 59 64 67	C_2	0.3719	0.0016	0.3735
134:2022	1 7 9 11 25 42 47 50 54 56 61 68	C_1	0.3646	0.0013	0.3659
134:13844	1 7 9 13 30 35 49 51 54 57 66 69	C_1	0.3732	0.0061	0.3793
134:13982	1 7 9 13 30 37 47 50 56 62 65 69	C_1	0.3647	0.0022	0.3669
134:13997	1 7 9 13 30 37 50 56 58 61 63 66	C_s	0.3582	0.0193	0.3775
134:14553	1 7 9 13 32 39 47 50 54 59 66 69	C_1	0.3481	0.0008	0.3489
136:1536	1 7 9 11 24 37 43 54 61 66 68 70	C_1	0.3597	0.0161	0.3757
136:2475	1 7 9 11 26 38 43 47 50 65 67 70	C_1	0.3666	0.0072	0.3738
136:2476	1 7 9 11 26 38 43 47 60 63 66 68	C_1	0.3668	0.0074	0.3472
136:2701	1 7 9 11 34 37 39 51 56 58 65 67	C_1	0.3476	0.0021	0.3497
136:2950	1 7 9 11 34 39 49 54 57 61 67 69	C_1	0.3494	0.0024	0.3518
136:2960	1 7 9 11 34 39 52 54 57 59 61 64	C_1	0.3553	0.0047	0.3600
136:2961	1 7 9 11 34 39 52 54 59 61 67 70	C_1	0.3579	0.0204	0.3783
136:3051	1 7 9 11 34 42 49 52 55 61 66 69	C_1	0.3611	0.0001	0.3612
136:25831	1 7 9 25 33 38 40 49 57 61 66 68	D_{2d}	0.2282	0.0020	0.2301
138:62	1 7 9 11 13 40 47 53 60 66 68 71	C_1	0.3601	0.0051	0.3652
138:2486	1 7 9 11 25 43 48 51 55 62 67 69	C_1	0.3515	0.0065	0.3580

Table 3.2. *(Continued)*

Isomer	Ring spiral	G	HOMO	LUMO	Band gap
138:3563	1 7 9 11 37 39 45 52 58 60 67 71	C_s	0.3429	0.0246	0.3675
138:21483	1 7 9 13 36 38 44 46 56 60 63 66	C_s	0.3389	0.0091	0.3480
140:88	1 7 9 11 13 41 48 59 61 65 67 72	C_1	0.3198	0.0228	0.3426
140:3481	1 7 9 11 34 38 53 57 59 61 65 67	C_2	0.3599	0.0018	0.3617
140:4308	1 7 9 11 40 43 48 51 54 58 63 71	C_1	0.3714	0.0019	0.3732
140:20857	1 7 9 13 23 42 50 54 56 60 69 71	C_1	0.3396	0.0064	0.3460
140:20874	1 7 9 13 23 42 54 56 59 62 64 71	C_1	0.3516	0.0038	0.3554
140:21608	1 7 9 13 30 35 49 51 57 64 67 72	C_1	0.3574	0.0055	0.3629
140:21857	1 7 9 13 30 37 50 56 58 63 68 71	C_1	0.3519	0.0050	0.3569
140:22396	1 7 9 13 32 37 50 56 58 63 66 71	C_1	0.3480	0.0048	0.3527
140:22661	1 7 9 13 32 39 50 52 57 59 69 72	C_1	0.3557	0.0171	0.3728
140:22903	1 7 9 13 32 47 50 52 55 60 66 68	C_1	0.3478	0.0083	0.3561
140:24687	1 7 9 13 35 48 50 55 58 60 67 69	C_1	0.3382	0.0022	0.3404
140:25127	1 7 9 13 39 42 48 51 57 61 63 69	C_1	0.3741	0.0003	0.3745

shells share a precarious status: in either case, a shift of the LUMO by a small fraction of β would lead to reclassification of the cluster as pseudo-closed. In the limit as gaps decrease even further, the energy difference between low-lying electronic states will become very small, and the distinction between leapfrog and other closed shells, between properly and pseudo-closed shells, and even between open and closed shells will be blurred.

3.7 Conclusion

Within the Hückel model, the relationship between the geometric and electronic structures of fullerenes is basically simple. Isolation of pentagons appears to be necessary but not sufficient for a closed shell. High symmetry is not necessary, since properly closed shells can be found in all symmetry groups permitted to fullerenes. Two precise geometric requirements that guarantee a closed shell are that the cluster should be either the leapfrog of a smaller fullerene, or a particular type of carbon cylinder, in each case with a predictable atom count. These rules are exhaustive in the range $20 <= n <= 100$. If other systematic series of closed shells do exist, they do not have members in this range. For example, the series of closed-shell clusters of D_2 symmetry obtained by expansion of C_{60}

with stoichiometry C_n $(n = 84 + 4m)$[33] turns out to be a subset of the known leapfrog series. Closed shells outside the two major series do eventually start to appear, as discussed in § 3.6, but with very small LUMO energies and small HOMO–LUMO gaps, so that the corresponding cages are unlikely to differ significantly in electronic properties from pseudo-closed-shell isomers of the same nuclearity. Even so, the first two fullerenes to be synthesized in the laboratory were respectively the first leapfrog (C_{60}) and the first carbon cylinder (C_{70}).

When attempting to predict or rationalize the stabilities of the higher fullerenes, it is necessary to consider effects beyond the HMO model. These effects can be lumped together under the general heading of steric strain, and will be considered in detail in the next chapter.

References and notes

[1] Hückel molecular orbital theory is discussed in detail in a number of chemistry textbooks. See, for example: A. Streitwieser, Molecular orbital theory for organic chemists (Wiley, New York, 1961).

[2] P. W. Fowler, P. Lazzeretti, and R. Zanasi, Chem. Phys. Lett. 165 (1990) 79.

[3] W. Byers Brown, Chem. Phys. Lett. 136 (1987) 128.

[4] J. M. Hawkins, A. Meyer, T. A. Lewis, S. Loren, and F. J. Hollander, Science 252 (1991) 312.

[5] P. W. Fowler, D. J. Collins, and S. J. Austin, J. Chem. Soc. Perkin Trans. 2 (1993) 275.

[6] P. J. Fagan, J. C. Calabrese, and B. Malone, Acc. Chem. Res. 25 (1992) 134.

[7] P. W. Fowler, Phil. Mag. Lett. 66 (1992) 277.

[8] For a review of conductivity and superconductivity in fulleride compounds see: R. C. Haddon, Acc. Chem. Res. 25 (1992) 127.

[9] D. E. Manolopoulos, J. C. May, and S. E. Down, Chem. Phys. Letters 181 (1991) 105.

[10] K. Wade, Electron deficient compounds (Nelson, London, 1971).

[11] D. M. P. Mingos and D. J. Wales, Introduction to cluster chemistry (Prentice-Hall, Englewood Cliffs, New Jersey, 1990).

[12] The dodecahedral C_{20} cage is an exception, with 19 bonding orbitals below a partly filled fivefold degenerate nonbonding HOMO, but this would appear from the data in Tables A.21 and A.22 to be the only C_n fullerene in the range with fewer than $n/2$ bonding orbitals. However, some very large tetrahedral fullerenes ($n > 600$) with open-shell configurations have recently been found to have marginally antibonding HOMOs (P. W. Fowler, J. Chem. Soc. Faraday Trans. 93 (1997) 1). Closed-shell

anions of these, or neutral fullerenes with filled antibonding HOMOs would be *meta*-closed in the terminology of ref. 30.

[13] R. G. Pearson, Inorg. Chim. Acta 198 (1992) 781.

[14] H. S. M. Coxeter, Virus macromolecules and geodesic Domes, in: A spectrum of mathematics, J.C. Butcher, ed. (Oxford University Press/Auckland University Press, Oxford/Auckland, 1971), p. 98.

[15] P. W. Fowler, J. E. Cremona, and J. I. Steer, Theor. Chim. Acta 73 (1988) 1.

[16] P. W. Fowler, Chem. Phys. Lett. 131 (1986) 444.

[17] M. Goldberg, Tôhoku Math. J. 43 (1937) 104.

[18] P. W. Fowler and J. I. Steer, J. Chem. Soc. Chem. Comm., (1987) 1403.

[19] An alternative version of the leapfrog transformation, in which the dual of the original fullerene is formed and then truncated on all vertices, can also be useful for some purposes. The intermediate in this process is different, but the end result is the same – a leapfrog fullerene with three times as many vertices. See: R. L. Johnston, J. Chem. Soc. Faraday Trans. 87 (1991) 3353.

[20] P. W. Fowler and D. B. Redmond, Theor. Chim. Acta 83 (1992) 367.

[21] P. W. Fowler, J. Chem. Soc. Perkin Trans. 2 (1992) 145.

[22] D. E. Manolopoulos, D. R. Woodall, and P. W. Fowler, J. Chem. Soc. Faraday Trans. 88 (1992) 2427.

[23] P. W. Fowler and V. Morvan, J. Chem. Soc. Faraday Trans. 88 (1992) 2631.

[24] A. Ceulemans and P. W. Fowler, Nature 353 (1991) 52.

[25] R. Taylor, J. Chem. Soc. Perkin Trans. 2 (1992) 3.

[26] The work of Fries is discussed in: M. Randić, J. Chem. Phys. 34 (1961) 693.

[27] E. Clar, Polycyclic hydrocarbons (Wiley, New York, 1964).

[28] E. Clar, The aromatic sextet (Wiley, New York, 1972).

[29] J. R. Dias, Chem. Phys. Lett. 204 (1993) 486.

[30] P. W. Fowler and T. Pisanski, J. Chem. Soc. Faraday Trans. 90 (1994) 2865.

[31] P. W. Fowler, S. J. Austin, and D. E. Manolopoulos, Competing factors in fullerene stability, in: Chemistry and physics of the fullerenes, K. Prassides, ed. (Kluwer Academic Press, Dordrecht, 1994)

[32] P. W. Fowler, J. Chem. Soc. Faraday Trans. 86 (1990) 1991.

[33] P. Labastie, R. L. Whetten, H.-P. Cheng, and K. Holczer, unpublished work.

4

STERIC STRAIN

The simple Hückel molecular orbital model discussed in the last chapter is known to give reliable predictions about the relative stabilities of planar aromatic hydrocarbons when it is combined with standard C–C and C–H σ bond energies.[1] However, it is not generally found to work so well for fullerenes. One apparent reason for this is that the curvature of a fullerene surface destroys the separability of σ and π orbitals upon which Hückel theory is based, making the theory less well justified for fullerenes than it is for planar hydrocarbons. Although the separation may be approximately maintained in C_{60} itself (§ 3.1), it can be lost entirely in less symmetrical cases. Moreover, the consequences of this breakdown in σ–π separability can vary dramatically from one fullerene isomer to another, because different fullerene isomers can have surfaces with very different local curvatures. Thus the energetic penalty associated with curvature can be an important factor in fullerene stability.

This chapter investigates the consequences of fullerene curvature in an elementary way which allows us to discuss the relative stabilities of different fullerene isomers without resorting to detailed electronic structure calculations. The theoretical basis of this investigation is discussed in §§ 4.1 and 4.2, where it is argued that the strain energy due to the curvature of a fullerene surface can be assessed on purely topological grounds by considering the arrangement of its 12 pentagons. In particular, it is argued that the strain energy will be minimized when the pentagons are as far apart as possible, in accordance with the celebrated isolated-pentagon rule.[2,3] § 4.3 proceeds along these lines by defining pentagon neighbour indices for fullerene isomers and using them to examine the implications of steric strain for each of the lower fullerenes $C_{n<=70}$.

§ 4.4 discusses the case of higher fullerenes $C_{n>70}$, for which hexagon neighbour indices are more appropriate.[4] These indices are used in § 4.5 to give an analysis of curvature-induced strain in leapfrog and carbon cylinder fullerenes which reveals that π electronic stability and steric strain are generally in opposition beyond the isolated-pentagon rule. Indeed the uniquely stable C_{60} and C_{70} structures are shown to be the only fullerenes with fewer than 180 atoms that are optimal with respect to both criteria.[5] Finally, § 4.6 discusses the competition between π electronic stability and steric strain in some of the (still comparatively few) higher fullerenes that have been characterized in experiments to date.

4.1 Steric strain and rehybridization

Although simple Hückel theory alone is too crude to give reliable predictions about fullerene stability, it is convenient to use it as a starting point. The stability of fullerenes can then be attributed to a balance between π electronic stability and *steric strain*, with the latter encompassing all the effects in the electronic structure of fullerenes that are neglected at the Hückel level.

There are of course many such effects, involving both the σ and the π electrons. In order to discuss them qualitatively it is convenient to assume that the valence orbitals of each carbon atom in a fullerene are hybridized in such a way that the resulting three σ orbitals point directly along the bonds to neighbouring atoms, as in the π orbital axis vector (POAV2) scheme of Haddon.[6] The direction of the remaining π orbital, and the s and p components of all four hybrid orbitals, are then uniquely determined by orthogonality contraints which ensure that local σ–π orthogonality is maintained within each atomic valence shell.[6]

The assumption that the σ orbitals point directly along the bonds is intuitively appealing, and Haddon has shown that it holds well even for quite large torsional distortions in *ab initio* calculations on twisted ethylene.[7] One limitation is that the association of bond-directed σ orbitals with sp^m hybrids breaks down for bond angles less than $90°$, because it is impossible to construct sp^m hybrid orbitals that are more closely aligned than the pure p orbitals themselves.[8] This mathematical breakdown has led Haddon to argue that the assumption of sp^m hybrid orbitals pointing along the bonds may also be unreliable in practice for bond angles less than $100°$. However, such small bond angles are unlikely to be found in the fullerenes.

In the case of an idealized planar carbon atom with $120°$ bond angles, perfect sp^2 hybridization is retained and no rehybridization is necessary. However, since 60 of the bond angles in a fullerene are closer to $108°$ than $120°$, with five bond

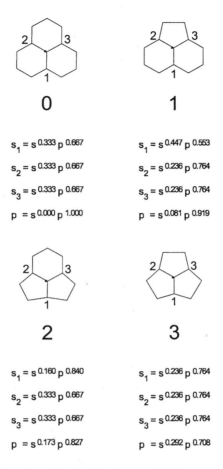

FIG. 4.1. Idealized fractional hybridizations of fullerene carbon atoms. The fractional s and p components of the σ and π hybrid orbitals on the central atoms have been calculated assuming bond-directed σ orbitals and idealized polygonal bond angles. Rehybridization away from sp^2 (i.e., $\pi = s^{0.000}p^{1.000}$) increases in the order **0→3**.

angles of approximately 108° in each of the 12 pentagons, the rehybridization in fullerenes is substantial. Figure 4.1 shows the expected hybridizations of fullerene carbon atoms in four different local environments, assuming regular hexagonal and pentagonal bond angles in each case. It is clear from this figure that the rehybridization away from the sp^2 ideal is least for atoms surrounded by hexagons and greatest for atoms surrounded by pentagons, and increases in the order **0→3**.

The natural temptation at this point is to attribute the steric strain in

fullerenes *directly* to rehybridization, and to argue that the least strained isomers of a given fullerene C_n will be those in which the total rehybridization is
minimized. While this does give results in agreement with more sophisticated
models, it is not entirely satisfactory. The valence orbital barycentre of each
carbon atom is actually preserved on rehybridization, with the three σ orbitals
moving up in energy and the π orbital moving down to compensate as the p
character of the σ orbitals increases. Therefore, assuming that each of the four
hybrid orbitals is singly occupied in the hypothetical 'valence state' that exists
before (or at least logically prior to) bond formation,[8] the rehybridization itself
is energetically neutral.

Whatever strain energy accompanies the rehybridization must therefore orig-
inate at the bonding stage, in which the σ hybrid orbitals on neighbouring
carbon atoms combine to give notionally localized σ bonds and the π orbitals
combine to give delocalized π bonds that sandwich the σ bonding framework.
Several possible sources of strain energy can be identified at this stage, involving both the σ and the π electrons. The most obvious, and arguably the most
important, is the reduced π orbital overlap and therefore weaker π bonding that
results from the imperfect alignment of neighbouring π hybrid orbitals. This effect was discussed by Schmalz et al.,[3] who termed it 'π-interaction diminution',
and Haddon describes it as 'steric inhibition of resonance'.[9] Since the effect is
not incorporated in simple Hückel theory, it comes under our general heading
of steric strain.

The qualitative features of π orbital misalignment are well illustrated by
two simple limiting examples. In the case of a planar sheet of graphite the p_π
orbitals on neighbouring carbon atoms are perfectly aligned by symmetry, as
shown in Fig. 4.2 (a). However, in the case of the C_{20} fullerene structure, the
π orbitals are approximately sp^3 hybridized and point outwards from the vertices of a dodecahedron, as shown in Fig. 4.2 (b). These hybrid orbitals clearly
overlap less than in the planar case, resulting in weaker π bonding and concomitant steric strain. An explicit calculation with standard Slater hybrid orbitals
at the graphite bond length of 1.42Å shows that the overlap integral between
neighbouring π hybrid orbitals in C_{20} is only 0.676 times the corresponding integral between neighbouring p_π orbitals in graphite, despite the perfect torsional
alignment of the C_{20} π orbitals that results from C_{3v} atomic site symmetry.
Since the overlap integral between bonded orbitals is often used as a measure of

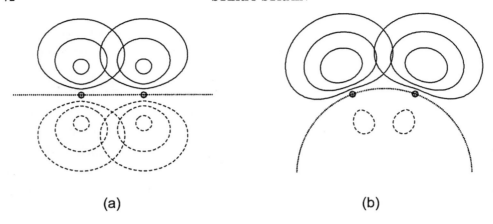

FIG. 4.2. Neighbouring POAV2 π hybrid orbitals in (a) graphite and (b) C_{20}. The hybrids are constructed from carbon 2s and 2p Slater orbitals with the same exponent, and the positive (solid line) and negative (dashed line) contours correspond to the same numerical values in both cases.

bond strength in qualitative molecular orbital theory, this is a clear indication of the importance of the π orbital misalignment effect.

The rehybridization away from sp^2 is clearly maximized among the fullerenes by C_{20}, where all the σ bond angles are fixed by symmetry at exactly 108°. Other fullerenes have more diverse carbon atom environments with varying degrees of rehybridization, and will thus exhibit intermediate degrees of π orbital misalignment between the graphite and C_{20} extremes. This therefore establishes at least one *indirect* way in which steric strain can be attributed to rehybridization.

Within the assumption of bond-directed σ hybrid orbitals, the directional properties of the σ orbitals are always perfect for bonding and there is no analogue of π orbital misalignment in the σ framework. Additional effects which come under our general heading of steric strain can nevertheless be identified for both σ and π electrons. For example, it is known from elementary molecular orbital theory that any difference in the energy of the two orbitals that combine to form a bond also reduces the overall strength of the bond. [8] In the planar graphite limit the σ bonds are composed solely of carbon sp^2 hybrid orbitals and the π bonds of pure carbon p orbitals, and no such difference arises. In contrast, energetic mismatch can occur in the fullerenes in both σ and π systems as a result of the different hybridizations of neighbouring atoms. Since this effect is again ignored in simple Hückel theory it is another possible source of steric strain. In this case the strain originates from *differences* in rehybridization between neighbouring sites and will be minimized when the local environments of neighbouring atoms are as similar as possible.

Many other effects in the electronic structure of fullerenes are also ignored at the simple Hückel level, but these effects, such as variations in orbital overlap with bond length and variations in electron repulsion at different sites, are also present in planar aromatic hydrocarbons where Hückel theory is comparatively reliable.[1] The two effects considered above – the directional misalignment of neighbouring π hybrid orbitals and the energetic mismatch of neighbouring σ and π orbitals – are additional consequences of the curvature of fullerene surfaces and the concomitant rehybridization away from sp^2. These are therefore likely to be among the effects which dominate steric strain in the fullerenes, with the directional misalignment probably being the more important of the two.

Some indication that this is indeed the case is provided by the calculations of Bakowies *et al.*,[10,11] who have shown that the energy ordering of a number of fullerene isomers within the semi-empirical modified-neglect-of-differential-overlap (MNDO) model of electronic structure is explained by simple Hückel theory once the resonance integrals of each isomer are scaled to account for reduced π orbital overlap at the optimized MNDO geometry.

4.2 The isolated-pentagon rule

The single most important consequence of steric strain in the fullerenes is the isolated-pentagon rule (IPR), which says that the most stable fullerenes are those in which all the pentagons are isolated. As mentioned in § 1.1, this rule was first proposed by Kroto in 1987, in a paper on the C_{24}, C_{28}, C_{32}, C_{36}, C_{50}, C_{60}, and C_{70} magic-number ion signals in early graphite laser vaporization experiments.[2] A more theoretical discussion of the rule was given soon afterwards by Schmalz *et al.* as part of their comprehensive 1988 paper in support of the fullerene hypothesis.[3] Since then the IPR has stood the test of time, and has proved particularly valuable in helping to unravel the experimental structures of the higher fullerenes $C_{n>70}$.[12,13,14]

The IPR can be justified very easily using the rehybridization ideas of § 4.1. In particular, it has been argued on the basis of the graphite and C_{20} limits that the dominant π orbital misalignment contribution to the strain energy will increase with increasing local rehybridization away from the sp^2 ideal. The strain energy will therefore increase in the order **1→3** in the last three fragments shown in Fig. 4.1, and since these fragments contain respectively 0, 1, and 3 fused pentagon pairs the IPR follows automatically. Indeed the same argument suggests more generally that when the IPR ideal cannot be realized, the least strained isomers of a given fullerene C_n will be those containing the smallest number of fused pentagon pairs. The implications of this more general pentagon isolation criterion will be examined in § 4.3.

This justification for the IPR is based on steric strain. However, this is not the only factor in fullerene stability according to the definition in § 4.1. The π electronic resonance energies of different fullerene isomers can vary, and these variations should also be taken into account when deriving a topological rule. It is therefore especially significant that π electronic and steric strain effects both support the isolation of pentagons.[3,14] This follows because two fused pentagons have a cycle of length eight around their periphery which can be assumed, on the basis of the Hückel $4n$ rule, to have a net destabilizing (antiaromatic) effect on the overall π electronic structure. Moreover, the more general pentagon isolation criterion mentioned above is also supported by this π electronic argument, because the last three fragments shown in Fig. 4.1 contain respectively 0, 1/2, and 1 antiaromatic eight-cycles per pentagon in the order **1→3**.

This discussion has inevitably been rather qualitative, and it would be useful to have a more quantitative indication of the energy penalty incurred on bringing two pentagons together in a fullerene surface. Such an estimate is provided by the calculations of Zhang *et al.*,[15] who have used a tight-binding Hamiltonian fit to the results of more sophisticated local density functional calculations to investigate the effect of a Stone–Wales pyracylene transformation on C_{60}. As discussed in § 6.5, the Stone–Wales transformation converts the I_h isomer of C_{60} into a C_{2v} isomer with two fused pentagon pairs. The energy difference between the two isomers is calculated by Zhang *et al.* to be 1.4 eV, giving a very significant energy penalty of 0.7 eV (or 68 kJ/mol) for each of the fused pentagon pairs. If this energy penalty is the same order of magnitude in other circumstances, the validity of the IPR can be in little doubt. Calculations of the relative stabilities of all 40 fullerene isomers of C_{40}[16] using the semi-empirical QCFF/PI model[17,18] give an average penalty of 0.9 eV (or 87 kJ/mol) per pentagon pair.

4.3 Pentagon indices for lower fullerenes

The implications of steric strain in the lower fullerenes can be explored in more detail by introducing a set of topological indices for each fullerene isomer which summarizes how the pentagons are arranged on its surface. One such set of indices has been proposed by Raghavachari,[4] who defines the *neighbour index* of each pentagon in a fullerene as the number of hexagons to which it is adjacent. We shall modify this definition here so that the neighbour index of each pentagon is defined as the number of other pentagons to which it is adjacent. The aim of the present section is to show how these pentagon neighbour indices can be combined with the pentagon isolation criterion to select the most likely experimental structures for each of the lower fullerenes $C_{n<=70}$.

The arrangement of the pentagons in a fullerene can be characterized within the neighbour index scheme by a signature of the form $(p_0, p_1, p_2, p_3, p_4, p_5)$, where p_k denotes the number of pentagons with neighbour index k. For example, all 12 pentagons in the I_h C_{20} fullerene have neighbour index 5, corresponding to the signature $(0, 0, 0, 0, 0, 12)$, whereas all 12 pentagons in I_h C_{60} have neighbour index 0, corresponding to the signature $(12, 0, 0, 0, 0, 0)$. More generally, the pentagons in a fullerene surface can be in different local environments. For example, the pentagon neighbour index signature of the unique D_{3h} C_{26} fullerene is $(0, 0, 0, 6, 6, 0)$. (The sum of the six entries in the signature is always 12, which is simply the total number of pentagons.) The pentagon neighbour index signatures of all C_n fullerene isomers in the range $n = 20$ to 50 are given in Tables A.1 to A.9.

One useful property of these signatures is that they give the total number of fused pentagon pairs in an isomer as

$$N_p = \sum_{k=1}^{5} k\, p_k / 2. \tag{4.1}$$

As discussed above, each fused pentagon pair can be associated with a significant strain energy. This immediately suggests the number N_p as a qualitative filter for selecting stable isomers, with the least strained isomers of a given fullerene C_n being those with the minimum value of N_p. In the case of the higher fullerenes $C_{n>70}$, this filter reduces to the standard IPR, with the least strained C_n isomers all having isolated pentagons and realizing the $N_p = 0$ ideal. However, for the lower fullerenes $C_{n<=70}$, the filter becomes the more general pentagon isolation criterion described in § 4.2.

Table 4.1. Pentagon neighbour index signatures of the least strained lower fullerene isomers between C_{20} and C_{70}. The point groups in this table are idealized, as in the main isomer tables, and so are higher than the actual symmetry in cases where Jahn–Teller distortion occurs. The calculations by Zhang *et al.*[15] include these distortions.

Isomer	Ring spiral												Point group	Pentagon indices	N_p	$(\star)$
20:1	1 2 3	4	5	6	7	8	9 10 11 12						I_h	0 0 0 0 0 12	30	$\star$
24:1	1 2 3	4	5	7	8 10 11 12 13 14								D_{6d}	0 0 0 0 12 0	24	$\star$
26:1	1 2 3	4	5	7	9 11 12 13 14 15								D_{3h}	0 0 0 6 6 0	21	$\star$
28:2	1 2 3	5	7	9 10 11 12 13 14 15									T_d	0 0 0 12 0 0	18	$\star$
30:3	1 2 3	4	7 10 11 12 13 14 15 16										C_{2v}	0 0 2 10 0 0	17	$\star$
32:6	1 2 3	5	7	9 10 12 14 16 17 18									D_3	0 0 6 6 0 0	15	$\star$
34:5	1 2 3	5	7 10 11 13 15 16 18 19										C_2	0 0 8 4 0 0	14	$\star$
36:14	1 2 4	7	9 10 12 13 14 16 18 20										D_{2d}	0 0 12 0 0 0	12	$\star$
36:15	1 2 4	8	9 10 12 13 14 15 18 20										D_{6h}	0 0 12 0 0 0	12	
38:17	1 2 4	7	9 10 12 13 15 18 20 21										C_2	0 2 10 0 0 0	11	$\star$
40:38	1 2 4	7	9 11 13 15 17 19 20 22										D_2	0 4 8 0 0 0	10	$\star$
40:39	1 2 4	7	9 11 13 15 18 19 21 22										D_{5d}	2 0 10 0 0 0	10	
42:45	1 2 4	7	9 12 13 16 18 19 21 22										D_3	0 6 6 0 0 0	9	$\star$
44:75	1 2 4	7	9 12 13 16 18 21 23 24										D_2	0 8 4 0 0 0	8	$\star$
44:89	1 2 4	9 12 13 14 16 17 19 22 24											D_2	0 8 4 0 0 0	8	
46:99	1 2 4	7	9 11 13 19 21 22 23 24										C_s	2 4 6 0 0 0	8	
46:103	1 2 4	7	9 12 13 18 21 22 23 24										C_1	1 6 5 0 0 0	8	
46:107	1 2 4	7	9 12 15 18 19 21 23 25										C_s	1 6 5 0 0 0	8	
46:108	1 2 4	7	9 12 15 18 20 21 23 24										C_s	1 6 5 0 0 0	8	
46:109	1 2 4	7	9 12 16 18 20 21 22 24										C_2	0 8 4 0 0 0	8	$\star$
46:114	1 2 4	8 12 13 15 17 18 19 22 25											C_1	0 8 4 0 0 0	8	
46:116	1 2 4	9 12 13 15 16 18 20 22 24											C_2	0 8 4 0 0 0	8	
48:171	1 2 4	7	9 13 18 19 20 22 23 24										C_2	2 6 4 0 0 0	7	
48:196	1 2 4	9 12 13 15 16 19 22 24 25											C_1	1 8 3 0 0 0	7	
48:197	1 2 4	9 12 13 15 17 19 22 24 26											C_s	1 8 3 0 0 0	7	
48:199	1 2 4	9 12 14 15 17 19 21 23 25											C_2	0 10 2 0 0 0	7	$\star$
50:271	1 2 9 10 12 14 15 17 20 22 24 26												D_{5h}	2 10 0 0 0 0	5	$\star$
52:422	1 2 4	9 12 14 16 19 21 23 25 28											C_2	4 6 2 0 0 0	5	$\star$
54:540	1 2 4	9 12 14 16 19 23 25 27 28											C_{2v}	6 4 2 0 0 0	4	$\star$
56:843	1 2 4	9 12 16 19 22 23 24 26 28											C_2	6 4 2 0 0 0	4	
56:864	1 2 4	9 13 16 19 21 23 24 26 28											C_s	5 6 1 0 0 0	4	
56:913	1 2 9 10 12 14 16 20 22 24 26 30												C_{2v}	4 8 0 0 0 0	4	
56:916	1 2 9 10 12 14 17 22 24 25 27 29												D_2	4 8 0 0 0 0	4	$\star$
58:1205	1 2 9 11 13 15 17 20 23 25 27 29												C_{3v}	6 6 0 0 0 0	3	$\star$

Table 4.1. *(Continued)*

Isomer	Ring spiral	Point group	Pentagon indices	N_p	$(\star)$
60:1812	1 7 9 11 13 15 18 20 22 24 26 32	I_h	12 0 0 0 0 0	0	$\star$
62:2194	1 2 4 12 15 17 20 23 26 28 30 32	C_1	7 4 1 0 0 0	3	
62:2377	1 2 9 11 13 15 17 21 25 27 30 33	C_1	6 6 0 0 0 0	3	
62:2378	1 2 9 11 13 15 17 21 25 28 30 32	C_2	6 6 0 0 0 0	3	$\star$
64:3451	1 2 9 12 14 17 20 23 25 28 30 33	D_2	8 4 0 0 0 0	2	$\star$
64:3452	1 2 9 12 14 17 20 23 26 28 30 32	C_s	8 4 0 0 0 0	2	
64:2378	1 2 9 12 14 18 21 23 25 27 29 33	C_2	8 4 0 0 0 0	2	
66:4169	1 2 4 13 17 19 22 24 26 28 30 33	C_s	9 2 1 0 0 0	2	
66:4348	1 2 9 10 13 16 21 24 27 29 31 33	C_{2v}	8 4 0 0 0 0	2	
66:4466	1 2 9 11 15 17 22 24 26 29 30 32	C_2	8 4 0 0 0 0	2	$\star$
68:6073	1 2 9 11 13 15 17 28 30 32 34 36	C_{2v}	8 4 0 0 0 0	2	
68:6094	1 2 9 11 13 16 20 24 27 31 33 35	C_s	8 4 0 0 0 0	2	
68:6146	1 2 9 11 14 17 22 25 28 30 32 34	C_2	8 4 0 0 0 0	2	
68:6148	1 2 9 11 14 17 22 26 28 30 32 33	C_1	8 4 0 0 0 0	2	
68:6149	1 2 9 11 14 17 22 26 28 31 33 36	C_2	8 4 0 0 0 0	2	
68:6195	1 2 9 11 15 17 22 24 28 30 32 34	C_2	8 4 0 0 0 0	2	
68:6198	1 2 9 11 15 18 22 24 26 30 32 35	C_1	8 4 0 0 0 0	2	
68:6269	1 2 9 12 14 17 21 23 26 28 35 36	D_2	8 4 0 0 0 0	2	
68:6270	1 2 9 12 14 17 21 23 26 29 34 36	C_1	8 4 0 0 0 0	2	
68:6290	1 2 9 12 14 20 23 25 27 29 31 35	C_2	8 4 0 0 0 0	2	
68:6328	1 2 9 13 17 20 22 25 26 29 31 33	C_2	8 4 0 0 0 0	2	$\star$
70:8149	1 7 9 11 13 15 27 29 31 33 35 37	D_{5h}	12 0 0 0 0 0	0	$\star$

Table 4.1 shows the pentagon neighbour index signatures of all the sterically optimal lower fullerene isomers that are selected by this pentagon isolation criterion. The isomer entries in this table are labelled as $n{:}m$, where the m means that the isomer is the mth generated by the unrestricted spiral algorithm for the given value of n. The same labels are used in Tables A.1 to A.9, where all C_n fullerene isomers with $n <= 50$ are illustrated. The optimal isomers with $n = 60$ and 70 are the familiar C_{60} and C_{70} structures, and the remaining isomers with n between 52 and 68 in Table 4.1 are illustrated in Fig. 4.3.

The first thing to notice about Table 4.1 is that it contains only a minute fraction of the available C_n isomers in the range $n = 20$ to 70. The total number of fullerene isomers found by the spiral algorithm in this range is 30 579, and yet the table contains only 56 isomer entries. The pentagon isolation criterion

FIG. 4.3. Sterically optimum C_n fullerene cages in the range $n = 52$ to 68. The labels $n{:}m$ are the same as in Table 4.1, and the optimum C_{60} cage has the familiar icosahedral structure (not shown here).

therefore provides an extremely effective topological filter throughout the lower fullerene range. As can be seen from the table, the criterion selects a single optimal isomer for each of the lower fullerenes C_{20}, C_{24}, C_{26}, C_{28}, C_{30}, C_{32}, C_{34}, C_{38}, C_{42}, C_{50}, C_{52}, C_{54}, C_{58}, C_{60}, and C_{70}. Moreover, the maximum number of lower fullerene isomers that survive the filter is 11 for C_{68}. Although it would obviously not be feasible to perform detailed electronic structure calculations on all 6 332 C_{68} fullerene isomers, these 11 candidate structures could be studied comprehensively.

It can also be seen that one isomer entry in Table 4.1 has been starred for each given value of n. These starred isomers are the minimum energy C_n structures found by Zhang et al. in their simulated annealing study of the lower fullerenes.[15] As mentioned in § 4.2, the tight-binding Hamiltonian used by these authors reproduces the results of more sophisticated local density functional calculations on C_{60}. It should therefore give a realistic description of the relative stabilities of the lower fullerenes, and the fact that it agrees with Table 4.1 in every case provides convincing support for the pentagon isolation criterion.

This criterion has been based so far on the total number of fused pentagon pairs N_p in Eq. (4.1), which is clearly only a rather averaged property of the pentagon neighbour index signature. However, the starred minimum energy isomers in Table 4.1 suggest an even more specific steric rule for lower fullerenes. In particular, whenever two or more C_n isomers have the same value of N_p but different pentagon neighbour index signatures, the starred minimum energy isomer is invariably one in which the *maximum* pentagon neighbour index is *minimized*. This agrees with the expectation of increasing steric strain in the order $1 \rightarrow 3$ in the last three fragments in Fig. 4.1, and suggests further that the strain energy increases *more* rapidly than the number of fused pentagon pairs in each fragment. It also shows that the pentagon neighbour index signatures are more useful than N_p alone.

One further possibility for reduction of strain should be mentioned here. The generalized pentagon isolation criterion predicts the best structures within the fullerene class. It has been pointed out [19] that it is possible to reduce the number of pentagon adjacencies still further, by introducing four-membered rings into the cage structure; a trivalent polyhedron with s, p, and h rings of size 4, 5, and 6 would have (from Euler's theorem (Eq. (2.1))) $p = 12 - 2s$ pentagons and $h = n/2 - 10 + s$ hexagons where s can take values from 0 (the fullerene limit) to 6 (a cage without pentagonal faces). Not all combinations of n and s are possible, but in general, the introduction of a square (quadrilateral) face into a fullerene eliminates two pentagons and adds a new hexagon. It is conceivable that the penalty associated with the square face would be outweighed by the benefits of the reduction in N_p, despite the severe rehybridization implied by the idealized $90°$ bond angles in the square (§ 4.1). Explicit calculations on a small number of cages with one or two square faces show some of these 'generalized fullerenes' to be of lower total energy than some conventional fullerene isomers.[19] However, systematic comparisons of the *best* isomers of each type have yet to be carried out.

4.4 Hexagon indices for higher fullerenes

The pentagon isolation criterion reduces to the standard IPR for the higher fullerenes $C_{n>70}$, each of which admits at least one isomer with isolated pentagons according to Table 2.2. The problem of steric strain for higher fullerenes might therefore appear to be quite straightforward, with the least strained isomers of a given higher fullerene C_n being those with isolated pentagons. However, the number of C_n isomers with isolated pentagons increases very rapidly beyond $n = 70$, and soon becomes practically unmanageable. For instance, there are already 450 different isolated-pentagon isomers of C_{100}; it would clearly be desirable to have some more specific rule for steric strain in higher fullerenes in order to eliminate some of these possibilities.

Such a rule has been proposed by Raghavachari,[4] who argues that steric strain will be minimized beyond the IPR when the pentagon-induced curvature is distributed as *uniformly* as possible over the fullerene surface. This even distribution of curvature is expected to minimize the contribution from orbital energy mismatch to the strain energy (§ 4.1), and it may also reduce the contribution from π orbital misalignment when the bond angles have relaxed from their idealized values. However, the assumption of idealized polygonal bond angles in Fig. 4.1 is too crude to differentiate between isolated-pentagon C_n isomers, all of which have exactly $n - 60$ atoms in sites of type 0, and 60 atoms in sites of type 1. Variations in steric strain beyond the first-order IPR will therefore be comparatively subtle, and will be associated with smaller energetic penalties.

In order to quantify the uniform curvature rule, Raghavachari defines the neighbour index of each *hexagon* in a fullerene as the number of other hexagons to which it is adjacent.[4] These new indices are necessary because the pentagon neighbour indices used for the lower fullerenes become redundant for fullerenes obeying the IPR, all isolated-pentagon C_n isomers having the pentagon signature $(12, 0, 0, 0, 0, 0)$ and realizing the $N_p = 0$ ideal. The hexagon neighbour indices can, however, vary for different isolated-pentagon fullerenes, and these variations allow one to analyse the implications of steric strain in the higher fullerenes.

The analysis of hexagon neighbour indices is especially straightforward for fullerenes with isolated pentagons. Every fullerene isomer can be characterized by a signature of the form $(h_0, h_1, h_2, h_3, h_4, h_5, h_6)$, where h_k is the number of hexagons with neighbour index k. However, in an isolated-pentagon fullerene, every hexagon is adjacent to a minimum of three others. We can therefore dispense with h_0, h_1, and h_2, all three of which are zero, and write the hexagon neighbour index signature of an isolated-pentagon fullerene as (h_3, h_4, h_5, h_6). These abbreviated signatures are given for all isolated-pentagon isomers of the higher fullerenes up to C_{100} in Tables A.10 to A.20.

As in the case of the pentagon neighbour index signature, the sum of the entries in the hexagon neighbour index signature is simply the total number of hexagons ($n/2 - 10$ in a fullerene C_n). Therefore, for an isolated-pentagon fullerene, we have

$$h_3 + h_4 + h_5 + h_6 = n/2 - 10. \qquad (4.2)$$

Furthermore, since these $n/2 - 10$ hexagons have $3n - 60$ edges, of which 60 are shared with pentagons in fullerenes with isolated pentagons, we have

$$3h_3 + 4h_4 + 5h_5 + 6h_6 = 3n - 120. \qquad (4.3)$$

Eliminating n from the two equations leaves

$$3h_3 + 2h_4 + h_5 = 60, \qquad (4.4)$$

which simply counts the number of edges that are shared between pentagons and hexagons.

All isolated-pentagon C_n fullerenes have hexagon neighbour index signatures which satisfy these equations, but what are the *optimum* solutions? In order to minimize steric strain, according to Raghavachari's argument, the neighbour indices of all the hexagons in the structure are to be as similar to one another as possible. The optimum solutions are therefore those in which all the hexagon neighbour indices are equal. This is possible only for three very special values of n because of the constraints in the above equations. These three optimum solutions follow immediately from Eq. (4.4):

$$(h_3, h_4, h_5, h_6) = \begin{cases} (20, 0, 0, 0), & \text{if } n = 60; \\ (0, 30, 0, 0), & \text{if } n = 80; \\ (0, 0, 60, 0), & \text{if } n = 140. \end{cases} \tag{4.5}$$

The above equations thus allow optimum solutions in which all the hexagons have neighbour index 3, 4, or 5, but there is no optimum solution in which all the hexagons have neighbour index 6. This also follows immediately from Eq. (4.4), and simply reflects the fact that some of the hexagons in a fullerene must be adjacent to pentagons.

The three special values of n in Eq. (4.5) are not simply accidental: they correspond to the first three cases in which it is possible to realize maximum (icosahedral) fullerene symmetry within the isolated-pentagon constraint. For example, C_{60} is the smallest I_h fullerene with isolated pentagons, and all 20 of its hexagons are equivalent with neighbour index 3. The next smallest I_h fullerene is an isolated-pentagon isomer of C_{80}, in which all 30 of the hexagons are equivalent with neighbour index 4. (In fact there is also another isolated-pentagon C_{80} isomer, with D_{5h} symmetry, all of the hexagons in which have neighbour index 4. This is similar in many respects to the I_h isomer, but not all of its hexagons are equivalent by symmetry. Thus the fact that it too enjoys an optimum hexagon neighbour index signature might indeed be regarded as accidental.) The next smallest icosahedral fullerene is a chiral isolated-pentagon isomer of C_{140} with I symmetry, all 60 hexagons of which are equivalent with neighbour index 5.

The next question is what happens between these very special values of n. Clearly, since it is impossible for all the hexagons to have the same neighbour index when n is not 60, 80, or 140, the least strained isomers with intermediate values of n will have two different hexagon neighbour indices. Furthermore, in order for the local environments of the hexagons to be as similar as possible, these two neighbour indices will be consecutive. Combining Eqs (4.2) and (4.3) with these new constraints gives the following optimum signatures, which

Table 4.2. Enumeration of isolated-pentagon C_n fullerene isomers found by the spiral algorithm in the range $n = 70$ to 140 with optimum hexagon neighbour index signatures.

n	Isomers	n	Isomers	n	Isomers	n	Isomers
70	1	88	1	106	33	124	1
72	0	90	6	108	34	126	0
74	1	92	28	110	21	128	0
76	1	94	15	112	19	130	1
78	1	96	17	114	6	132	0
80	2	98	42	116	15	134	0
82	1	100	38	118	3	136	0
84	3	102	38	120	4	138	0
86	1	104	44	122	3	140	1

include the exceptional signatures of Eq. (4.5) as special cases:

$$(h_3, h_4, h_5, h_6) = \begin{cases} (80 - n, 3n/2 - 90, 0, 0), & \text{if } 60 <= n <= 80; \\ (0, 70 - n/2, n - 80, 0), & \text{if } 80 <= n <= 140; \\ (0, 0, 60, n/2 - 70), & \text{if } 140 <= n. \end{cases} \quad (4.6)$$

According to Raghavachari's argument,[4] the least strained isolated-pentagon isomers of the higher fullerenes C_n will be those in which these optimum hexagon neighbour index signatures are realized. Table 4.2 shows that there is at least one such isomer for each value of n in the range $n = 70$ to 124, with the sole exception of $n = 72$. However, optimum signatures are more elusive again beyond $n = 124$.

The exception at $n = 72$ in Table 4.2 is particularly interesting because the (unique) isolated-pentagon isomer of C_{72} is the leapfrog of the D_{6d} C_{24} fullerene. As discussed in § 3.4, this leapfrog has a large band gap and π bonding resonance energy in Hückel theory, and thus has a stable π electronic structure. However, the C_{72} mass peak is conspicuously absent from experimental reports on the higher fullerenes. The reason for this is evidently steric strain, which by our definition includes everything that the simple Hückel approximation neglects. In fact, the hexagon neighbour index signature of the C_{72} leapfrog is $(12, 12, 0, 2)$, and this differs markedly from the (unrealizable) $(8,18,0,0)$ ideal in Eq. (4.6). Thus the hexagon neighbour index signature does seem, at least in this case, to quantify steric strain in a way which is consistent with experiment.

Finally, it is useful for some purposes to reduce the four potentially non-zero numbers in an abbreviated hexagon neighbour index signature to a single parameter that quantifies steric strain in a manner similar to N_p for lower fullerenes. However, since the strain is minimized in this case when the hexagon neighbour indices are all as similar to one another as possible, the direct analogue of N_p in Eq. (4.1) is inappropriate. A more suitable steric parameter is the standard deviation σ_h of the hexagon neighbour index distribution:[5]

$$\sigma_h = \sqrt{\langle k^2 \rangle - \langle k \rangle^2},\qquad(4.7)$$

where

$$\langle k \rangle = \sum_{k=3}^{6} k h_k \Big/ \sum_{k=3}^{6} h_k,\qquad(4.8)$$

and

$$\langle k^2 \rangle = \sum_{k=3}^{6} k^2 h_k \Big/ \sum_{k=3}^{6} h_k.\qquad(4.9)$$

The strain parameter σ_h can be calculated from these equations for any chosen isolated-pentagon fullerene isomer from the information that is given in the main tables. The ideal value of the parameter is zero, which can be realized only when $n = 60$, 80, or 140. We shall use σ_h as a measure of steric strain in the examples in § 4.6.

4.5 Steric strain in leapfrogs and carbon cylinders

One striking fact that emerges from a comparison between early theoretical discussions of the higher fullerenes and the subsequent experiments is that C_{60} is the only leapfrog fullerene, and C_{70} the only carbon cylinder fullerene, to have been isolated so far. As we have seen in § 3.4 and 3.5, these two classes of fullerene are known to have closed-shell electronic structures, with bonding HOMOs and antibonding (or nonbonding) LUMOs in Hückel theory. No analogous result has yet been proved for any other class of fullerene, and indeed the vast majority of fullerenes found by the spiral algorithm have at best pseudo-closed shell electronic structures. Why, then, of the several higher fullerenes that have now been identified in experiments, is one and only one a leapfrog and one and only one a carbon cylinder?

The hexagon neighbour index signatures of leapfrog and carbon cylinder fullerenes provide a simple answer to this question which generalizes the C_{72}

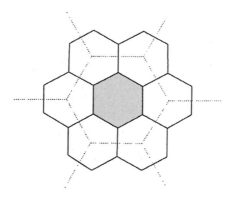

FIG. 4.4. During the leapfrog construction, each hexagonal face in the (dashed) $C_{n/3}$ parent becomes a hexagon with neighbour index 6 in the (solid) C_n leapfrog.

analysis given above. Recall that a leapfrog fullerene C_n is derived from its parent $C_{n/3}$ by a geometric construction, as described in § 3.4. During this construction each hexagon in the $C_{n/3}$ parent becomes a hexagon surrounded by other hexagons in the leapfrog, as shown in Fig. 4.4. Since the parent has a total of $n/6 - 10$ hexagons this implies a *minimum* of $n/6 - 10$ hexagons in the leapfrog having the maximum hexagon neighbour index of 6. All leapfrogs except I_h C_{60} therefore have $h_6 > 0$, and it follows immediately from Eq. (4.6) that no leapfrog fullerene in the range $60 < n <= 140$ can have an optimum hexagon neighbour index signature.

In fact this range can be extended. As shown in Fig. 4.5, each pair of adjacent pentagons in the parent gives rise to at least two hexagons with neighbour index 3 or 4 in the leapfrog, the precise number depending on how many other neighbouring pentagons are present at that location in the parent. In any event it is clear that all leapfrog fullerenes in the range $60 <= n < 180$ (which must be leapfrogs of parents with at least one pair of fused pentagons) have either $h_3 > 0$ or $h_4 > 0$ or both. Inspection of Eq. (4.6) therefore extends the range over which the optimum hexagon neighbour index signature cannot be realized by a leapfrog to $60 < n < 180$.

The limits of this extended range are now precise. C_{60}, the smallest leapfrog of all, obviously has an optimum hexagon neighbour index signature. Moreover the I_h C_{180} leapfrog of C_{60} also has an optimum signature, with $h_5 = 60$ and $h_6 = 20$. These two exceptional structures, the single and double leapfrogs of the

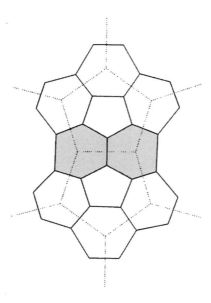

FIG. 4.5. During the leapfrog construction, each pair of adjacent pentagons in the (dashed) $C_{n/3}$ parent gives rise to at least two hexagons with neighbour index 3 or 4 in the (solid) C_n leapfrog.

dodecahedron, are thus the first two examples where the optimum π electronic stability inferred from the leapfrog rule is reinforced by a concomitant minimization of steric strain. Every leapfrog fullerene between C_{60} and C_{180}, and hence every leapfrog fullerene in the experimentally important higher fullerene range, has a hexagon neighbour index signature which departs markedly from the ideal in Eq. (4.6). These intermediate leapfrogs are therefore comparatively strained, and are unlikely to be observed experimentally.

A similar analysis applies to carbon cylinders. First consider the series of C_n carbon cylinders with fivefold axial symmetry, for which $n = 70 + 30k$. Each of these cylinders is characterized by two polar caps of the type shown in Fig. 4.6, the caps being joined together by an additional equatorial belt of $15k + 5$ hexagons. Two distinct cases arise. When $k = 0$ (corresponding to the experimentally observed C_{70} structure) the five additional hexagons in the equatorial belt are directly adjacent to one pentagon in each cap and therefore have neighbour index 4. In this case it is easy to see from Fig. 4.6 that the hexagon neighbour index signature of the cage as a whole is (10,15,0,0), which is precisely the optimum signature in Eq. (4.6). Therefore, like C_{60}, the

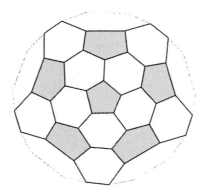

FIG. 4.6. A generic polar cap, containing 6 pentagons and 10 hexagons, in a carbon cylinder fullerene with $n = 70+3k$ atoms and fivefold axial symmetry. The five dashed hexagons form part of the equatorial belt.

experimentally observed C_{70} carbon cylinder has both a closed-shell electronic structure and minimum steric strain for the number of atoms it contains.

In the more general case, in which $k > 0$, Fig. 4.6 shows that ten of the additional $15k + 5$ equatorial hexagons are adjacent to a pentagon in one or other of the polar caps, and therefore have neighbour index 5. The remaining $15k - 5$ hexagons in the equatorial belt are adjacent only to other hexagons, and therefore have neighbour index 6. Hence the hexagon neighbour index signature of a general fivefold carbon cylinder with $n = 70 + 30k$ and $k > 0$ is $(10, 10, 10, n/2 - 40)$. Since this is far from the optimum solution in Eq. (4.6), the higher fivefold carbon cylinders beyond C_{70} are likely to be comparatively strained, despite their pronounced π electronic stability.

The cylinders with $n = 84 + 36k$ follow a similar pattern. In this case the polar caps have sixfold symmetry, as shown in Fig. 4.7, but the analysis runs along the same lines. The first sixfold cylinder, with $k = 0$, is a D_{6h} isomer of C_{84} with hexagon neighbour index signature $(0,30,0,2)$. While this is not ideal for $n = 84$, it is not far from the optimum C_{84} signature of $(0,28,4,0)$. However, as in the fivefold case, the higher members of the sixfold cylinder series are more strained. The hexagon neighbour index signature of a general carbon cylinder with $n = 84 + 36k$ and $k > 0$ is $(0, 24, 12, n/2 - 46)$, and this again differs markedly from the optimum signature in Eq. (4.6).

To summarize, it has been shown here by analysing hexagon neighbour index signatures that, once the first-order IPR has been satisfied, the remaining π

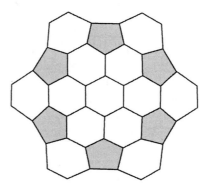

FIG. 4.7. A polar cap containing six pentagons and 13 hexagons in a carbon cylinder fullerene with $n = 84+36k$ atoms and sixfold axial symmetry. The six dashed hexagons form part of the equatorial belt.

electronic and steric strain effects in fullerene stability are generally in *opposite* directions. This contrasts strongly with the situation found in § 4.2 for lower fullerenes, where the two effects cooperate to push the pentagons apart. Realization of maximum π electronic stability via the leapfrog and carbon cylinder rules inevitably results in high steric strain, except in the very special cases of C_{60}, C_{70}, and possibly C_{84}.[5]

4.6 Selected higher fullerene examples

The idea that steric strain and π electronic effects are in competition beyond the IPR has been justified in the last section by considering the hexagon neighbour index signatures of leapfrog and carbon cylinder fullerenes, which provide extreme examples of both effects. This section shows that the available experimental results on the higher fullerenes $C_{n>70}$ provide further support for the idea by considering the contrasting cases of C_{76},[20] C_{78},[21,22] and C_{84},[22,23] which between them illustrate the two effects very well.

C_{76}

The two isolated-pentagon isomers of C_{76} are shown in Table A.10, and their π electronic and steric parameters are summarized in Table 4.3. One isomer has chiral D_2 symmetry and the other cubic T_d symmetry. The D_2 isomer is more strained than the T_d isomer, which has an optimum hexagon neighbour

Table 4.3. π electronic and steric parameters of the two isolated-pentagon isomers of C_{76} (see also Table A.10). The band gaps and resonance energies per electron are given in units of the Hückel resonance integral β. The steric strain parameter σ_h is defined in terms of the hexagon neighbour indices in Eq. (4.7).

Isomer	Point group	Band gap	Resonance energy	Hexagon indices	σ_h
76:1	D_2	0.3436	0.5551	8 16 4 0	0.6389
76:2	T_d	0.0000	0.5482	4 24 0 0	0.3500

index signature. However, the D_2 isomer has a more stable π electronic structure than the T_d isomer, which has a comparatively small π bonding resonance energy. It is therefore not obvious which will have the lower energy, because the two qualitative effects in fullerene stability are operating in different directions. However, the fact that the T_d isomer is actually open-shell in the Hückel approximation strongly favours the D_2 isomer,[12] and more sophisticated (HF/double zeta) electronic structure calculations which allow the T_d isomer to undergo Jahn–Teller distortion to D_{2d} symmetry confirm this expectation with a very significant energy difference of 180 kJ/mol between the two isomers.[24]

The experimental structure of C_{76} is thus the chiral D_2 isomer, which gives rise to 19 ^{13}C NMR lines of equal intensity[20] and has enantiomers which can be resolved by osmylation.[25] The case of C_{76} therefore provides an example in which differences in π electronic stability between competing isolated-pentagon isomers are more important than differences in steric strain.

C_{78}

The five isolated-pentagon isomers of C_{78} are shown in Table A.10, and their π electronic and steric parameters are summarized in Table 4.4. The five isomers have symmetries $C_{78}{:}1$ (D_3), $C_{78}{:}2$ (C_{2v}), $C_{78}{:}3$ (C_{2v}), $C_{78}{:}4$ (D_{3h}) and $C_{78}{:}5$ (D_{3h}), respectively. Isomer $C_{78}{:}4$ is the leapfrog of the unique C_{26} fullerene, and can be seen from Table 4.4 to have a significantly more stable π electronic structure than any of the remaining isomers. Indeed the band gaps and π bonding resonance energies in the table reveal the following π electronic stability sequence for the five isomers:[13]

$$C_{78}{:}4\ (D_{3h}) > C_{78}{:}2\ (C_{2v}) > C_{78}{:}1\ (D_3) > C_{78}{:}3\ (C_{2v}) > C_{78}{:}5\ (D_{3h}).$$

Table 4.4. π electronic and steric parameters of the five isolated-pentagon isomers of C_{78} (see also Table A.10).

Isomer	Point group	Band gap	Resonance energy	Hexagon indices	σ_h
78:1	D_3	0.2532	0.5552	8 15 6 0	0.6914
78:2	C_{2v}	0.3481	0.5557	6 20 2 1	0.6396
78:3	C_{2v}	0.1802	0.5528	4 23 2 0	0.4496
78:4	D_{3h}	0.6333	0.5578	8 18 0 3	0.8276
78:5	D_{3h}	0.0730	0.5490	2 27 0 0	0.2534

However, the standard deviations σ_h of the hexagon neighbour index signatures reveal that steric strain decreases in much the same direction.[5,26] The leapfrog isomer $C_{78}:4$ (D_{3h}) is therefore also the most strained of the isolated-pentagon possibilities, as expected from the discussion of leapfrog fullerenes in § 4.5. The least strained isomer is $C_{78}:5$ (D_{3h}), which has the optimum hexagon neighbour index signature for C_{78}. The two effects in fullerene stability are again in direct competition, and it is not at all clear from qualitative arguments exactly which C_{78} isomer should be the most stable. Five isomers are not, however, too many for expensive (HF/6-31G*) *ab initio* calculations,[26] which have been found to give the following relative isomer energies (in kJ/mol):

$$C_{78}:3\ (C_{2v}) \quad C_{78}:1\ (D_3) \quad C_{78}:2\ (C_{2v}) \quad C_{78}:5\ (D_{3h}) \quad C_{78}:4\ (D_{3h})$$
$$0 \qquad\qquad 14 \qquad\qquad 17 \qquad\qquad 30 \qquad\qquad 85$$

The ^{13}C NMR spectrum of the experimental C_{78} product is consistent with a mixture of the $C_{78}:1$ (D_3), $C_{78}:2$ (C_{2v}), and $C_{78}:3$ (C_{2v}) isomers, with the relative proportions depending on the precise experimental conditions.[21,22] The observed mixtures are clearly consistent with both the above discussion and the *ab initio* energy ordering. Each D_{3h} isomer fails badly on one criterion and is optimal on the other, whereas the three remaining candidates have intermediate π electronic stability and steric strain and are observed in the experimental product. The case of C_{78} therefore provides an example in which differences in π electronic stability and steric strain between competing isolated-pentagon isomers are of comparable importance.

Table 4.5. π electronic and steric parameters of the 24 isolated-pentagon isomers of C_{84} (see also Table A.12).

Isomer	Point group	Band gap	Resonance energy	Hexagon indices	σ_h
84:1	D_2	0.6143	0.5590	8 16 4 4	0.9270
84:2	C_2	0.3523	0.5570	6 18 6 2	0.7806
84:3	C_s	0.0191	0.5530	3 22 7 0	0.5449
84:4	D_{2d}	0.3519	0.5564	4 20 8 0	0.5995
84:5	D_2	0.2403	0.5546	4 20 8 0	0.5995
84:6	C_{2v}	0.1892	0.5555	4 20 8 0	0.5995
84:7	C_{2v}	0.1892	0.5527	2 24 6 0	0.4841
84:8	C_2	0.1776	0.5541	2 24 6 0	0.4841
84:9	C_2	0.0556	0.5531	2 24 6 0	0.4841
84:10	C_s	0.0916	0.5523	1 26 5 0	0.4146
84:11	C_2	0.2540	0.5553	2 24 6 0	0.4841
84:12	C_1	0.2164	0.5540	1 26 5 0	0.4146
84:13	C_2	0.0988	0.5526	2 24 6 0	0.4841
84:14	C_s	0.4054	0.5566	3 23 5 1	0.5995
84:15	C_s	0.2191	0.5540	1 26 5 0	0.4146
84:16	C_s	0.3369	0.5559	1 27 3 1	0.4841
84:17	C_{2v}	0.1745	0.5546	2 25 4 1	0.5449
84:18	C_{2v}	0.3285	0.5567	2 26 2 2	0.5995
84:19	D_{3d}	0.1861	0.5552	2 24 6 0	0.4841
84:20	T_d	0.6962	0.5588	4 24 0 4	0.7806
84:21	D_2	0.1381	0.5522	0 28 4 0	0.3307
84:22	D_2	0.3449	0.5549	0 28 4 0	0.3307
84:23	D_{2d}	0.3449	0.5546	0 28 4 0	0.3307
84:24	D_{6h}	0.5293	0.5565	0 30 0 2	0.4841

C_{84}

The 24 isolated-pentagon isomers of C_{84} are shown in Table A.12, and their π electronic and steric parameters are summarized in Table 4.5. Isomers $C_{84}:1$ (D_2) and $C_{84}:20$ (T_d) in these tables are leapfrogs of the two different C_{28} fullerenes, and $C_{84}:24$ (D_{6h}) is the unique C_{84} carbon cylinder.[27]

As expected, the two leapfrogs have both the best π electronic stability and the worst steric strain in the set. The carbon cylinder $C_{84}:24$ (D_{6h}) also has a very stable π electronic structure, and it is significantly less strained than the leapfrogs for reasons discussed in § 4.5. However, the least strained isomers in the table are C_{84}:21 (D_2), C_{84}:22 (D_2), and C_{84}:23 (D_{2d}), all of which have optimum hexagon neighbour index signatures. Of these, C_{84}:22 (D_2) and C_{84}:23 (D_{2d}) are more stable than C_{84}:21 (D_2) on π electronic grounds. Overall, however, it is clear that π electronic and steric strain effects are again in competition for C_{84}, with π electronic stability favouring the leapfrogs C_{84}:1 (D_2) and C_{84}:20 (T_d) and steric strain favouring C_{84}:21 (D_2), C_{84}:22 (D_2), and C_{84}:23 (D_{2d}). Thus it is again hard to decide which isomer(s) will be the most stable on the basis of qualitative arguments alone.[14]

Detailed electronic structure calculations have not yet been performed on all 24 isomers at an *ab initio* level of theory. However, Zhang *et al.*[28] have performed semi-empirical calculations on the complete list using the tight-binding Hamiltonian discussed in § 4.3. These calculations find C_{84}:22 (D_2) and C_{84}:23 (D_{2d}) to be essentially isoenergetic at the bottom of the isomer energy ladder, and the same conclusion is also reached with other semi-empirical methods.[4,11] The similarity in energy of isomers C_{84}:22 (D_2) and C_{84}:23 (D_{2d}) can be can be rationalised from the similarity of their π electronic and steric parameters in Table 4.5. However, the fact that they are also the most stable of the 24 isomers would be more difficult to predict from this table without assuming steric strain to be the dominant effect.

All the indications are that the experimental structure of C_{84} is indeed dominated by a 2:1 mixture of the C_{84}:22 (D_2) and C_{84}:23 (D_{2d}) isomers, which gives a ^{13}C NMR spectrum showing 31 lines of equal intensity and one line of half intensity.[22,23] This assignment is actually quite subtle, because there are also three other D_2 isomers in Table 4.5 which would give the same ^{13}C NMR count, but the fact that the observed 2:1 ratio implies an equilibrium mixture of *isoenergetic* D_2 and D_{2d} isomers points unambiguously to C_{84}:22 (D_2) and C_{84}:23 (D_{2d}).[23] The case of C_{84} therefore provides a final contrasting example in which differences in steric strain between competing isolated-pentagon isomers are more important than differences in π electronic stability.

Taking these three examples together, it can be seen that steric strain effects become progressively more important from C_{76} to C_{84}. It is therefore tempting to argue that steric strain will be the deciding factor in the stability of even higher fullerenes. However, the qualitative nature of our arguments and the delicacy of the energy differences involved both warn against this extrapolation. At present, the best way to predict the experimental structures of the higher fullerenes beyond C_{84} is to perform explicit semi-empirical calculations on all isolated-pentagon isomers and select the most likely candidate structures for calculations at progressively higher levels of electronic structure theory. Simple topological arguments are useful for deriving general rules such as the high steric strain in leapfrog and carbon cylinder fullerenes (§ 4.5), and for rationalizing known results of *ab initio* calculations and experiments. However, they do not appear at present to be sophisticated enough to have *predictive* power for higher fullerenes. This contrasts sharply with the situation found in § 4.3, where simple topological arguments were shown to predict uniquely stable candidate structures for more than half of the lower fullerenes. The higher fullerene problem involves a competition of effects beyond the first-order IPR and is evidently more subtle.

References and notes

[1] J. N. Murrell, S. F. A. Kettle, and J. M. Tedder, The chemical bond (Wiley, New York, 1978).

[2] H. W. Kroto, Nature 329 (1987) 529.

[3] T. G. Schmalz, W. A. Seitz, D. J. Klein, and G. E. Hite, J. Am. Chem. Soc. 110 (1988) 1113.

[4] K. Raghavachari, Chem. Phys. Lett. 190 (1992) 397.

[5] P. W. Fowler, S. J. Austin, and D. E. Manolopoulos, Competing factors in fullerene stability, in: Chemistry and physics of the fullerenes, K. Prassides, ed. (Kluwer, Dordrecht, 1994).

[6] R. C. Haddon, J. Am. Chem. Soc. 108 (1986) 2837.

[7] R. C. Haddon, Chem. Phys. Lett. 125 (1986) 231.

[8] C. A. Coulson, Valence (Clarendon, Oxford, 1952).

[9] R. C. Haddon, Science 261 (1993) 1545.

[10] D. Bakowies, A. Gelessus, and W. Thiel, Chem. Phys. Lett. 197 (1992) 324.

[11] D. Bakowies, M. Kolb, W. Thiel, S. Richard, R. Ahlrichs, and M. M. Kappes, Chem. Phys. Lett. 200 (1992) 411.

[12] D. E. Manolopoulos, J. Chem. Soc. Faraday Trans. 87 (1991) 2861.

[13] P. W. Fowler, R. C. Batten, and D. E. Manolopoulos, J. Chem. Soc. Faraday Trans. 87 (1991) 3103.

[14] D. E. Manolopoulos and P. W. Fowler, J. Chem. Phys. 96 (1992) 7603.

[15] B. L. Zhang, C. Z. Wang, K. M. Ho, C. H. Xu, and C. T. Chan, J. Chem. Phys. 97 (1992) 5007.

[16] P. W. Fowler, D. E. Manolopoulos, G. Orlandi, and F. Zerbetto, J. Chem. Soc. Faraday 91 (1995) 1421.

[17] A. Warshel and M. Karplus, J. Am. Chem. Soc. 94 (1972) 5612.

[18] F. Negri, G. Orlandi, and F. Zerbetto, Chem. Phys. Lett. 144 (1988) 31.

[19] Y.-D. Gao and W. C. Herndon, J. Am. Chem. Soc. 115 (1993) 8459.

[20] R. Ettl, I. Chao, F. Diederich, and R. L. Whetten, Nature 353 (1991) 149.

[21] F. Diederich, R. L. Whetten, C. Thilgen, R. Ettl, I. Chao, and M. M. Alvarez, Science 254 (1991) 1768.

[22] K. Kikuchi, N. Nakahara, T. Wakabayashi, S. Suzuki, H. Shiromaru, Y. Miyake, K. Saito, I. Ikemoto, M. Kainosho, and Y. Achiba, Nature 357 (1992) 142.

[23] D. E. Manolopoulos, P. W. Fowler, R. Taylor, H. W. Kroto, and D. R. M. Walton, J. Chem. Soc. Faraday Trans. 88 (1992) 3117.

[24] J. R. Colt and G. E. Scuseria, J. Phys. Chem. 96 (1992) 10265.

[25] J. M. Hawkins and A. Meyer, Science 260 (1993) 1918.

[26] K. Raghavachari and C. M. Rohlfing, Chem. Phys. Lett. 208 (1993) 436.

[27] P. W. Fowler, J. Chem. Soc. Faraday Trans. 87 (1991) 1945.

[28] B. L. Zhang, C. Z. Wang, and K. M. Ho, J. Chem. Phys. 96 (1992) 7183.

5

SYMMETRY AND SPECTROSCOPY

The presence of symmetry elements in a molecule can have important theoretical and practical consequences because the spectroscopy of a molecular species is governed by symmetry-based rules. Given the point group[1] of a molecular structure, symmetry arguments tell us whether the molecule is chiral and so could be optically active, whether it may have a permanent dipole moment and therefore an allowed pure rotational spectrum, and whether its infrared and Raman spectra may have peaks in common. Molecular symmetry also determines the number of signals to be expected in the NMR spectrum. As ^{13}C NMR spectroscopy is the main tool used so far to probe the structures of the higher fullerenes (§§ 1.2 and 4.6), it is clearly important to know the possible symmetries of fullerene structures.

Fullerene topology is compatible with only a few types of symmetry element, and all the infinitely many possible isomers belong to a small finite set of point groups. The present chapter shows that there are just 28 fullerene point groups, and describes how the idealized symmetry and spectroscopic signature of each fullerene isomer can be assigned automatically using the information contained in its adjacency matrix or molecular graph.

5.1 The fullerene point groups

As we have seen in earlier chapters, a fullerene C_n is defined by a pseudospherical polyhedral cage of n vertices, each connected to three others, arranged in 12 pentagonal and $n/2 - 10$ hexagonal faces. To determine the overall symmetries available to such an object, it is useful to consider how symmetry operations affect its components. The $3n + 2$ *special points* on a fullerene polyhedron are the n vertices, $3n/2$ edge midpoints and $n/2 + 2$ face centres. Each point is characterized by a site symmetry — the group of operations that leave the surrounding environment unchanged as seen from that particular site — which is determined by the symmetry elements of the whole object that pass through

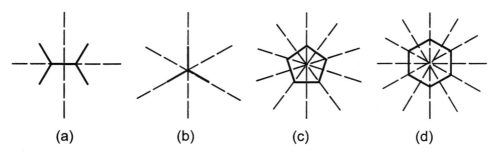

FIG. 5.1. Site symmetries of special points of a fullerene polyhedron. The dotted lines show possible mirror planes perpendicular to the page and passing through (a) an edge, (b) a vertex, (c) the centre of a pentagonal face, and (d) the centre of a hexagonal face. In each case there may also be an axis of proper rotation normal to the plane of the page; this would be C_2 through an edge, C_3 through a vertex, C_5 through a pentagonal face, and C_6, C_3, or C_2 through a hexagonal face. The maximal site symmetry groups are thus C_{2v}, C_{3v}, C_{5v}, and C_{6v}, respectively.

the site. For example, Fig. 5.1 shows that a trivalent vertex can lie on a threefold rotational axis and in three mirror planes. The maximum site symmetry for a fullerene vertex, if the rest of the structure is appropriately arranged, is therefore C_{3v}. If the rotation axis alone is present the site group is C_3, if only one mirror plane passes through the vertex the site the group is C_s, and if no axis or plane passes through the vertex the site group is the trivial C_1. In short, a fullerene vertex has site symmetry C_{3v} or a subgroup. The maximal site symmetries of edge midpoints and face centres can be deduced similarly from Fig. 5.1, giving the full list of site symmetries for the special points as:[2]

$$C_{3v}, C_3, C_s, C_1 \text{ for vertices;}$$

$$C_{2v}, C_2, C_s, C_1 \text{ for edge centres;}$$

$$C_{5v}, C_5, C_s, C_1 \text{ for pentagon centres;}$$

$$C_{6v}, C_6, C_{3v}, C_3, C_{2v}, C_2, C_s, C_1 \text{ for hexagon centres.}$$

In addition, there is at least one point within the cage (e.g., the centre of mass) that lies on all symmetry elements and therefore has the full point group of the cage as its site group. All positions on the polyhedron surface, apart from the special points, have either C_s or C_1 site symmetry.

All axes and planes of symmetry of the fullerene must pass through special points, and so fullerene point groups are limited to those that admit *only* site groups from the above list for surface sites. A secondary logical requirement is that a fullerene point group should admit at least one site group for each type of component, but this is trivially satisfied since C_1 is available in every point group as the site symmetry of the general position.

Before listing the groups that satisfy these criteria, it is useful to introduce the concept of an *orbit*. In a structure belonging to any nontrivial point group, various sets of *equivalent positions* can be defined, and each set is a realization of an *orbit* of the group. Examples are the eight vertices of a cube (O_8 in O_h), the midpoints of the six edges of a regular tetrahedron (O_6 in T_d), and the 12 pentagon centres of a regular dodecahedron (O_{12} in I_h). Equivalent positions are in general permuted by operations of the group, and are characterized by a common site symmetry. Multiple copies of an orbit may be present in a structure, and all will be characterized by the same site symmetry. For example, in the group C_s generated by a single mirror plane, any point in the plane is an example of the O_1 orbit (site symmetry C_s), and any pair of points exchanged by reflection is an example of the O_2 orbit (site symmetry C_1). The number of points in an orbit and the number of operations in the point group are related by [3]

$$m_A \times |G_A| = |G|, \tag{5.1}$$

where m_A is the size of the orbit, $|G_A|$ is the order of the site symmetry group and $|G|$ is the order of the full point group.

Every group contains the *regular orbit* O_{reg} of points at general positions, with one point for every operation of the group and with site symmetry C_1. Every group also contains an orbit O_1 consisting of a single point lying on all symmetry elements for which the site symmetry is identical with the point group. O_{reg} and O_1 are, of course, one and the same for the C_1 group. All orbits apart from O_1 have site symmetry C_{mv} or lower. Lists of orbits and site symmetries are available for all the point groups important in chemistry. [3]

Returning to the specific problem of fullerene symmetry, it is easily proved that the site groups listed earlier for the orbits of vertices, edge midpoints, and face centres are compatible with exactly 36 point groups. [4] Apart from the low symmetries C_1, C_i and C_s, and the high symmetries I_h, I, T_d, T_h, and T, the groups that satisfy the site criterion are D_{nh}, D_{nd}, D_n, S_{2n}, C_{nh}, C_{nv}, C_n with $n = 2, 3, 5$, and 6. However, more detailed consideration using the spiral algorithm (in a situation where it can be proved to be complete) shows that whenever a C_5 or C_6 rotational axis is present, the geometrical closure of the fullerene cage forces a perpendicular twofold rotational axis.[2] This means that C_5, C_{5v}, C_{5h}, and S_{10} symmetries occur only as subgroups of operations for fivefold dihedral or icosahedral fullerenes, and likewise C_6, C_{6v}, C_{6h}, and S_{12} symmetries occur only as subgroups of sixfold dihedral fullerene groups. The final list of 28 fullerene point groups is therefore: I_h, I (icosahedral); T_h, T_d, T (tetrahedral); D_{6h}, D_{6d}, D_6, D_{5h}, D_{5d}, D_5, D_{3h}, D_{3d}, D_3, D_{2h}, D_{2d}, D_2 (cylindrical); S_6, S_4, C_{3h}, C_{2h} (skewed); C_{3v}, C_3, C_{2v}, C_2 (pyramidal); C_s, C_i, C_1 (low symmetry).

All of the allowed groups are subgroups of I_h, T_d, or $D_{\infty h}$, and all but five of them (I_h, T_d, D_{5h}, D_{6h}, D_{6d}) are proper subgroups of another group in the list. The relationships between the fullerene groups are listed in Table 5.1. All but one of these groups (I) can be realized by fullerene isomers with adjacent pentagons and Fig. 5.2 shows the smallest example of each symmetry. Since any fullerene produces an isolated-pentagon isomer of the same point group on leapfrogging, all 28 groups can be realized by fullerenes with isolated pentagons. The groups I and I_h are unusual in that apart from C_{20} (I_h), they can be realized *only* by isolated-pentagon fullerenes; this is a consequence of the fact that I symmetry fixes the positions of all 12 pentagon centres.

The subgroup/supergroup relationships in Table 5.1 are important, for example when counting cages by the Coxeter tessellation method described in § 2.3. The construction of all Coxeter-type nets that satisfy the constraints of T symmetry generates T_h, T_d, I, and I_h fullerenes as special cases, for example, and I and I_h also arise if the working symmetry is D_5 (when all D_{5h} and D_{5d} nets are also generated).

Two important global characteristics of the fullerenes are also apparent from the list of point groups. Fullerenes may be *chiral* and they may be *polar*. A chiral structure belongs to a pure rotational point group (i.e., one without improper operations) and the nine chiral fullerene groups are therefore I, T, D_6, D_5, D_3, D_2, C_3, C_2, and C_1. Permanent dipole moments are excluded in point groups with multiple rotational axes, improper rotational axes, or a centre of inversion. A polar fullerene must therefore belong to one of the six groups C_{3v}, C_3, C_{2v}, C_2, C_s, or C_1.

Fullerenes may be spherical tops (I_h, I, T_h, T_d, T), symmetric tops (D_{6h}, D_{6d}, D_6, D_{5h}, D_{5d}, D_5, D_{3h}, D_{3d}, D_3, S_6, S_4, C_{3h}, C_{3v}, C_3), or asymmetric tops (D_{2h}, D_2, C_{2h}, C_{2v}, C_2, C_s, C_i, C_1). A pure rotational spectrum is allowed

Table 5.1. Subgroup relations within the 28 fullerene groups. A fullerene group is a possible *maximal* point group of a fullerene polyhedron. Each row of the table lists one such group (as a Schönflies symbol), its order, and the set of its immediate subgroups that are themselves fullerene groups. Chains of subgroups can be followed through the table: for example, I_h (row 1) has a subgroup I (row 2) that has a subgroup T (row 5) ... and so the chain $C_1 \subset C_3 \subset T \subset I \subset I_h$ (amongst others) can be realized entirely within the fullerene family. The carbon frameworks of fullerene compounds are also classified within the same list of 28 maximal groups, but when decorations of the skeleton are taken into account the list is extended to 36. For example, a hypothetical $C_{60}H_5^+$ cation in which H has been added to every member of one pentagonal ring of icosahedral C_{60} has maximal group I_h when only the framework is considered, but the probable 'physical' group is C_{5v}.

Group	Order	Subgroups	Group	Order	Subgroups
I_h	120	I, T_h, D_{5d}, D_{3d}	D_{2h}	8	C_{2v}, C_{2h}, D_2
I	60	T, D_5, D_3	D_{2d}	8	C_{2v}, D_2, S_4
T_d	24	T, D_{2d}, C_{3v}	D_2	4	C_2
T_h	24	T, D_{2h}, S_6	S_6	6	C_3, C_i
T	12	D_2, C_3	S_4	4	C_2
D_{6h}	24	$D_{3d}, D_{3h}, D_6, D_{2h}$	C_{3h}	6	C_3, C_s
D_{6d}	24	D_6, D_{2d}	C_{3v}	6	C_3, C_s
D_6	12	D_3, D_2	C_3	3	C_1
D_{5h}	20	D_5, C_{2v}	C_{2h}	4	C_2, C_s, C_i
D_{5d}	20	D_5, C_{2h}	C_{2v}	4	C_2, C_s
D_5	10	C_2	C_2	2	C_1
D_{3h}	12	$C_{3v}, C_{3h}, D_3, C_{2v}$	C_s	2	C_1
D_{3d}	12	C_{3v}, D_3, S_6, C_{2h}	C_i	2	C_1
D_3	6	C_3, C_2	C_1	1	–

only for the six polar fullerene groups, but a forbidden spectrum induced by a centrifugal distortion mechanism [5] is possible in these six and an additional ten groups ($T_d, T, D_6, D_5, D_3, D_2, D_{3h}, D_{2d}, S_4$, and C_{3h}). Geometric information is in principle available from the rotational structure of infra-red transitions for fullerenes of any point group, but the very small rotational constants will make it difficult to resolve the spectra.

In order to make use of all this symmetry information we need a way of assigning point groups to fullerene isomers. Practical considerations dictate that this must be capable of computerization. The next two sections describe an approach to this problem that is an extension of the way that a human being

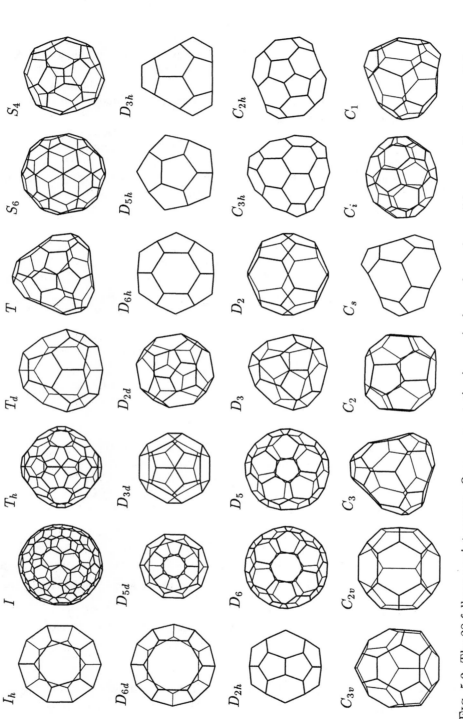

FIG. 5.2. The 28 fullerene point groups. One representative isomer is shown for each possible group. In each case it is the first of that particular symmetry to be found by the spiral algorithm. The atom counts are: C_{20} (I_h), C_{24} (D_{6d}), C_{26} (D_{3h}), C_{28} (D_2, T_d), C_{30} (D_{5h}, C_{2v}), C_{32} (D_2, D_{3d}, D_3), C_{34} (C_s, C_{3v}), C_{36} (C_1, D_{2d}, D_{6h}), C_{40} (D_{5d}, C_3, D_{2h}), C_{44} (T, S_4), C_{48} (C_{2h}), C_{56} (C_i), C_{60} (D_5), C_{62} (C_{3h}), C_{68} (S_6), C_{72} (D_6), C_{92} (T_h), C_{140} (I).

would assign symmetry, by constructing a model or picture and inspecting it for symmetry elements. In the computerized method we construct a set of Cartesian coordinates (the 'topological coordinates') for the atoms in the isomer and use a program to catalogue all symmetry operations consistent with them.[6] Pictures can also be constructed from the topological coordinates, and indeed all the fullerene drawings in this book have been produced in this way, but pictures are not required for the assignment of the point group or the idealized spectroscopic signature.

5.2 Topological coordinates

Given the adjacency matrix of a fullerene isomer, an appropriate set of Cartesian coordinates for its atoms can be obtained in a number of ways. One conventional approach is to feed the bonding connectivity into a molecular mechanics program, which then minimizes some parameterized potential energy function involving local bond-length and bond-angle terms. This approach is widely used, and has the advantage that the resulting coordinates are in some sense physically faithful and therefore suitable as an initial guess for an *ab initio* calculation. However, there is an even simpler approach that is equally valid for the present topological purposes and is entirely parameter-free.

Fullerenes are pseudospherical clusters, and the eigenvectors of a fullerene adjacency matrix bear traces of its descent from a hypothetical spherical ancestor. In particular, the symmetries spanned in the finite point group by these eigenvectors (i.e., by the Hückel molecular orbitals of the surface π system) match those of the spherical harmonics.[7,8,9] The largest eigenvalue corresponds to an eigenvector with equal coefficients $1/\sqrt{n}$ on all n atoms, for example, which is a sampling of a giant S function spread over the whole cluster. Amongst the next few eigenvalues are invariably found three belonging to eigenvectors with the symmetry of giant P functions.

In the notation of Stone's tensor surface harmonic theory,[7,8] the set of n eigenvectors spans the $S^\sigma, P^\sigma, D^\sigma \ldots$ scalar spherical harmonics and, roughly speaking, the higher the L value, the higher the energy of the orbital (and therefore the lower the corresponding eigenvalue of the adjacency matrix). Our recipe for topological coordinates takes this spherical analogy seriously: since the P^σ eigenvectors on a sphere are proportional to the X, Y, and Z coordinates we can interpret the coefficients in these eigenvectors directly as scaled coordinates.

Suppose that $\mathbf{A}$ is a fullerene adjacency matrix, with elements $A_{ij} = 1$ if atom i is bonded to atom j and $A_{ij} = 0$ otherwise. Let $\mathbf{C}_k$ be the kth orthonormal eigenvector of $\mathbf{A}$, and x_k the corresponding eigenvalue (§ 3.1):

$$\sum_j A_{ij} C_{jk} = C_{ik} x_k, \qquad (5.2)$$

and

$$\sum_j C_{jk} C_{jl} = \delta_{kl}. \qquad (5.3)$$

The eigenvalues x_k are conventionally arranged in descending order, so that $\mathbf{C}_1$ is the totally symmetric S^σ eigenvector with $x_1 = +3$. The topological coordinates (X_j, Y_j, Z_j) of atom j are simply

$$\begin{aligned}
X_j &= C_{jk_X} \, s_X, \\
Y_j &= C_{jk_Y} \, s_Y, \\
Z_j &= C_{jk_Z} \, s_Z,
\end{aligned} \qquad (5.4)$$

where s_X, s_Y, and s_Z are axis scaling factors and k_X, k_Y, and k_Z are the indices that label the three P^σ eigenvectors.

The P^σ eigenvectors are the only Hückel molecular orbitals with a single nodal plane, or equivalently one positive and one negative lobe, and they can be identified as such in practice using the connectivity information in $\mathbf{A}$.[6] For a given eigenvector $\mathbf{C}_k$ ($k > 1$) the atoms are labelled black (positive), white (negative), or grey (zero) according to the size and sign of the coefficient C_{jk}. If $\mathbf{C}_k$ is a P^σ eigenvector then

(1) every pair of black atoms is connected by a bonding path passing only through black atoms, and similarly every pair of white atoms is connected through white atoms;

(2) similar paths can still be found when atoms directly bonded to neighbours of the opposite colour are removed from each set.

Together these two criteria serve to find three and only three P^σ eigenvectors in all cases tested so far. The existence of precisely three such eigenvectors is a consequence of the fact, conjectured in our earlier work,[6] but now proved,[10] that fullerene graphs have a unique realization in three-dimensional space. The three P^σ molecular orbitals nearly always correspond to the eigenvalues x_2, x_3, and x_4, and the rare exceptions can be rationalized on considerations of cage asymmetry. In a long thin cage, a Hückel molecular orbital with two nodal planes along the long axis may be more bonding than a P^σ molecular orbital with a single nodal plane cutting the short axis, just as certain components of the D^σ manifold may drop below the highest component of the P^σ manifold for a particle in an eccentric box. An example of this kind, with $k_Z = 2$ but $k_X = 4$ and $k_Y = 5$, is shown in Fig. 5.3.

The topological coordinates defined in Eq. (5.4) have a number of useful properties, which follow immediately from the orthogonality relationships in Eq. (5.3) and the fact that the totally symmetric S^σ eigenvector $\mathbf{C}_1$ has equal coefficients on all atoms. The centre of mass of an isotopically pure fullerene is at the topological coordinate origin, for example, and the topological coordinate axes are the principal axes of rotation. Topological moments of inertia calculated from these coordinates[6] give a classification of fullerenes into spherical, prolate, oblate, and asymmetric types.

Although these properties hold for any choice of the axis scaling factors s_X, s_Y, and s_Z in Eq. (5.4), the choice of scaling factors is not entirely arbitrary. In particular, they must be chosen in such a way as to reflect any inherent axis degeneracies, so that, for example, $s_X = s_Y$ whenever $x_{k_X} = x_{k_Y}$. One convenient choice of scaling factors which satisfy this degeneracy requirement is[6]

$$
\begin{aligned}
s_X &= 1/\sqrt{x_1 - x_{k_X}}, \\
s_Y &= 1/\sqrt{x_1 - x_{k_Y}}, \\
s_Z &= 1/\sqrt{x_1 - x_{k_Z}},
\end{aligned}
\tag{5.5}
$$

and this particular choice was used to draw the pictures in this book. It has the advantage that, for the most part, it places each atom closest in three-dimensional space to the three atoms to which it is bonded. This geometrical faithfulness is not required for the computer assignment of symmetry, but it is clearly an aid to visualization. It *is* necessary for the coordinates to be topologically faithful, so that equivalent vertices of the graph lead to atoms at equivalent spatial positions, and this requirement is satisfied by Eq. (5.5).

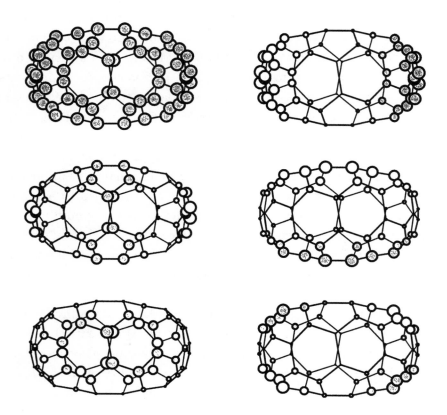

FIG. 5.3. An 'awkward' fullerene in which the molecular orbitals taken purely in order of energy would not determine the topological Cartesian coordinates. The fullerene illustrated is a prolate D_2 isomer of C_{60} with a spiral code containing 12 5s and 20 6s: 55555665656666666666666666656655555. In the diagrams of the six lowest molecular orbitals, shaded and unshaded circles denote the contributions of local atomic orbitals to the MO; shaded for an orbital with its positive lobe on the outside of the cage, unshaded on the inside. (The corresponding atomic orbitals are *exo*-pointing sp^m hybrid orbitals as described in § 4.1.) The overall sign of the eigenvector is arbitrary, but the relative phases are not. In ascending order of energy the Hückel eigenvectors (and eigenvalues) are (i) S^σ ($\epsilon = \alpha + 3\beta$), (ii) P_z^σ ($\epsilon = \alpha + 2.8989\beta$), (iii) $D_{z^2}^\sigma$ ($\epsilon = \alpha + 2.6109\beta$), (iv) P_x^σ ($\epsilon = \alpha + 2.6031\beta$), (v) P_y^σ ($\epsilon = \alpha + 2.5716\beta$), (vi) D_{xz}^σ ($\epsilon = \alpha + 2.4170\beta$). This energy order is easy to rationalize on the particle-in-a-box model: for a long thin cage, the energy of the wavefunction with two nodal planes cutting the long axis can drop below that of each function with only a single nodal plane bisecting a short axis. Only two of the 1812 isomers of C_{60} show this behaviour, and both of these are highly strained. Once detected the problem is easily solved as described in the text.

5.3 Symmetry assignment

Once the topological coordinates have been calculated the computer assignment of a symmetry group is straightforward. The set of atomic coordinates is first used to generate explicit values of the coordinates of all $3n + 2$ special points. Vectors from the topological coordinate origin to the special points include all possible rotation axes; all possible reflection planes pass through the coordinate origin and either include at least one bond or bisect at least one bond at right angles. A catalogue of symmetry operations can therefore be constructed, and a decision tree similar to the usual one in symmetry textbooks employed to find the appropriate point group. Provided that sufficient care is taken to trap the accumulation of numerical errors in testing for a symmetry operation, this is a robust procedure. [6,11]

Table 5.2 shows a breakdown of the isomers of fullerenes C_{20} to C_{100} into symmetry groups as calculated by the procedure. Each entry for a chiral group stands for a pair of enantiomers and so, for example, a purely random choice of an isomer of C_{70} has an overwhelming ($\sim 99\%$) chance of leading to a chiral structure and a very high ($\sim 95\%$) chance of leading to a structure with no symmetry at all. The isolated-pentagon isomers from C_{60} to C_{140} (Table 5.3) show a similar dominance of chiral and low-symmetry isomers. All higher fullerenes that have been experimentally characterized so far show some symmetry.

Having produced a point group assignment from what is effectively the molecular graph of the fullerene, it is necessary to enquire how this group relates to the symmetry of a real molecular framework. In fact, the group obtained with the aid of the topological coordinates is the *highest possible* point symmetry group compatible with the graph. It does not refer to any particular electronic configuration or electron count for the fullerene or one of its ions, and it neglects the possibility of any geometric distortion in the three-dimensional structure driven by electronic effects.

The most important of these electronic effects is the first-order Jahn–Teller distortion. If a calculation of the electronic structure of a neutral fullerene in the maximal symmetry group indicates a spatially degenerate ground state (i.e., an open shell), then Jahn–Teller distortion to lower symmetry is to be expected. Distortion takes place in accordance with the epikernel principle, [12] which states that molecular symmetry will descend through a chain of subgroups, stopping at the largest group in which the pertinent degeneracy is resolved (the epikernel

Table 5.2. Symmetry groups of the fullerene isomers for C_{20} to C_{100}. All isomers represented by spiral codes starting on pentagons are listed, counting only one member of each enantiomeric pair. C_{100} has a T_d isomer for which all spirals start on a hexagonal face and this has been added to the appropriate entry. See Table 2.1 for the total isomer count at each carbon number.

Fullerene	I_h	I	T_h	T_d	T	D_{6d}	D_{5d}	D_{3d}	D_{2d}	D_{6h}	D_{5h}	D_{3h}	D_{2h}	D_6	D_5	D_3	D_2	S_6	S_4	C_{3h}	C_{2h}	C_{3v}	C_{2v}	C_3	C_2	C_s	C_i	C_1
C_{20}	1	0	0	0	0	0	0	0	0	0	0	0	0	0	0	0	0	0	0	0	0	0	0	0	0	0	0	0
C_{24}	0	0	0	0	0	1	0	0	0	0	0	0	0	0	0	0	0	0	0	0	0	0	0	0	0	0	0	0
C_{26}	0	0	0	0	0	0	0	0	0	0	0	1	0	0	0	0	0	0	0	0	0	0	0	0	0	0	0	0
C_{28}	0	0	0	1	0	0	0	0	0	0	0	0	0	0	0	0	1	0	0	0	0	0	0	0	0	0	0	0
C_{30}	0	0	0	0	0	0	0	0	0	0	1	0	0	0	0	0	0	0	0	0	0	0	2	0	0	0	0	0
C_{32}	0	0	0	0	0	0	0	1	0	0	0	1	0	0	0	1	1	0	0	0	0	1	0	0	2	0	0	0
C_{34}	0	0	0	0	0	0	0	0	0	0	0	0	0	0	0	0	0	0	0	0	0	1	0	0	3	2	0	2
C_{36}	0	0	0	0	0	0	0	0	2	0	0	1	0	0	0	0	2	0	0	0	0	0	1	0	4	2	0	2
C_{38}	0	0	0	0	0	0	0	0	0	0	0	1	0	0	0	1	0	0	0	0	0	1	2	0	5	0	0	7
C_{40}	0	0	0	1	0	0	2	0	0	0	0	1	1	0	0	0	3	0	0	0	0	1	2	1	14	7	0	8
C_{42}	0	0	0	0	0	0	0	0	0	0	0	0	0	0	0	1	0	0	0	0	0	0	4	0	11	6	0	23
C_{44}	0	0	0	0	1	0	0	3	0	0	0	0	0	0	0	2	6	0	1	0	0	0	3	0	22	7	0	42
C_{46}	0	0	0	0	0	0	0	0	0	0	0	0	0	0	0	0	0	0	0	0	0	0	4	2	22	19	0	69
C_{48}	0	0	0	0	0	0	0	0	5	0	0	0	2	0	0	1	0	0	0	0	1	0	3	0	52	16	0	117
C_{50}	0	0	0	0	0	0	1	0	0	0	0	0	0	0	2	0	5	0	0	0	0	0	6	0	37	25	0	195
C_{52}	0	0	0	0	0	0	0	0	0	0	0	0	0	0	0	2	0	0	0	0	0	1	3	2	78	26	0	307
C_{54}	0	0	0	0	0	0	0	2	0	0	0	1	0	0	0	0	9	0	0	0	0	2	8	3	62	38	1	470
C_{56}	0	0	0	1	0	0	0	0	1	0	0	0	1	0	0	6	10	0	0	0	2	0	13	3	135	49	1	700
C_{58}	0	0	0	0	0	0	0	0	0	0	0	0	0	0	0	0	0	0	0	0	0	2	6	4	98	58	0	1037
C_{60}	1	0	0	0	0	0	1	0	0	0	0	0	0	0	0	3	0	0	2	0	4	1	9	1	189	67	0	1508
C_{62}	0	0	0	0	0	0	0	0	0	0	0	2	0	0	0	4	0	0	0	1	0	1	16	4	142	80	0	2135
C_{64}	0	0	0	0	0	0	0	0	0	0	0	0	1	0	0	0	17	0	0	0	4	4	4	8	316	118	2	2990
C_{66}	0	0	0	0	0	0	0	0	0	0	0	0	0	0	0	2	0	1	0	0	0	1	18	0	211	112	0	4134
C_{68}	0	0	0	1	1	0	0	3	2	0	0	1	2	0	0	10	28	1	2	1	7	0	21	5	411	122	0	5714
C_{70}	0	0	0	0	0	0	0	0	0	0	2	0	0	0	2	0	0	0	0	0	0	5	14	8	300	186	0	7634

Table 5.2. (*Continued*)

Fullerene	I_h	I	T_h	T_d	T	D_{6d}	D_{5d}	D_{3d}	D_{2d}	D_{6h}	D_{5h}	D_{3h}	D_{2h}	D_6	D_5	D_3	D_2	S_6	S_4	C_{3h}	C_{2h}	C_{3v}	C_{2v}	C_3	C_2	C_s	C_i	C_1
C_{72}	0	0	0	0	0	2	0	0	4	0	0	0	5	1	0	3	24	0	0	0	7	1	26	1	619	190	3	10304
C_{74}	0	0	0	0	0	0	0	0	0	0	0	3	0	0	0	6	0	0	0	1	0	1	18	9	414	237	0	13557
C_{76}	0	0	0	0	1	0	0	0	3	0	0	0	0	0	0	0	45	0	4	0	11	5	14	14	800	246	2	18005
C_{78}	0	0	0	0	0	0	0	0	0	0	0	2	0	0	0	3	0	0	0	0	0	0	35	2	557	312	0	23197
C_{80}	1	0	0	0	0	0	2	2	0	0	2	2	0	0	0	12	39	0	1	0	16	2	28	15	1146	371	5	30280
C_{82}	0	0	0	0	0	0	0	0	0	0	0	0	0	0	0	0	0	0	0	0	0	5	28	15	742	380	0	38548
C_{84}	0	0	0	1	0	0	0	1	9	0	0	1	6	0	0	6	59	0	3	0	4	1	29	1	1436	434	9	49590
C_{86}	0	0	0	0	0	0	0	0	0	0	0	3	0	0	0	9	0	0	0	0	0	5	36	15	976	505	0	62212
C_{88}	0	0	0	0	0	0	0	0	4	0	0	0	5	0	0	0	52	0	1	0	16	8	43	24	1945	596	10	79033
C_{90}	0	0	0	0	0	0	0	0	0	0	2	1	0	1	0	4	0	0	0	0	0	0	50	3	1266	655	0	97936
C_{92}	0	0	1	1	1	0	0	5	4	0	0	3	4	0	0	19	80	0	5	2	13	2	38	20	2412	646	12	123141
C_{94}	0	0	0	0	0	0	0	0	0	0	0	0	0	0	0	0	0	0	0	0	0	4	42	26	1603	879	0	150939
C_{96}	0	0	0	0	0	3	0	1	3	0	0	1	3	0	0	9	70	0	0	0	16	2	28	4	3200	972	20	187505
C_{98}	0	0	0	0	0	0	0	1	0	0	0	2	0	0	0	13	0	0	0	1	0	1	58	27	2029	952	0	227934
C_{100}	0	0	0	1	2	0	2	0	5	0	0	0	2	2	2	0	114	0	9	0	28	5	66	40	3801	1093	14	280730

Table 5.3. Symmetry groups of the isolated-pentagon isomers for C_{60} to C_{140}. Only one member of each enantiomeric pair is counted. Above C_{100} the counts refer to spirals starting on pentagons. See Table 2.2 for the total isomer counts.

Fullerene	I_h	I	T_h	T_d	T	D_{6d}	D_{5d}	D_{3d}	D_{2d}	D_{6h}	D_{5h}	D_{3h}	D_{2h}	D_6	D_5	D_3	D_2	S_6	S_4	C_{3h}	C_{2h}	C_{3v}	C_{2v}	C_3	C_2	C_s	C_i	C_1
C_{60}	1	0	0	0	0	0	0	0	0	0	0	0	0	0	0	0	0	0	0	0	0	0	0	0	0	0	0	0
C_{70}	0	0	0	0	0	0	0	0	0	0	1	0	0	0	0	0	0	0	0	0	0	0	0	0	0	0	0	0
C_{72}	0	0	0	0	0	1	0	0	0	0	0	0	0	0	0	0	0	0	0	0	0	0	0	0	0	0	0	0
C_{74}	0	0	0	0	0	0	0	0	0	0	0	1	0	0	0	0	0	0	0	0	0	0	0	0	0	0	0	0
C_{76}	0	0	0	1	0	0	0	0	0	0	0	0	0	0	0	0	1	0	0	0	0	0	0	0	0	0	0	0
C_{78}	0	0	0	0	0	0	0	0	0	0	0	2	0	0	0	1	0	0	0	0	0	0	2	0	0	0	0	0
C_{80}	1	0	0	0	0	0	1	0	0	0	0	0	0	0	0	2	1	0	0	0	0	0	2	0	0	0	0	0
C_{82}	0	0	0	0	0	0	0	0	0	0	0	0	0	0	0	0	0	0	0	0	0	2	1	0	3	3	0	0
C_{84}	0	0	0	1	0	0	0	1	2	0	0	0	0	0	0	2	4	0	0	0	0	0	4	0	5	3	0	1
C_{86}	0	0	0	0	0	0	0	0	0	0	0	0	0	0	0	0	0	0	0	0	0	0	2	0	6	5	0	6
C_{88}	0	0	0	0	1	0	0	0	0	0	0	0	0	0	0	1	2	0	0	0	0	0	3	0	7	3	0	11
C_{90}	0	0	0	0	0	0	0	0	0	0	0	0	0	0	0	0	0	0	0	0	0	0	7	0	16	11	0	16
C_{92}	0	0	0	0	1	0	0	0	0	1	0	0	1	0	0	0	4	0	0	0	0	0	2	0	26	6	0	38
C_{94}	0	0	0	0	0	0	0	0	1	0	0	0	1	0	0	5	8	0	0	0	0	1	2	0	26	8	0	89
C_{96}	0	0	0	0	0	2	0	1	0	1	0	0	0	0	0	0	0	0	0	0	0	0	3	1	43	13	0	108
C_{98}	0	0	0	0	0	0	1	0	1	1	0	1	0	0	0	3	9	0	0	0	0	1	5	0	49	14	0	169
C_{100}	0	0	0	0	1	0	0	0	0	0	0	0	0	1	0	3	0	0	0	0	0	0	5	1	62	30	0	336
C_{102}	0	0	0	0	0	0	0	1	0	0	0	0	0	0	0	0	11	0	0	0	0	0	8	1	73	31	0	488
C_{104}	0	0	0	0	0	0	0	0	1	0	0	1	2	0	0	0	0	0	1	0	3	3	8	0	123	44	0	644
C_{106}	0	0	0	0	0	0	0	0	2	0	0	0	0	0	0	4	0	0	0	0	0	3	8	0	105	26	0	1054
C_{108}	0	0	0	0	0	0	0	1	0	0	0	2	1	0	0	0	21	0	1	0	0	1	9	4	201	54	0	1479
C_{110}	0	0	0	0	0	0	0	0	0	0	1	0	0	0	1	6	0	0	0	0	0	0	8	6	168	72	0	2111
C_{112}	0	0	1	0	0	0	0	0	0	0	0	0	0	0	0	0	20	0	0	0	2	2	5	0	250	56	0	2950
C_{114}	0	0	0	0	0	0	0	0	0	0	0	1	0	0	0	0	0	0	0	0	0	2	18	1	258	94	0	4089
C_{116}	0	0	1	0	1	0	0	0	4	0	0	0	1	0	0	10	22	0	0	2	2	3	7	7	386	112	0	5508

Table 5.3. (Continued)

Fullerene	I_h	I	T_h	T_d	T	D_{6d}	D_{5d}	D_{3d}	D_{2d}	D_{6h}	D_{5h}	D_{3h}	D_{2h}	D_6	D_5	D_3	D_2	S_6	S_4	C_{3h}	C_{2h}	C_{3v}	C_{2v}	C_3	C_2	C_s	C_i	C_1
C_{118}	0	0	0	0	0	0	0	0	0	0	0	0	0	0	0	0	0	0	0	0	0	3	13	11	303	148	0	7670
C_{120}	0	0	0	1	0	1	3	0	1	0	1	0	3	1	1	4	29	0	0	0	5	1	25	1	612	180	1	9904
C_{122}	0	0	0	0	0	0	0	0	0	0	0	0	0	0	0	7	0	0	0	0	0	0	13	12	472	178	0	13295
C_{124}	0	0	0	2	0	0	0	0	3	0	0	0	1	0	0	0	41	0	3	0	5	2	10	15	735	200	1	17751
C_{126}	0	0	0	0	0	0	0	0	0	0	0	0	0	0	0	5	0	0	0	0	0	0	29	2	625	242	0	22686
C_{128}	0	0	0	0	0	0	0	2	1	0	0	4	0	0	0	11	36	1	0	0	4	0	13	12	1052	269	3	29275
C_{130}	0	0	0	0	0	0	0	0	0	0	1	0	0	0	1	0	0	0	0	0	0	6	29	24	773	274	0	38285
C_{132}	0	0	0	0	1	0	0	5	3	0	0	3	2	1	0	8	71	0	6	0	10	2	35	2	1369	346	0	48013
C_{134}	0	0	0	0	0	0	0	0	0	0	0	1	0	0	0	13	0	0	0	1	0	1	26	13	1117	433	0	60767
C_{136}	0	0	0	0	1	0	0	0	3	0	0	0	3	0	0	0	60	0	2	0	12	2	32	23	1740	408	4	77072
C_{138}	0	0	0	0	0	0	0	0	0	0	0	0	0	0	0	5	0	0	0	0	0	2	31	6	1382	517	0	96598
C_{140}	0	1	0	0	2	0	1	0	0	0	0	0	0	0	1	17	71	0	4	1	7	0	26	24	2335	535	8	118321

group). In a large group there may be several possible resolutions of a given degeneracy and the choice of subgroup chain will depend on the specific energetics of the case, but in an important sense the distortion and the loss of symmetry are guaranteed to be minimal. All epikernels of the fullerene point groups that allow degeneracy are themselves fullerene point groups[2] and so Jahn–Teller distortion cannot lead to a group outside the original list. It is important to note, however, that when a neutral fullerene is assigned to a degenerate point group and predicted in the main tables to have a zero band gap, this is a danger sign. In such a case, the spectroscopic signatures must be interpreted with some care.

A more subtle symmetry-breaking effect can occur via second-order Jahn–Teller distortion. If the HOMO and LUMO of a fullerene, calculated in the maximal symmetry group, are sufficiently close in energy *and* if the product of their symmetries matches one of the normal modes of vibration then it *may* be energetically favourable for the molecule to distort.[13] This does not require an open-shell configuration and would be an analogue of the bending of 14-electron AH_2 species predicted by Walsh's rules.[14] It is another possibility to be kept in mind when scanning the main tables, but again cannot lead to a symmetry outside the list of 28 groups. Both types of Jahn–Teller distortion are beyond the scope of the nearest-neighbour simple Hückel Hamiltonian with all α and β parameters set equal.

A limited list of point groups also describes compounds of fullerenes that preserve the basic carbon skeleton. Addition of functional groups to atoms of a fullerene cage either has no effect on the symmetry (when all members of an equivalent set are attacked by ligands whose symmetry matches the site group) or causes symmetry lowering (e.g., when addition is to part of a set). Similar remarks apply to the η^2 addition of functional groups, atoms, or ligands across equivalent edges of the cage, or to the hypothetical η^5 or η^6 addition of groups to faces. Endohedral metallofullerenes, *exo-* or *endo-* functionalized C_nX_m molecules, heterofullerenes, and organometallic complexes with a single fullerene ligand are therefore all described by subgroups of fullerene groups. The full list of fullerene point groups and subgroups takes us back to the original list of 36 groups proposed for fullerenes themselves,[4] namely the 28 fullerene groups plus C_5, C_{5v}, C_{5h}, S_{10}, C_6, C_{6v}, C_{6h}, and S_{12}, all of which are possible for fullerene derivatives. A substituted or functionalized fullerene must therefore belong to one of 36 point groups, and a fullerene itself to one of only 28.

5.4 ^{13}C NMR spectra

In natural abundance, only one in every hundred carbon nuclei is ^{13}C, and in ideal circumstances the ^{13}C NMR spectrum of a fullerene isomer will be very simple. It will consist of a number of peaks, one for each equivalent set of atomic sites, with intensities proportional to the number of sites in each set. In the language of orbits, every atomic orbit will give rise to a peak, and the peak height will be proportional to the size of the orbit and therefore (from Eq. (5.1)) inversely proportional to the order of the site group (i.e., with relative heights of 1 for C_{3v}, 2 for C_3, 3 for C_s, and 6 for C_1 sites). A symmetry analysis can therefore predict the numbers and relative intensities of peaks in the hypothetical ^{13}C NMR spectrum of each fullerene isomer.

The orbits can be counted easily using the topological coordinates. A topological radius is assigned to each vertex by computing $r_j = (x_j^2 + y_j^2 + z_j^2)^{1/2}$, and the vertices are sorted into sets of equal radius. Errors caused by exact coincidences of radii for distinct orbits are rare and can be trapped by checking the symmetry elements passing through each point, or by checking that all atoms at a given radius are exchanged by symmetry operations. The site symmetry follows immediately from the size of each orbit and the order of the point group, again from Eq. (6.1). Results for each possible isomer are given in the main tables as a list of peak counts at each possible height.

Symmetry places one very useful limitation on the ^{13}C NMR spectrum. Only four site groups are available to a fullerene atom and so no more than four different peak heights are possible.[6] However, the C_{3v} and C_3 site groups are mutually exclusive, because no point group has two distinct sets of C_3 axes. The idealized stick spectrum of a pure fullerene isomer may therefore contain at most *three* different peak heights.[2] If a larger number is observed, the sample *must* contain a mixture of isomers. The 28 fullerene point groups give NMR patterns as follows:

C_1 sites only: $C_1, C_i, C_2, S_4, D_2, D_5, D_6$ all heights equal;

C_1 and C_s sites: $C_s, C_{2h}, C_{2v}, D_{2h}, D_{5h}, D_{6h}, D_{2d}, D_{5d}, D_{6d}$ at most 2 heights (2:1);

C_1 and C_3 sites: C_3, S_6, D_3, T, I at most 2 heights (3:1);

$C_1, C_s,$ and C_3 sites: C_{3h}, T_h at most 3 heights (3:2:1);

$C_1, C_s,$ and C_{3v} sites: $C_{3v}, D_{3h}, D_{3d}, T_d, I_h$ at most 3 heights (6:2:1).

Various more specialized results can also be derived by thinking about the number of copies of an orbit that can be present in a structure consisting of a single pseudospherical shell. The C_3 and C_{3v} orbits consist *either* of a single point on each C_3 axis *or* of a point and its opposite pole on each C_3 axis, depending on the overall symmetry. In the groups D_{3h}, D_{3d}, and I_h there is at most one C_{3v} orbit of atoms, whereas in C_{3v} and T_d there may be up to two orbits. Likewise in S_6, D_3, C_{3h}, T_h, and I there is at most one C_3 orbit, but up to two in C_3 and T. Thus in favourable cases the NMR spectrum gives a strong hint about the symmetry of the isomer.

In some groups only one type of orbit is possible; the seven groups that will always give spectra consisting of peaks of equal height are C_1 to D_6 as listed above. Some trivial mathematical consequences are that all D_2 and S_4 fullerenes contain multiples of four atoms, whereas D_5 and D_6 fullerenes have multiples of 10 and 12 atoms, respectively.

Symmetry alone gives no information on ^{13}C NMR chemical shifts, and it is possible that nuclei in chemically similar but inequivalent sites may give rise to overlapping peaks. The effective point group of the molecule may also be lower than the maximal group, as discussed earlier, so that the idealized pattern predicted in the main tables may be further split. All the above rules still apply, however, to the molecule in its distorted configuration.

It is worth emphasizing that the number of peaks in the NMR spectrum can serve to eliminate candidate structures for a fullerene, but does not always uniquely predict a structure. Perhaps the best example of this is the famous 'one-line proof' of the structure of C_{60} itself.[15] The ^{13}C NMR spectrum of C_{60} has a single peak at ~ 14 ppm downfield of the benzene signal. Amongst the 1812 fullerene isomers, only one can have such a simple spectrum — the correct icosahedral structure. However, without the strong presumption that a fullerene structure is involved, a number of alternative arrangements of 60 equivalent atoms could be proposed – a ring of 60 atoms (in D_{60h} or D_{30h} point groups), two rings of 30 atoms (in D_{30h}, D_{30d}, or D_{30}) or a truncated dodecahedron (Fig. 5.4) The latter structure has full I_h symmetry and exactly the same NMR signature (and the same IR and Raman spectroscopic signatures, see later) as the truncated icosahedral fullerene C_{60}. The two structures are in fact realizations of the same orbit of I_h, but with different values of the two independent structural parameters. Hückel calculations indicate a highly de-

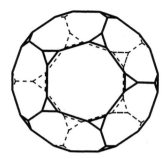

FIG. 5.4. The truncated dodecahedron. This polyhedron has 60 equivalent vertices and a carbon cluster of this shape would be indistinguishable on the basis of the numbers of peaks in its NMR, IR, and Raman spectra from the icosahedral C_{60} fullerene.

generate nonbonding HOMO for truncated dodecahedral C_{60}, and the presence of 20 triangular rings rules it out as a chemically plausible structure for C_{60}, but the NMR evidence alone does not do so.

More significantly, equal peak counts can be predicted for different isomers within the fullerene constraint. For example, amongst the 40 isomers of C_{40} are 14 with signatures of 20 lines of equal height, and 8 with 40 lines of equal height. Stability considerations may cut down the possibilities. For C_{76} there are 50 isomers (45 of D_2, 4 of S_4, and 1 of C_{2h} symmetry) which give rise to 19 peaks in maximal symmetry (as observed in experiment), but of these only one D_2 isomer has isolated pentagons and a favourable electronic structure. Nevertheless, addition of a few more carbon atoms to give the C_{82} cage is sufficient to give isolated-pentagon isomers that are indistinguishable on spectroscopic signature alone.

5.5 IR and Raman spectra

Once its point group is known, the vibrations of a molecule can be analysed and the IR-active, Raman-active, polarized Raman, and IR/Raman coincident modes counted, fully characterizing the spectrum. A fullerene C_n has $3n - 6$ vibrations and their symmetries are given by Γ_{vib}, found by subtracting the translational and rotational symmetries from the representation of the $3n$ Cartesian atomic displacement coordinates. Reduction of the representation can be done systematically using σ and π representations for the various orbits,[3] or with the help of the Brester tables,[16] which give an orbit-by-orbit breakdown of vibrational symmetry. If the atoms of a fullerene C_n are sorted into m_1, m_s, m_3, and m_{3v} orbits of each site symmetry (C_1, C_s, C_3, and C_{3v}) then

$$m_1 + 2\,m_s + 3\,m_3 + 6\,m_{3v} = n, \qquad (5.6)$$

and all the various counts (with one exception discussed below) are functions of these four numbers.

A fundamental transition is infra-red-active if the normal mode involved belongs to the same representation of the point group as one or more components of the dipole moment, and Raman-active if the mode shares a representation with one or more components of the polarizability tensor.[16] A mode can in general be IR-active, Raman-active, or neither or both. Simultaneously IR- and Raman-active modes are impossible in point groups with a centre of symmetry (the rule of mutual exclusion), and in groups with certain combinations of other elements (D_{5h}, D_{6d} and I amongst the fullerene groups). Raman lines are characterized by a depolarization ratio[16] ρ that describes the ratio of scattered intensity polarized parallel and perpendicular to the electric vector of the incident light. Different ratios are defined depending on whether the incident light is plane (linearly) polarized (ρ_l) or natural, unpolarized (ρ_n). Only totally symmetric vibrations can give depolarization ratios $\rho_n < \frac{6}{7}$ and $\rho_l < \frac{3}{4}$, and these are termed 'polarized' modes. They include all the totally symmetric vibrations, because at least one component of the polarizability tensor is totally symmetric in every point group. In the absence of Coriolis perturbations, vibrations of other symmetries give $\rho_n = \frac{6}{7}$ and $\rho_l = \frac{3}{4}$.

Table 5.4 gives expressions for the numbers of IR-active (N_I), Raman-active (N_R), polarized Raman (N_P), and coincident (N_C) lines in terms of the orbit counts within the 28 fullerene groups. In one case, that of D_{3h}, it is necessary to analyse the m_s count more carefully as a sum $m_h + m_v$ according to whether the atoms with C_s site symmetry lie in the unique horizontal mirror plane (m_h) or in one of the vertical mirror planes (m_v). Even in this group the sum determines all but the coincidence count N_C. Results computed using these expressions are listed in the main tables for all fullerene isomers considered. N_P is not listed there explicitly but it can be worked out in each case as follows.

Since the polarized Raman vibrations are totally symmetric, their displacement coordinates preserve the site symmetry of each vibrating atom. Each atom in a C_1 site therefore contributes three, each atom in a C_s site contributes

Table 5.4. Vibrational modes of fullerenes. The four rows associated with each of the 28 fullerene point groups give respectively the counts for IR-active, Raman-active, polarized Raman-active, and IR/Raman coincident modes. Each count is expressed in terms of the numbers m_1, m_s, m_3, and m_{3v} of orbits with site symmetries C_1, C_s, C_3, and C_{3v}. For example, the 60 atoms in I_h C_{60} comprise a single orbit O_{60} with site symmetry C_s, giving 4 IR-active, 10 Raman-active, 2 polarized Raman-active, and 0 IR/Raman coincident modes. In the D_{3h} point group it is necessary to resolve m_s into horizontal (m_h) and vertical (m_v) orbit counts as $m_s = m_h + m_v$.

Group	Vibrational modes	Group	Vibrational modes
C_1	$3m_1 - 6$	C_i	$3m_1 - 3$
	$3m_1 - 6$		$3m_1 - 3$
	$3m_1 - 6$		$3m_1 - 3$
	$3m_1 - 6$		0
C_s	$3m_s + 6m_1 - 6$	C_2	$6m_1 - 6$
	$3m_s + 6m_1 - 6$		$6m_1 - 6$
	$2m_s + 3m_1 - 3$		$3m_1 - 2$
	$3m_s + 6m_1 - 6$		$6m_1 - 6$
C_3	$2m_3 + 6m_1 - 4$	C_{2h}	$3m_s + 6m_1 - 3$
	$2m_3 + 6m_1 - 4$		$3m_s + 6m_1 - 3$
	$m_3 + 3m_1 - 2$		$2m_s + 3m_1 - 1$
	$2m_3 + 6m_1 - 4$		0
C_{3h}	$2m_3 + 3m_s + 6m_1 - 2$	C_{2v}	$5m_s + 9m_1 - 5$
	$3m_3 + 5m_s + 9m_1 - 3$		$6m_s + 12m_1 - 6$
	$m_3 + 2m_s + 3m_1 - 1$		$2m_s + 3m_1 - 1$
	$m_3 + 2m_s + 3m_1 - 1$		$5m_s + 9m_1 - 5$
C_{3v}	$2m_{3v} + 5m_s + 9m_1 - 3$	S_4	$6m_1 - 3$
	$2m_{3v} + 5m_s + 9m_1 - 3$		$9m_1 - 4$
	$m_{3v} + 2m_s + 3m_1 - 1$		$3m_1 - 1$
	$2m_{3v} + 5m_s + 9m_1 - 3$		$6m_1 - 3$
S_6	$2m_3 + 6m_1 - 2$	D_2	$9m_1 - 6$
	$2m_3 + 6m_1 - 2$		$12m_1 - 6$
	$m_3 + 3m_1 - 1$		$3m_1$
	0		$9m_1 - 6$
D_3	$3m_3 + 9m_1 - 4$	D_5	$9m_1 - 4$
	$3m_3 + 9m_1 - 2$		$15m_1 - 2$
	$m_3 + 3m_1$		$3m_1$
	$2m_3 + 6m_1 - 2$		$6m_1 - 2$

Table 5.4. *(Continued)*

Group	Vibrational modes	Group	Vibrational modes
D_6	$9m_1 - 4$	D_{2h}	$5m_s + 9m_1 - 3$
	$15m_1 - 2$		$6m_s + 12m_1 - 3$
	$3m_1$		$2m_s + 3m_1$
	$6m_1 - 2$		0
D_{3h}	$2m_{3v} + 5m_h + 5m_v + 9m_1 - 2$	D_{5h}	$5m_s + 9m_1 - 2$
	$3m_{3v} + 8m_h + 8m_v + 15m_1 - 2$		$8m_s + 15m_1 - 1$
	$m_{3v} + 2m_h + 2m_v + 3m_1$		$2m_s + 3m_1$
	$m_{3v} + 4m_h + 3m_v + 6m_1 - 1$		0
D_{6h}	$5m_s + 9m_1 - 2$	D_{2d}	$5m_s + 9m_1 - 3$
	$8m_s + 15m_1 - 1$		$8m_s + 15m_1 - 3$
	$2m_s + 3m_1$		$2m_s + 3m_1$
	0		$5m_s + 9m_1 - 3$
D_{3d}	$2m_{3v} + 5m_s + 9m_1 - 2$	D_{5d}	$5m_s + 9m_1 - 2$
	$2m_{3v} + 5m_s + 9m_1 - 1$		$8m_s + 15m_1 - 1$
	$m_{3v} + 2m_s + 3m_1$		$2m_s + 3m_1$
	0		0
D_{6d}	$5m_s + 9m_1 - 2$	T	$3m_3 + 9m_1 - 2$
	$8m_s + 15m_1 - 1$		$5m_3 + 15m_1 - 2$
	$2m_s + 3m_1$		$m_3 + 3m_1$
	0		$3m_3 + 9m_1 - 2$
T_h	$3m_3 + 5m_s + 9m_1 - 1$	T_d	$2m_{3v} + 5m_s + 9m_1 - 1$
	$5m_3 + 8m_s + 15m_1 - 1$		$4m_{3v} + 10m_s + 18m_1 - 1$
	$m_3 + 2m_s + 3m_1$		$m_{3v} + 2m_s + 3m_1$
	0		$2m_{3v} + 5m_s + 9m_1 - 1$
I	$3m_3 + 9m_1 - 2$	I_h	$2m_{3v} + 5m_s + 9m_1 - 1$
	$6m_3 + 18m_1$		$4m_{3v} + 10m_s + 18m_1$
	$m_3 + 3m_1$		$m_{3v} + 2m_s + 3m_1$
	0		0

two, and each atom in a C_3 or C_{3v} site contributes one to N_P (because, for example, each atom in a C_{3v} site must remain on the C_3 axis and therefore only has one degree of freedom). However, the sum of these contributions includes the totally symmetric rotations and translations which do not appear in Γ_{vib}. Subtracting these latter motions accordingly leaves

$$N_P = 3\,m_1 + 2\,m_s + m_3 + m_{3v} - \Delta, \tag{5.7}$$

where Δ is a number characteristic of the point group ($\Delta = 6$ for C_1, 3 for C_i and C_s, 2 for C_2 and C_3, 1 for C_{2h}, C_{3h}, C_{2v}, C_{3v}, S_4, and S_6, and 0 for the other 17 fullerene groups).

The trick that is required to evaluate Eq. (5.7) from the information in the main tables is to extract m_1, m_s, m_3, and m_{3v} from the tabulated ^{13}C NMR signature, and this can be done using Eq. (5.1) and a knowledge of the various point group orders in Table 5.1. For example, icosahedral C_{60} has the NMR signature 1×60, corresponding to a single orbit of 60 atoms. Equation (5.1) gives the order of the site symmetry group of this orbit as $120/60 = 2$, and the site symmetry is therefore C_s. (This assignment is unambiguous, because the four site symmetry groups available to a fullerene atom all have different orders: 6 for C_{3v}, 3 for C_3, 2 for C_s, and 1 for C_1.) Icosahedral C_{60} therefore has $m_s = 1$ and $m_{3v} = m_3 = m_1 = 0$, and since $\Delta = 0$ for I_h Eq. (5.7) gives $N_P = 2$. A similar analysis shows that a C_1 C_{60} isomer with NMR signature 60×1 would have $N_P = 174$. These numbers are of interest to quantum chemists as much as to spectroscopists, because the number of totally symmetric vibrations N_P is equal to the number of independent structural parameters in an isomer – i.e., the number of parameters that must be varied independently in an *ab initio* geometry optimization calculation.

Since the three remaining vibrational mode counts are also functions of the orbit counts m_{3v}, m_3, m_s, and m_1, except in the special case of the D_{3h} coincidence count N_C, these counts too can be computed from the ^{13}C NMR signature. However, because the computation in these cases requires reference to Table 5.4, N_I, N_R and N_C are given explicitly in the main tables for convenience. An important corollary of the fact that these counts are also ultimately contained in the NMR pattern is that any two fullerene isomers with the same point group symmetry and the same ^{13}C NMR signature will also have identical IR/Raman signatures (excepting the D_{3h} case noted above), and so will be difficult to distinguish spectroscopically unless their chemical shifts and/or vibrational frequencies are calculated to be markedly different.

References and notes

[1] Point group theory and the uses of symmetry in chemistry are treated in many textbooks, e.g., F. A. Cotton, Chemical applications of group theory (Wiley, New York, 1990); D. M. Bishop, Group theory and chemistry (Oxford University Press, Oxford, 1973); S. F. Kettle, Symmetry and structure (Wiley, New York, 1985); J. S. Griffith, The theory of transition metal ions (Cambridge University Press, Cambridge, 1961).

[2] P. W. Fowler, D. E. Manolopoulos, D. B. Redmond, and R. P. Ryan, Chem. Phys. Lett. 202 (1993) 371.

[3] P. W. Fowler and C. M. Quinn, Theor. Chim. Acta 70 (1986) 333.

[4] P. W. Fowler, J. Cremona, and J. I. Steer, Theor. Chim. Acta 73 (1988) 1.

[5] J. K. G. Watson, J. Mol. Spec. 40 (1971) 536.

[6] D. E. Manolopoulos and P. W. Fowler, J. Chem. Phys. 96 (1992) 7603.

[7] A. J. Stone, Mol. Phys. 41 (1980) 1339.

[8] A. J. Stone, Inorg. Chem. 20 (1981) 563.

[9] P. W. Fowler and J. Woolrich, Chem. Phys. Lett. 127 (1986) 78.

[10] More precisely, a fullerene dual graph has an essentially unique realization in three-dimensional space as a convex deltahedron with every edge tangent to a unit sphere. This is discussed in C.-H. Sah, A generalized leapfrog for fullerene structures, to be published in Croat. Chem. Acta (1994), where the proof is attributed to W. P. Thurston, Shapes of polyhedra, Research Report GCG 7, Geometry Centre, University of Minnesota (1991)).

[11] An alternative procedure based on a complete set of spiral unwindings of the fullerene surface is implemented in the computer program in the Appendix (see also § 2.5). This has the advantage that it can be coded entirely in integer arithmetic, and is therefore immune to rounding errors. However, both procedures give the same results in practice, and the topological coordinate approach has the advantage that it is less abstract and so is easier to describe.

[12] A. Ceulemans and L. G. Vanquickenborne, Structure and Bonding 71 (1989) 125.

[13] An example is C_{44}:73, which has maximal T symmetry but is predicted by various semi-empirical models to distort to a structure of D_2 symmetry with a wider HOMO–LUMO gap (P. W. Fowler and J. P. B. Sandall, J. Chem. Soc. Perkin 2 (1994) 1917).

[14] T. A. Albright, J. K. Burdett and M. H. Whangbo, Orbital interactions in chemistry (Wiley, New York, 1985).

[15] R. Taylor, J. P. Hare, A. K. Abdul-Sada, and H. W. Kroto, J. Chem. Soc. Chem. Comm. (1990) 1423.

[16] G. Herzberg, Molecular spectra and structure II: infrared and raman spectra of polyatomic molecules (Van Nostrand, Princeton, 1945). See especially Chap. II,4.

6

FULLERENE ISOMERIZATION

A striking characteristic of the fullerenes as a class, and an important reason for making the compilation in this book, is the huge number of isomers corresponding to the fullerene definition that can be described by the single formula C_n. Knowing the extent of this isomerism, one might find the experimental observations puzzling. On the one hand, a random process of self-assembly should lead to formation of many if not all of the conceivable isomers, and on the other it is known that the experimental products contain no more than a few isomers at each n.[1-6] It seems likely that under the violent conditions of graphite vaporization the isomers can interconvert, allowing conversion of thermodynamically unfavourable cages to the few favoured structures. Since the essential difference between fullerene isomers lies in the distribution of the pentagons, this hypothesis implies that there should be a mechanism for motion of pentagons on the fullerene surface. One such mechanism is described here and is used in the tables to separate the isomers into conversion classes. This is the so-called Stone–Wales or pyracylene rearrangement.

6.1 The Stone–Wales rearrangment

The pyracylene rearrangement was first proposed in the context of fullerenes in a paper by Stone and Wales[7] on possible non-icosahedral isomers of C_{60}, but is not restricted to this application, and in fact it provides a general hypothetical mechanism for interconversion between fullerene isomers.

To perform the transformation it is first necessary to find a place on the surface of the fullerene polyhedron where two hexagons and two pentagons meet in a pyracylene/pyracene patch (**I** in Fig. 6.1). If such a patch exists, the central bond (the Stone–Wales or SW bond) can be twisted, breaking two of its neighbouring bonds and making two more, thereby swapping pentagons and hexagons, and producing a 'rotated' patch within the same perimeter of 12 atoms and 12 bonds. The product is still a fullerene and is in general different

(a)   (b)

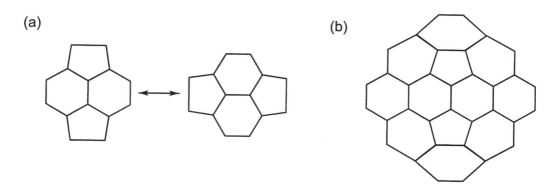

FIG. 6.1. Patches of the fullerene surface involved in (a) the Stone–Wales transformation between general fullerene isomers (type **I**), and (b) between isolated-pentagon fullerenes (type **II**).

from the starting isomer, though it is possible for the initial and final fullerenes to be enantiomeric or even conceivably identical. The 'sense' of the twist is immaterial because rotations of $\pm(\pi/2)$ lead to the same product, though possibly via different transition states. Any single SW transformation is reversible and so links isomers in a well-defined way. Given the complete set of fullerene isomers of a particular C_n it is straightforward to generate all possible SW interconversions and so to construct a map of isomerizations allowed under this mechanism. The smallest fullerene to support a Stone–Wales transformation is C_{28}:1 (D_2), which consists of two type **I** patches separated by a strip of eight pentagons.

When only isolated-pentagon isomers are under consideration, it is useful to define a restricted version of the transformation for which both initial and final fullerenes have isolated pentagons. If these isomers are indeed more stable than those with adjacent pentagons, for the reasons discussed in Chapters 3 and 4, then such restricted SW transformations may have greater chemical relevance than more general SW isomerizations. A necessary and sufficient condition for an isolate-to-isolate transformation is the presence of an appropriate 12-ring patch (**II** in Fig. 6.1) in the initial isolated-pentagon fullerene. Rotation of the central SW bond within the 20-atom perimeter of this patch introduces no new pentagon adjacency. The smallest isolated-pentagon fullerenes with type **II** patches are four of the five isomers of C_{78} in Table A.10.

The SW transformation is expected to be a high-energy process. It proceeds

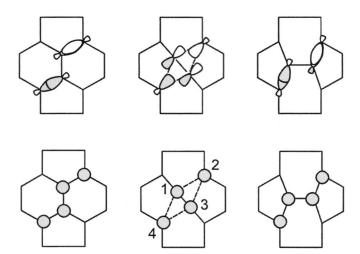

FIG. 6.2. Orbital changes during a (concerted) Stone–Wales transformation. The top and bottom rows show changes in the surface (σ) and radial (π) systems, respectively. As the rearrangement proceeds, two old bonds are broken and two new bonds are formed in each system. The topological reaction coordinate r defines the Hückel integrals for the participating bonds via $\beta_{12} = \beta_{34} = (1/2)(1 - r)\beta$ and $\beta_{23} = \beta_{41} = (1/2)(1+r)\beta$. The three snapshots (left to right) correspond to r values of -1, 0, and $+1$, respectively. At $r = \pm 1$ all four atoms contribute sp^2 hybrids to the σ bonding, but at $r = 0$ the hybridization has changed to sp (on atoms 2 and 4) and p_σ (on atoms 1 and 3).

via an antiaromatic 4-electron-4-centre transition state and so is thermally forbidden under the Woodward–Hoffman rules[8] for concerted reactions. Figure 6.2 shows the changes in the σ and π systems of orbitals as the SW bond is rotated. To a first approximation, the σ and π changes are independent and the total σ bond order is conserved throughout, with two bonds breaking as two others form. A crude topological model of the energy profile of the process can be obtained[9] by representing the Hückel resonance integrals of the four active bonds as linear functions of a reaction coordinate r (see Fig. 6.2). The π energy is then calculated as a function of r by diagonalizing the interpolated Hückel Hamiltonian at values of r between 0 and 1. This π-only model probably underestimates the energetic cost of the isomerization, since it neglects the distortion and strain in the σ framework, but it is a simple way of generating qualitatively reasonable reaction profiles.

Within the topological model the SW conversion of one enantiomer of $C_{28}{:}1$

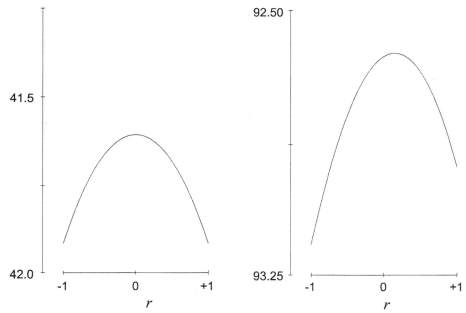

FIG. 6.3. Model reaction profiles for Stone–Wales transformations of (a) C_{28}:1 (D_2) into its enantiomer, (b) C_{60}:1812 (I_h) into its nearest neighbour on the Stone–Wales map. The ordinate in each case is the total Hückel π energy in units of β.

(D_2) to its mirror image goes via a C_{2v} transition state that is $\sim 0.31|\beta|$ higher in π energy than the isoenergetic reactant and product (Fig. 6.3a and 6.4a). More detailed calculations that include both σ and π electrons predict a barrier of 525 kJ/mol (in the semi-empirical PM3 method) or 665 kJ/mol (in the *ab initio* SCF method with a minimal basis), and give a local energy minimum for the C_{2v} cage with a nearby asymmetric (C_1) transition state lying a few kJ/mol higher in energy.

Rotation of a SW bond in icosahedral C_{60} gives a C_{2v} isomer with two fused pentagon pairs; the simple π electronic model predicts an activation energy of $\sim 0.33|\beta|$ for the conversion back to icosahedral C_{60} (Fig. 6.3b and 6.4b), and similar energies for the cascade of isolate-to-isolate transformations C_{78}:5 (D_{3h}) $\rightarrow$ C_{78}:3 (C_{2v}) $\rightarrow$ C_{78}:2 (C_{2v}) $\rightarrow$ C_{78}:4 (D_{3h}).[9] Calculations using a density functional approach[10] suggest that the barrier for the conversion of 'defect' C_{60} (i.e., the C_{2v} isomer) to icosahedral C_{60} should be 520 kJ/mol, confirming prediction of the topological model that transformation of one SW bond costs roughly the same in C_{28} and C_{60}.

Large activation energies are not unexpected for a process in which σ bonds

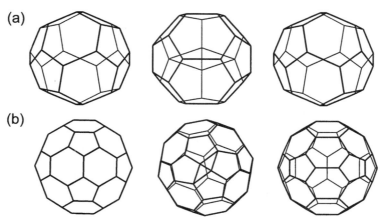

FIG. 6.4. Steps along the idealized Stone–Wales reaction coordinates for $C_{28}(D_2)$ and $C_{60}(I_h)$. (a) Left- and right-handed enantiomers of C_{28} (D_2) are interconverted by a single Stone–Wales step. The central C_{2v} polyhedron corresponds to the transition state in the simplest topological model of the process and has 2 hexagonal, 14 pentagonal, and 2 triangular faces. (b) Left to right: the isolated-pentagon C_{60} isomer with full icosahedral symmetry ($r = -1$), a symmetrized transition state (C_2) with two extra pentagonal and two triangular faces ($r = 0$), and the C_{2v} product fullerene isomer with two fused pentagon pairs. ($r = +1$).

are made and broken, and this is obviously unavoidable for a fullerene isomerization mechanism. Attempts to find isomerization processes with lower barriers have not produced promising alternatives to the Stone–Wales transformation, though they do suggest a modification to the assumed concerted pathway. In a number of model systems and some fullerenes themselves, calculations by various methods favour sequential $1, 2$ shifts (involving an sp^3–sp intermediate) over the concerted route in Fig. 6.2.[11] Activation energies are still of the order of $500 - 600$ kJ/mol, and since the two processes convert the same reactant to the same product, the details of the path are largely irrelevant.

A different sort of generalization of the Stone–Wales transformation can also be imagined. An edge of any trivalent polyhedron, unless it forms part of a triangular face, defines a patch of four rings in contact that can be regarded as the central bond of a Stone–Wales-like rotation. The rings may be labelled by their size as p, q, r, and s, where p and q are rings sharing the edge and r and s each contain only one of its vertices (Fig. 6.5). The rotation leads to a patch in which rings p and q have contracted and rings r and s have expanded by one vertex each. The normal SW transformation has $p = q = 6$

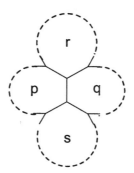

FIG. 6.5. The active patch of four rings for the generalized Stone–Wales transformation of a trivalent polyhedron. p, q, r, and s denote the ring sizes. After the transformation the ring sizes are $p - 1$, $q - 1$, $r + 1$, and $s + 1$, respectively.

and $r = s = 5$, but in principle a bond in a four-hexagon patch on a fullerene (with $p = q = r = s = 6$) could be rotated to give two extra pentagons and two heptagons, or a four-pentagon patch (with $p = q = r = s = 5$) could be converted to two squares and two hexagons. Rotation of one of the vertical edges of a trigonal prism (with $p = q = 4$ and $r = s = 3$) would turn this polyhedron into itself; rotation of one edge of a cube (with $p = q = r = s = 4$) produces a strange C_{2v} object with two triangular, two quadrilateral, and two pentagonal faces. These more general transformations take a carbon cage out of the fullerene series and have as yet no direct chemical relevance but they could be useful in a mathematical context for constructing and counting trivalent polyhedra. It is also possible that fullerene-like cages with some square and/or heptagonal faces could be more stable than fullerenes of the same nuclearity in cases where the fullerenes would be forced to have pentagon adjacencies (§ 4.3).

6.2 Symmetry aspects

We have seen that a general fullerene may belong to one of only 28 point groups (§ 5.1). In the present context, molecular symmetry also restricts the isomeric fullerenes that may be linked by a single Stone–Wales transformation, and so should be considered when drawing interconversion maps.

The key symmetry restriction is that the site symmetry of the active SW bond is preserved in the transformation.[9] An edge in a fullerene polyhedron has at most C_{2v} site symmetry, and the central bond of any SW transformation is therefore in a site of C_{2v}, C_2, C_s, or C_1 symmetry. Though the overall molecular symmetry is usually different for initial and final fullerenes, and though the site symmetry at a general point along the reaction coordinate is at best C_2 (see Fig. 6.2), the subset of symmetry elements that leave the active SW bond and its environment unshifted is common to both initial and final fullerene. Preservation of site symmetry is thus a selection rule for one-step SW transformations that applies equally to the general and isolate-to-isolate processes. (Symmetry and site symmetry are used here in the sense of the maximal point group compatible with the molecular graph, as described in § 5.3.)

As with other structural components, the SW bonds of a given fullerene can be classified into orbits (§ 5.1), with all members of a single orbit sharing a common site symmetry. Since the site symmetries are limited to C_{2v} and its subgroups, the possible numbers of equivalent SW bonds within a set are $|G|$ (for C_1 sites), $|G|/2$ (for C_2 and C_s sites), or $|G|/4$ (for C_{2v} sites), where $|G|$ is the order of the molecular point group. These are also the numbers of SW patches (**I** or **II**), though in counting patches it should be remembered that they may overlap so that some rings, but not the central SW bond, may be common to more than one patch. For example, each hexagon in icosahedral C_{60} is simultaneously part of three SW patches. The number of distinct orbits also gives an upper bound on the number of exits from a given fullerene on an isomerization map when no distinction is made between enantiomers (it is only an upper bound because a fullerene can transform into its own enantiomer, or conceivably into itself).

A necessary (but not sufficient condition) for an allowed SW transformation between two isomers is that they should belong to point groups that have at least one SW site symmetry in common. Table 6.1 lists the possible orbits of edges in fullerenes belonging to the 28 point groups. All equivalent sets of SW bonds are covered by this list, but not all entries in the table can correspond to SW bonds. Some site symmetries cannot be realized in particular point groups because of limits on the total number of SW bonds in a fullerene, or because the orbit can only occur in particular isomers where the pentagons are correctly spaced. For example, the requirement for a shared orbit prohibits transformations between I_h and C_s or C_1 isomers because an I_h fullerene has neither C_s nor C_1 SW bonds. It is not, however, necessary for two SW-related fullerenes to belong to a point group and one of its subgroups, because symmetry elements other than those passing through the SW bond itself can be either created or destroyed

Table 6.1. Sets of equivalent bonds in the fullerene point groups. Orbits O_n of each allowed group G are collected together according to site symmetry and labelled by size (n) and, where necessary, by the generating symmetry element of the site group (′ for C_2', x for C_{2x}, d for σ_d, xy for σ_{xy}, and so on). Not every bond orbit can be a set of Stone–Wales bonds because of limitations on the number of simultaneous SW patches (30 for type **I** patches and 12 for type **II**) and because of geometric factors that allow some orbits to occur only in specific clusters. For example, O_{30} of the I_h group describes SW bonds only in C_{60} itself. No I cluster has any SW bonds.

| | | Site group | | |
G	C_{2v}	C_2	C_s	C_1
I_h	O_{30}	–	O_{60}	O_{120}
I	–	O_{30}	–	O_{60}
T_h	O_6	–	O_{12}	O_{24}
T_d	O_6	–	O_{12}	O_{24}
T	–	O_6	–	O_{12}
D_{6d}	–	O_{12}	O_{12d}	O_{24}
D_{5d}	–	O_{10}	O_{10d}	O_{20}
D_{3d}	–	O_6	O_{6d}	O_{12}
D_{2d}	O_2	O_4	O_{4d}	O_8
D_{6h}	$O_6, O_{6'}$	–	$O_{12h}, O_{12v}, O_{12d}$	O_{24}
D_{5h}	O_5	–	O_{10h}, O_{10v}	O_{20}
D_{3h}	O_3	–	O_{6h}, O_{6v}	O_{12}
D_{2h}	O_{2x}, O_{2y}, O_{2z}	–	$O_{4xy}, O_{4xz}, O_{4yz}$	O_8
D_6	–	$O_6, O_{6'}$	–	O_{12}
D_5	–	O_5	–	O_{10}
D_3	–	O_3	–	O_6
D_2	–	O_{2x}, O_{2y}, O_{2z}	–	O_4
S_6	–	–	–	O_6
S_4	–	O_2	–	O_4
C_{3h}	–	–	O_3	O_6
C_{2h}	–	O_2	O_{2h}	O_4
C_{3v}	–	–	O_3	O_6
C_{2v}	O_1	–	O_{2xz}, O_{2yz}	O_4
C_3	–	–	–	O_3
C_2	–	O_1	–	O_2
C_s	–	–	O_1	O_2
C_i	–	–	–	O_2
C_1	–	–	–	O_1

during the transformation. Thus, for example, a one-step interconversion of D_{2d} and D_{2h} isomers is possible.[9]

The total number of SW bonds that may be present simultaneously in a fullerene is 30, if all belong to type **I** patches, or 12 if all belong to patches of type **II**. These limits follow from the facts that a pentagon may belong simultaneously to at most five (two) type **I** (type **II**) patches, each patch involves two pentagons, and the total number of pentagons is 12. The limit of 30 type **I** patches is realized in just one fullerene, the icosahedral isomer of C_{60} where all 30 SW bonds are equivalent and belong to the O_{30} orbit which has C_{2v} site symmetry. In fact, C_{60}:1 is the only icosahedral fullerene (I or I_h) with any SW bonds at all; C_{20}:1 has no hexagonal face, and in C_{80}:7 and beyond, all pentagon pairs are separated by more than one bond. Every I fullerene is therefore an island on the SW interconversion map. The limit of 12 patches of type **II** is realized by C_{84}:24, the D_{6h} isolated-pentagon isomer, where the patches occur in rings of 6 at the north and south poles of the cylinder. Cylindrical fullerenes formed by inserting equatorial belts of six hexagons into this C_{84} structure also have 12 type **II** patches suitable for isolate-to-isolate transformations. Tetrahedral clusters cannot have more than six type **II** patches,[9] but may have 12 of type **I** (and the T_d isolated-pentagon isomer C_{84}:20 achieves 12 type **I** and six type **II** patches simultaneously). Although site symmetry of the active bond is preserved during the transformation, the total number of SW bonds may change. Rotation of one type **I** bond can remove as many as eight others (Fig. 6.6).

The operations of leapfrogging and SW transformation are connected. Leapfrog fullerenes may contain SW bonds, and each one arises from the leapfrogging of a pair of adjacent pentagons. Since no leapfrog itself contains adjacent pentagons, it follows that repetition of the leapfrog operation removes all SW bonds. It can also be shown that no two leapfrog isomers of a C_n cage can be directly connected by a single SW transformation,[9] though as they may be indirectly connected by a sequence of transformations they can still occur on the same isomerization map.

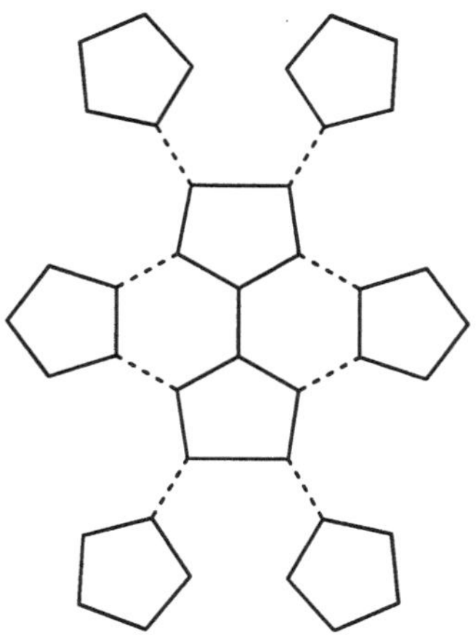

FIG. 6.6. Twisting a single Stone–Wales bond can change the total number of patches available for further transformation. All eight dotted lines are SW bonds when the central bond is in the vertical position but none remain SW bonds when the central bond is rotated to the horizontal.

6.3 Chirality and the Stone–Wales transformation

All sufficiently large fullerenes have both chiral and achiral isomers and these will often be interconvertible by a sequence of SW transformations. General selection rules based on site symmetry can be formulated to describe the interaction of handedness and the Stone–Wales transformation. If A and B are fullerenes connected by a Stone–Wales transformation then there are three distinct cases: (a) neither A nor B is chiral, (b) one of A and B is chiral (B, say), (c) A and B are both chiral. In cases (b) and (c) the site symmetry of the active bond is either C_2 or C_1; in (a) it may be any subgroup of C_{2v}.

In case (b) the active bond has a chiral site symmetry in an overall achiral molecule. It must therefore belong to an orbit consisting of an even number of equivalent members, and because transformations starting from an achiral point group cannot show a bias towards a particular handedness, independent rotations of one half of the bonds in the orbit will lead to the left-handed enantiomer of B and rotations of the other half will lead to its mirror image. The general rule[9] is that any two members of the orbit that are exchanged by *proper* operations will lead to the *same* enantiomer of B, and any two that are exchanged by *improper* operations will lead to opposite enantiomers (Fig. 6.7). In case (a) all members of the orbit will give the same SW product because B

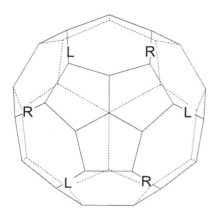 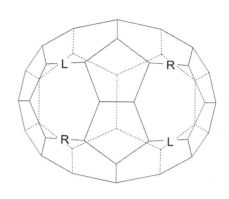

FIG. 6.7. Transformations between chiral and achiral fullerenes. Rotation of equivalent SW bonds in an achiral isomer may produce either left or right forms (arbitrarily assigned labels L and R) of a chiral product. In the diagrams each bond is labelled with the chirality of the product that would be obtained by rotating that bond. In general, SW bonds producing the same enantiomer are exchanged by proper operations, those producing opposite enantiomers are exchanged by improper operations. The examples shown are (a) C_{38}:16 (C_{3v}), which has six equivalent SW bonds that yield L and R enantiomers of C_{38}:14 (C_1), and (b) C_{40}:8 (C_{2v}), which has four equivalent SW bonds yielding L and R enantiomers of C_{40}:6 (C_1).

and its mirror image are then indistinguishable.

In case (c) a chiral reactant is converted to a chiral product. All patches of a given orbit in A are exchanged by proper operations alone, and so all must lead to the same enantiomer of B. If A and B are distinct (neither identical nor enantiomers of one another) then the transformation exists in two forms, one linking A to B and one linking A* to B* where the star signifies a mirror image.

These rules may affect prospects for the resolution of optically active fullerenes, since if an isomerization map includes an achiral isomer then all chiral fullerenes on the map will be in racemic equilibrium under conditions where the SW transformations are facile. Only at temperatures where the transformations are frozen out would it be possible to maintain a pure enantiomer once separated. The recent resolution of C_{76}:1 by chemical means[12] suggests that SW transformations to neighbours with adjacent pentagons do not occur at a significant rate at room temperature once the fullerene has been formed.

6.4 Isomerization maps

It is straightforward to design a computer program to determine all possible
Stone–Wales transformations within a set of fullerene isomers and the information so obtained can be presented in a variety of ways. Tables A.1 to A.9
simply list in the final column of the entry for each isomer all those partners
that can be reached in a single SW step, each followed by a number in brackets
to indicate how many patches lead to the same or enantiomorphic products.
(The bracketed number is omitted when, as is most often the case, the patch is
unique.) It is usually possible to deduce the site symmetry of the patch from
this number and the point group of the initial isomer. The exception is for C_2
and C_s patches when the initial isomer belongs to one of the groups C_{2h} or D_{nd}
($n = 2, 3, 5, 6$), because these point groups admit SW bonds at both C_2 and C_s
sites and the orders of these site groups are the same. If the product isomer does
not belong to one of these groups then the assignment can be made by considering the reverse transformation, but if both initial and final isomers belong to
these groups then more detailed consideration is needed. Tables A.10 to A.20
give equivalent information for isolate-to-isolate conversion via the restricted
SW transformation.

Tables 6.2 and 6.3 collate the possible transformations and give lists of the
families of interconverting isomers at each n. Table 6.2 lists all SW transformations for general isomers in the range C_{28} to C_{50}; Table 6.3 lists all direct
isolate-to-isolate transformations for isomers in the range C_{60} to C_{100}. When
the families are small enough it is useful to represent the transformations on
an isomerization map which gives an idea of the 'distance' between isomers.
Examples of such maps are shown in Fig. 6.8 and 6.9, where each (reversible)
transformation is represented by an arrow labelled with the site symmetry of
the transforming patch. Given the rules for conversion of chiral isomers in § 6.3
it is not necessary to distinguish enantiomers on the maps.

The most obvious feature of these maps is that they fall into disconnected
regions. From a given starting isomer it is not always possible to reach a given
product, no matter how many SW transformations are made. This is analogous to the 'factorization' observed for rearrangement mechanisms in other
fields of chemistry. In boranes and carboranes, for example, the diamond-
square-diamond mechanism [13] gives a way of changing the connectivity of the
vertices of a deltahedral cage and is in fact the dual of a generalized Stone–Wales

Table 6.2. Connected transformation maps for fullerenes in the range C_{20} to C_{50}

Fullerene	Map	Isomer(s)	Fullerene	Map	Isomer(s)
C_{20}	1	1	C_{44}	1	2
				2	3
C_{24}	1	1		3	13
				4	38
C_{26}	1	1		5	85
				6	86
C_{28}	1	1		7	all 83 other isomers
	2	2			
			C_{46}	1	1
C_{30}	1	1, 2, 3		2	20
				3	113
C_{32}	1	5		4	all 113 other isomers
	2	all 5 other isomers			
			C_{48}	1	1
C_{34}	1	1, 3, 6		2	2
	2	2, 4, 5		3	16
				4	186
C_{36}	1	5		5	189
	2	all 14 other isomers		6	all 194 other isomers
C_{38}	1	2	C_{50}	1	1
	2	12		2	2
	3	all 15 other isomers		3	3
				4	10
C_{40}	1	1		5	27
	2	3		6	33
	3	all 38 other isomers		7	125
				8	157
C_{42}	1	1		9	16, 28
	2	all 44 other isomers		10	all 261 other isomers

Table 6.3. Connected transformation maps for isolated-pentagon fullerenes in the range C_{60} to C_{100}

Fullerene	Map	Isomer(s)		Fullerene	Map	Isomer(s)
C_{60}	1	1		C_{88}	1	1
					2	3
C_{70}	1	1			3	4
					4	34
C_{72}	1	1			5	2, 7
					6	all 29 other isomers
C_{74}	1	1				
				C_{90}	1	1
C_{76}	1	1			2	4
	2	2			3	2, 3, 5, 8
					4	6, 7, 9, 10, 23
C_{78}	1	1			5	all 35 other isomers
	2	all 4 other isomers				
				C_{92}	1	1
C_{80}	1	1			2	4
	2	2			3	23
	3	4			4	24
	4	7			5	25
	5	3, 5, 6			6	28
					7	29
C_{82}	1	all 9 isomers			8	36
					9	5, 6
C_{84}	1	1, 2, 5			10	2, 3, 9
	2	all 21 other isomers			11	21, 22, 35, 45
					12	7, 8, 10, 11, 12, 13, 14, 15, 16, 26, 27, 34
C_{86}	1	3			13	all 57 other isomers
	2	9				
	3	all 17 other isomers				

Table 6.3. (*Continued*)

Fullerene	Map	Isomer(s)
C$_{94}$	1	3
	2	4
	3	32
	4	46, 52
	5	7, 8, 51
	6	1, 2, 5, 6, 9
	7	10, 11, 12, 13, 14, 15, 16, 17, 18, 19, 20, 21, 22, 23, 24, 31, 33, 34, 35, 36, 37, 39, 40, 42, 43, 44, 45
	8	all 94 other isomers
C$_{96}$	1	1
	2	2
	3	3
	4	9
	5	109
	6	111
	7	187
	8	49, 50
	9	4, 5, 6, 14
	10	7, 8, 10, 11
	11	12, 13, 15, 55
	12	86, 120, 122, 123
	13	33, 110, 148, 149, 184
	14	all 157 other isomers

Fullerene	Map	Isomer(s)
C$_{98}$	1	3
	2	8
	3	21
	4	47
	5	48
	6	64
	7	66
	8	85
	9	88
	10	148
	11	13, 62
	12	18, 23
	13	1, 2, 11
	14	9, 10, 72
	15	24, 61, 65
	16	4, 5, 6, 7
	17	12, 14, 15, 16, 17, 19, 22, 49
	18	20, 25, 26, 27, 28, 29, 30, 31, 32, 33, 34, 35, 36, 37, 38, 39, 40, 41, 42, 43, 44, 45, 46, 50, 51, 52, 53, 54, 55, 56, 57, 58, 59, 60, 63, 67, 68, 69, 70, 71, 74, 90, 115, 124, 125, 126, 127, 128, 129, 130, 131, 132, 133, 134, 135, 136, 137, 138, 139, 140, 141, 166
	19	all 162 other isomers

Table 6.3. (*Continued*)

Fullerene	Map	Isomer(s)	Fullerene	Map	Isomer(s)
C_{100}	1	1	C_{100}	17	3, 18, 37
	2	2		18	16, 17, 103
	3	4		19	55, 66, 92
	4	5		20	149, 150, 395
	5	14		21	25, 36, 64, 93, 94
	6	23		22	6, 7, 9, 10, 11, 12, 15, 38, 39
	7	34		23	40, 41, 42, 43, 44, 54, 65, 99, 100, 101, 102, 116, 151, 152, 154, 159, 160, 330
	8	35			
	9	67		24	19, 20, 21, 22, 24, 26, 27, 28, 29, 30, 31, 32, 33, 45, 47, 70, 91, 95, 96, 97
	10	98			
	11	115		25	46, 48, 49, 50, 51, 52, 53, 56, 57, 58, 59, 60, 61, 62, 68, 69, 71, 72, 73, 74, 75, 76, 77, 78, 79, 80, 81, 82, 83, 84, 85, 86, 87, 88, 89, 90, 128, 129, 146, 148, 153, 161, 162, 163, 164, 165, 166, 167, 168, 169, 319, 447
	12	212			
	13	321			
	14	450			
	15	8, 13			
	16	63, 170		26	all 316 other isomers

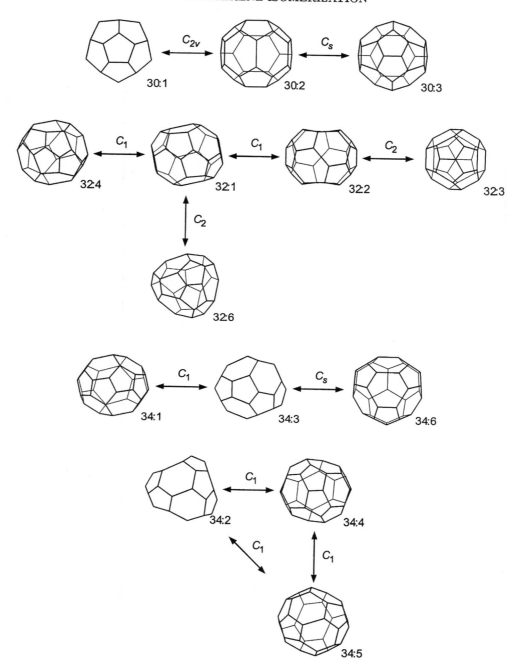

FIG. 6.8. Stone–Wales interconversion maps for fullerenes with 30 to 34 atoms. (a) The isomers of C_{30} form a single family, (b) C_{32} forms two families, but one consists only of isomer $C_{32}{:}5$ which is disconnected from the rest and so is not shown here, and (c) C_{34} forms two families. For the conventions used in this and the subsequent maps see Fig. 6.9.

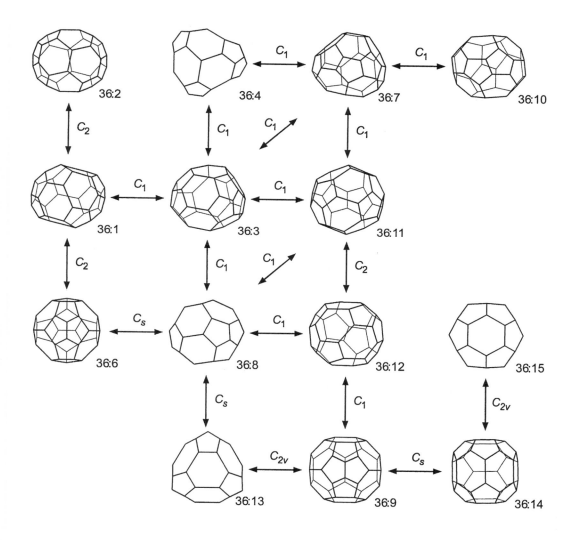

FIG. 6.9. The Stone–Wales interconversion map for C_{36}. Isomer C_{36}:5 is disconnected from the others and is not shown. The conventions used in Figs 6.8 to 6.12 are as follows: each transformation is labelled by the preserved site symmetry and each isomer by its position in the canonical spiral ordering of the main tables; enantiomers are not distinguished and transformations that have no effect on the eigenvalue spectrum (e.g., the transformation of a fullerene into itself or to its own enantiomer) are not shown, though they are listed in the main tables.

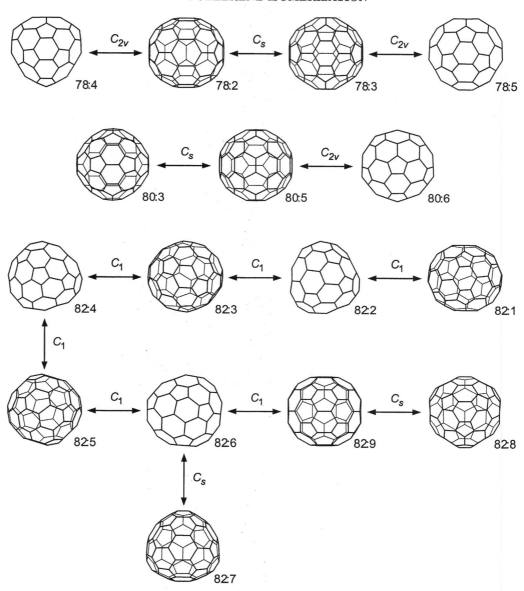

FIG. 6.10. Stone–Wales interconversion maps for isolated-pentagon isomers of (a) C_{78}, (b) C_{80}, and (c) C_{82}. Only direct isolate-to-isolate transformations are shown. Under this restriction, isomers C_{78}:1, C_{80}:1, C_{80}:2, C_{80}:4, and C_{80}:7 are disconnected from the others. All the isolated-pentagon isomers of C_{82} lie on a single map. Other transformations proceeding via intermediates with adjacent pentagons are possible, and may connect the different families.

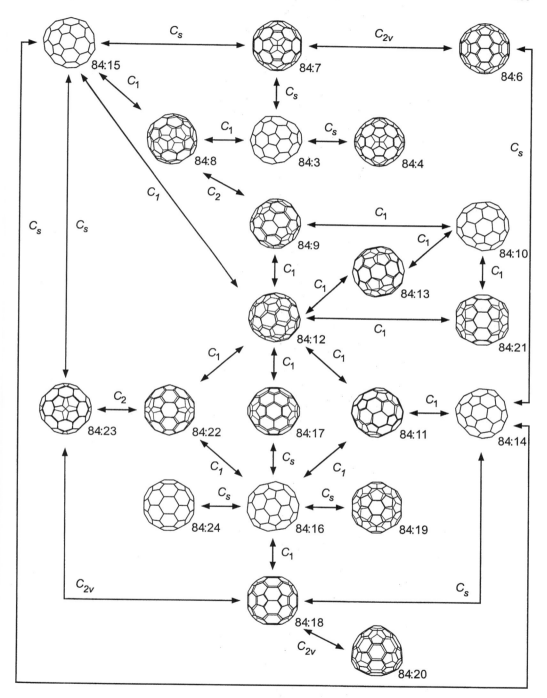

FIG. 6.11. Stone–Wales interconversion map for isolated-pentagon isomers of C_{84}.

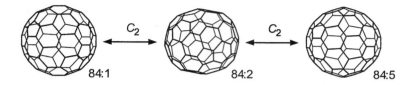

F IG . 6.12. The smaller Stone–Wales family of three interconverting, chiral, isolated-pentagon isomers of C_{84}. The other 21 of the 24 isolated-pentagon isomers form a single map, shown in Fig. 6.11.

transformation. Of the three possible isomers of $C_2B_{10}H_{12}$, two (the (1,2) and (1,7) forms) interconvert under a concerted multiple DSD rearrangement but the third (the (1,12) form) remains unchanged. The rearrangement involves a centrosymmetric cuboctahedral transition state,[13] and has the property that inversion partners in the initial cage remain inversion partners in the product; the (1,12) isomer in which the C atoms occupy opposite vertices cannot convert to a form in which each C is antipodal to B. The Stone–Wales families presumably also correspond to conservation of a graphical invariant of a less obvious kind.

The small selection of maps given in this chapter is already sufficient to show examples of self-racemizing fullerenes, homo-chiral pathways and closed loops. As a rule, the high-symmetry isomers have a few large orbits of patches and tend to lie at the ends of branches, whereas low-symmetry isomers are often at multiple junction points on the map. A notable example of this is icosahedral C_{60} with its 30 patches of type **I**, all of which lead to the same C_{2v} isomer (Fig. 6.4b) with two fused pentagon pairs; the map then fans out from this bottleneck as described in more detail below. If the process of formation of C_{60} involves the mediation of a cage with a nonideal distribution of pentagons and subsequent rearrangement by SW transformations, then all material ending up as icosahedral C_{60} must funnel through this C_{2v} isomer.

Maps for restricted SW transformations among the isolated-pentagon isomers of some example higher fullerenes are given in Fig. 6.10 to 6.12. The two-family map for isolated-pentagon isomers of C_{78} was the first to appear in the chemical literature. It was presented in the paper by Diederich *et al.* that described the first separation of C_{78} from arc-processed soot,[3] and the existence of two disjoint families was correlated with the observation of two isomers in the experiment. However, later work[5] has shown that at least three isomers can be

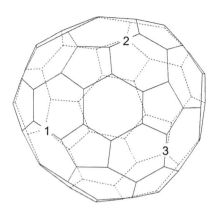

FIG. 6.13. Conversions between Stone–Wales families of isolated-pentagon isomers. C$_{78}$:1 (D_3) is not connected to the other isomers of C$_{78}$ by isolate-to-isolate transformations. Successive rotation of the equivalent bonds 1, 2, and 3 (part of an orbit of six SW bonds) leads to a D_{3h} member of the main family of IPR isomers, but only by passing through two fullerenes with adjacent pentagons.

obtained from the soot under some conditions and the picture is therefore more complicated. C$_{80}$ also has disjoint families of isomers with isolated pentagons, but C$_{82}$ is apparently unique in that all nine of its isolated-pentagon isomers belong to a single SW family. C$_{30}$ occupies an equivalent position for the type **I** transformation, which connects all three of its isomers. C$_{84}$ has two families, each containing one leapfrog isomer, but examples of leapfrog isomers within a single family can also be found (e.g., for C$_{96}$).

If the restriction to direct isolate-to-isolate transformations is relaxed, then many of the disjoint families of isolated-pentagon isomers can be linked up. For example, the D_3 isolated-pentagon isomer C$_{78}$:1 has no type **II** SW patch but does have 18 type **I** patches; consecutive rotation of three of these converts it to a D_{3h} isolated-pentagon isomer in the main SW family (Fig. 6.13). The intermediates have adjacent pentagons and are presumably energetically unfavourable.

6.5 The C$_{60}$ Stone–Wales map

If a chemist were to set out to find a rational synthesis of C$_{60}$, the brute-force method of graphite evaporation would surely not be the first to spring to mind. Nevertheless, this process works. It yields fullerenes and is relatively specific

Table 6.4. Near neighbours of icosahedral C_{60} on the Stone–Wales map. The isomer labels correspond to those in Fig. 6.14. d is the 'distance' of the isomer from I_h C_{60}; i.e., the minimum number of consecutive SW transformations required to convert the isomer back to 1812. All isomers up to $d = 3$ are listed.

Isomer	Ring spiral	Point group	d
60:1662	1 2 4 12 15 16 18 20 23 27 29 31	C_1	3
60:1714	1 2 4 12 16 18 20 23 24 26 29 32	C_1	3
60:1728	1 2 4 13 15 16 18 20 22 28 31 32	C_1	3
60:1752	1 2 4 13 16 17 20 22 24 27 29 32	C_1	3
60:1756	1 2 4 13 16 18 20 22 24 25 27 29	C_1	3
60:1757	1 2 4 13 16 18 20 22 24 26 29 32	C_s	2
60:1758	1 2 4 13 16 18 20 22 25 26 29 31	C_2	3
60:1760	1 2 4 13 16 18 20 23 26 29 31 32	C_s	3
60:1761	1 2 4 13 16 20 22 24 26 27 29 30	C_{3v}	3
60:1780	1 2 9 10 13 15 21 24 26 28 29 31	C_s	3
60:1786	1 2 9 11 12 14 17 24 25 27 29 30	C_1	3
60:1787	1 2 9 11 12 15 16 23 26 27 29 30	C_2	3
60:1789	1 2 9 11 12 15 17 23 25 27 29 30	C_2	2
60:1790	1 2 9 11 12 15 17 23 25 28 29 32	C_2	3
60:1791	1 2 9 11 12 15 18 23 25 26 29 30	C_1	3
60:1797	1 2 9 11 14 16 19 22 23 25 27 29	C_1	3
60:1802	1 2 9 12 13 17 20 21 24 25 27 32	C_1	3
60:1803	1 2 9 12 14 17 20 21 23 25 26 28	D_3	3
60:1804	1 2 9 12 14 17 20 21 23 25 27 32	C_s	2
60:1805	1 2 9 12 14 17 20 22 25 27 30 32	D_{2d}	3
60:1806	1 2 9 12 15 16 20 22 24 27 28 30	C_2	2
60:1807	1 2 9 12 15 17 19 22 24 26 29 30	C_2	2
60:1808	1 2 9 12 15 17 20 21 23 24 27 32	D_{2d}	3
60:1809	1 2 9 12 15 17 20 22 24 26 28 30	C_{2v}	1
60:1810	1 2 9 12 15 17 20 22 24 27 30 32	D_{2h}	2
60:1811	1 2 9 12 16 17 19 22 24 25 29 30	D_3	3
60:1812	1 7 9 11 13 15 18 20 22 24 26 32	I_h	0

to C_{60} (and C_{70}). Explanations of these features have been proposed[11] but a detailed growth mechanism for C_{60} has yet to be proved. The possible inter-conversions amongst isomers of C_{60} are of obvious interest in this connection, and the automated methods described earlier can be applied to generate the full Stone–Wales map for all 1812 isomers. Space does not permit a full listing of all the transformations but some general features of the map are worth quoting.

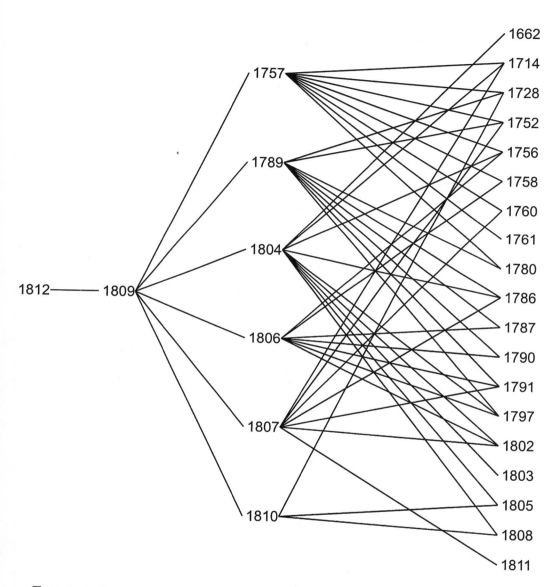

FIG. 6.14. Structure of the Stone–Wales map for C$_{60}$ in the vicinity of the favoured icosahedral isomer. The isomer labels denote the position in the full catalogue of 1812 spirals for C$_{60}$. Table 6.4 gives the spiral code for each isomer.

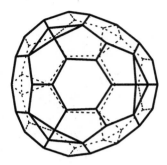

FIG. 6.15. A possible transition state for a hypothetical concerted rotation of six octahedrally disposed SW bonds of $C_{60}(I_h)$. This structure has T_h symmetry. The net result of the transformation would be to convert $C_{60}(I_h)$ back to itself, but with exchange of some atom labels.

The 1812 isomers of C_{60} (counting each enantiomeric pair as a single isomer) break down into 44 SW families, of which 33 consist of single isomers, 4 of pairs, 2 of triples and 1 each of 4, 6, 18 and 27 members, respectively. All the rest fall into one giant family of 1710 members, and this includes the stable isolated-pentagon cage. Thus the vast majority of adjacent-pentagon fullerenes with 60 atoms can be converted into icosahedral C_{60} by a sufficiently long sequence of SW transformations of type **I**.

The local structure of the map in the vicinity of I_h C_{60} was briefly described in § 6.4. Icosahedral C_{60} is on a spur of the main map. A single transformation based on any one of its 30 patches gives only one product, a second transformation produces one of six distinct products and a further SW step yields one of 19 third-generation isomers. Table 6.4 lists spiral codes for all isomers that can be reached from icosahedral C_{60} in up to three SW steps. Some of these 'defective' isomers have been studied before.[7,14,15] As Fig. 6.14 shows, all but four of the 19 third-generation products can be reached in a single step from at least two different second-generation reactants. This means that there are many possible pathways by which I_h C_{60} could be converted back into itself by a sequence of six consecutive SW transformations without any one step reversing its immediate predecessor. One particularly attractive sequence involves a set of six independent patches of the icosahedral cage formed by taking SW bonds whose centres lie on the vertices of an octahedron. Rotation of all six patches (simultaneously or consecutively in any order) gives the original structure back again unchanged apart from a trivial rotation of the molecule as a whole. Figure 6.15 shows a hypothetical transition state for a concerted rotation of all six SW bonds in the same sense. The activation energy of such a process would probably be high as the topological model described earlier in this chapter gives a barrier of $\sim 0.47|\beta|$ per participating SW bond ($\sim 2.8|\beta|$ in total).

6.6 Isomer distributions

As n increases, the number of disjoint families must be expected to grow rapidly. Very large higher fullerenes will have many isomers with well separated pentagons and therefore no SW patch of either type. These isomers are necessarily islands on the Stone–Wales map. This bears on the question of the expected product distributions for giant fullerenes, but even in the range of n where the maps are still fairly simple, pentagon rearrangement has implications for isomer distributions.

Prediction of the isomer distribution within the Stone–Wales regime depends on assumptions about relative stabilities and activation energies across an isomerization map. At the 'entropic' extreme, where all isomers in a family are of equal stability and all barriers are low in comparison to the available thermal energy, the equilibium mole fraction of each isomer will be inversely proportional to the order of its molecular point group.[9] The proof is straightforward. Suppose first that just two isomers are involved. The equilibrium constant for interconverting isomers A and B is given by elementary statistical mechanics[16] as

$$K_c = [B]/[A] = (q_B/q_A) \exp(-\Delta E_0/kT), \qquad (6.1)$$

where q_A and q_B are the molecular partition functions of A and B. On the hypothesis that A and B are isoenergetic, $\Delta E_0 = 0$. On the further assumptions (also reasonable) that both molecules have similar vibrational frequencies, rotational constants and electronic structures, the ratio of partition functions is simply $q_B/q_A = \sigma_A/\sigma_B$ where σ_A and σ_B are symmetry numbers. For a nonlinear molecule the symmetry number is the order of the rotational subgroup of the point group ($=$ the number of indistinguishable orientations of the molecule that can be reached by proper symmetry operations). Three cases must therefore be considered, depending on the chirality of A and B.

(1) If A and B are both achiral, then $\sigma_A = |G_A|/2$ and $\sigma_B = |G_B|/2$ giving $[B]/[A] = |G_A|/|G_B|$.

(2) If A is achiral and B is chiral with R and S forms, then $\sigma_A = |G_A|/2$ and $\sigma_B = |G_B|$ giving $[B_R]/[A] = [B_S]/[A] = |G_A|/2|G_B|$ and hence $[B]/[A] = ([B_R] + [B_S])/[A] = |G_A|/|G_B|$.

(3) If A and B are both chiral with R and S forms, then $\sigma_A = |G_A|$ and $\sigma_B = |G_B|$ giving $[B_R]/[A_R] = \ldots = [B_S]/[A_S] = |G_A|/|G_B|$ and hence $[B]/[A] = ([B_R] + [B_S])/([A_R] + [A_S]) = |G_A|/|G_B|$.

In each case $|G_X|$ is the order of the molecular point group of X (= the order of the rotational subgroup for a chiral molecule and twice the order of the rotational subgroup for an achiral molecule), and the final result is the same:

$$K_c = [B]/[A] = |G_A|/|G_B|. \tag{6.2}$$

This result was first obtained for isomers on the same SW map using kinetic arguments based on microscopic reversibility.[9] However, the present derivation is independent of the specific mechanism of the transformation between the two isomers, and generalizes to any number of interconverting species.

The opposite ('enthalpic') assumption is that isomers within a given map differ significantly in stability. A Boltzmann distribution of populations of all the members of an interconverting family would then be expected, and in the case where one isomer is much more stable than any other in the same family, the limit of one product isomer per family envisaged by Diederich et al.[3] would be realized.

In reality, the experimental picture is unlikely to be as clearcut as either extreme assumption would suggest. Experimental data for C_{78} and C_{84} were discussed in Chapter 4. The distribution of C_{78} isomers has already been mentioned: two C_{2v} isomers from the larger family and the lone D_3 isomer are found in the chromatographically separated extract of processed soot; this may indicate an equilibrium in and between the maps that became frozen in at a temperature where the energy differences within the large map were significant but the SW transformations were still facile. However, the composition of the C_{78} product mixture is sensitive to the method of preparation and separation.[3,5,6,17] In contrast, C_{84} mixtures as prepared by three different groups using different conditions have the same basic composition.[4–6,17] The NMR spectrum of the C_{84} chromatographic fraction has 31 lines of approximately equal intensity and 1 of half intensity, and hence (see Table A.12) is consistent with a 2 : 1 mixture of D_2 and D_{2d} isomers; this has been interpreted[4] as a statistical distribution of the two most stable isolated-pentagon isomers of the larger SW family C_{84}:22 (D_2) and C_{84}:23 (D_{2d}). Electronic structure calculations of various kinds [18-20] agree in placing these two isomers at almost equal energy and well below all other isolated-pentagon isomers of C_{84} (see also § 4.6). Thus it seems that in this case it is safe to interpret experimental product ratios in terms of thermodynamic stability, as was tacitly assumed in § 4.6. In other cases the observed ratios may represent a convolution of thermodynamic and kinetic factors.

A further complication in the interpretation is that, under conditions where graphite is vaporized, fullerenes are presumably able to convert not only to different isomers but to cages with different numbers of carbon atoms. Hypothetical processes of insertion and extrusion of carbon fragments will be considered in the next chapter.

References and notes

[1] R. Taylor, J. P. Hare, A. K. Abdul-Sada, and H. W. Kroto, J. Chem. Soc. Chem. Comm. (1990) 1423.

[2] R. Ettl, I. Chao, F. Diederich, and R. L. Whetten, Nature 353 (1991) 149.

[3] F. Diederich, R. L. Whetten, C. Thilgen, R. Ettl, I. Chao, and M. Alvarez, Science 254 (1991) 1768.

[4] D. E. Manolopoulos, P. W. Fowler, R. Taylor, H. W. Kroto, and D. R. M. Walton, J. Chem. Soc. Faraday Trans. 88 (1992) 3117.

[5] K. Kikuchi, T. Wakabayashi, N. Nakahara, S. Suzuki, H. Shiromaru, Y. Miyake, K. Saito, I. Ikemoto, M. Kaionsho, and Y. Achiba, Nature 357 (1992) 142.

[6] R. Taylor, G. J. Langley, A. G. Avent, T. J. S. Dennis, H. W. Kroto, and D. R. M. Walton, J. Chem. Soc. Perkin Trans. 2 (1993) 1029.

[7] A. J. Stone and D. J. Wales, Chem. Phys. Lett. 128 (1986) 501.

[8] R. B. Woodward and R. Hoffmann, The conservation of orbital symmetry (Verlag Chemie, Weinheim, 1970).

[9] P. W. Fowler, D. E. Manolopoulos, and R. P. Ryan, Carbon 30 (1992) 1235.

[10] J.-Y. Yi and J. Bernholc, J. Chem. Phys. 96 (1992) 8634.

[11] R. L. Murry, D. L. Strout, G. K. Odom, and G. E. Scuseria, Nature 366 (1993) 665.

[12] J. M. Hawkins and A. Meyer, Science 260 (1993) 1918.

[13] See e.g. F. A. Cotton and G. Wilkinson, Advanced inorganic chemistry, 3rd edition, Chapter 8 (Wiley, New York, 1972).

[14] C. Coulombeau and A. Rassat, J. Chim. Phys. 88 (1991) 173.

[15] C. Coulombeau and A. Rassat, J. Chim. Phys. 88 (1991) 665.

[16] See e.g. P. W. Atkins, Physical Chemistry, 4th edition, Chapter 20 (Oxford University Press, Oxford, 1990).

[17] F. Diederich and R. L. Whetten, Acc. Chem. Res. 25 (1992) 119.

[18] B. L. Zhang, C. L. Wang, and K. M. Ho, J. Chem. Phys. 96 (1992) 7183.

[19] K. Raghavachari, Chem. Phys. Lett. 190 (1992) 397.

[20] D. Bakowies and W. Thiel, J. Am. Chem. Soc. 113 (1991) 3704.

7

CARBON GAIN AND LOSS

Chapter 6 was concerned with processes in which one fullerene is converted to another with the same number of atoms. The present chapter deals with conversion processes in which the atom count changes. As several authors have remarked,[1,2] it comes as a shock to learn that the closed, highly symmetrical, 'low entropy' C_{60} molecule can form spontaneously during condensation of hot carbon vapour. The mechanism of fullerene growth is not yet well understood, though discussion has been extensive. It is clear that C_{60} cannot form by the enormously improbable simultaneous collision of 60 carbon atoms, though when made from isotopically enriched graphite, the fullerene does show isotopic scrambling.[3] The stable fullerenes must result from a process of accretion of smaller fragments, and several candidates for the pathway have been proposed. Some models invoke oligomerization of cyclic monomers,[4,5] and indeed C_{60} can be made by pyrolysis of naphthalene vapour.[6] Two contrasting pictures of the formation of C_{60} from small polycyclic fragments are reviewed by Curl;[1] they are the so-called pentagon and fullerene roads.

In the 'pentagon road' model,[7] small species are ingested by growing graphite sheets. Minimization of the number of dangling bonds in the sheet results in the formation of pentagonal defects that cause it to curl up, and rearrangements within the sheet remove some or all of the destabilizing pentagon adjacencies. In this picture, fullerene cages are formed when the open pseudo-graphite network happens to close. If the pentagonal faces appear at the wrong stage in the process or at the wrong place in the network, the structure may continue to grow, perhaps producing a multilayer spiral soot particle.[8]

In the 'fullerene road' model,[9] it is supposed that clusters of forty or so atoms are already fullerenes, and that the higher members of series such as C_{60} and C_{70} form from them by incorporation of C_2 or other small molecules, again[1] undergoing annealing to the most stable isomers by Stone–Wales or other rearrangements that reduce the number of pentagon adjacencies. In both models the dominance of C_{60} and C_{70} in the fullerene fraction of the product is a consequence of their extra stability: the process stumbles into a deep potential well and becomes trapped.

Both models have their problems in detail,[1,2] but there is evidence that under some circumstances, closed graphitic networks can grow and shrink by addition and loss of small fragments. Multiphoton fragmentation of mass-selected fullerenes leads to a series of clusters C_{n-2}, C_{n-4}, C_{n-6}, ..., all with even numbers of carbon atoms;[10] the implication is that the fullerene cages deflate by sequential loss of C_2 or other small molecules. In other experiments,[11] laser irradiation of solid C_{60} and C_{70} is found to produce a range of both larger and smaller clusters containing even numbers of atoms. This finding is consistent with the idea that fragments ejected by a target fullerene are ingested by another molecule, yielding cages $C_{n\pm2}$, $C_{n\pm4}$, and so on.

A similar mechanism on a larger scale has been invoked in the explanation of the formation and morphology of carbon nanotubes. Endo and Kroto[12] propose that ingestion of carbon takes place at the tips of these helically wound[13] tubes, laying down a spiral trail of atoms at the boundary between the pseudo-hemispherical cap and the cylinder wall, causing the cap to rotate by a full turn each time a coil of the helix is added. The tip is the place of maximum strain, because it contains six pentagons in close proximity, and it is likely to be the reactive region of the tube. Facile opening and resealing of nanotube tips may be steps in the mechanism of the reported filling of tubes with lead.[14]

Detailed discussion of all the proposed mechanisms for growth and fragmentation is well beyond the scope of the present treatment. Instead, we concentrate on the smallest possible fullerene-to-fullerene conversion: the insertion or extrusion of a C_2 molecule. As in Chapter 6, the aim is to study the qualitative topological and symmetry-related consequences of one particular hypothetical mechanism and to map out the allowed conversions between isomers of small fullerenes.

7.1 C_2 insertion and extrusion

Endo and Kroto[12] describe a hypothetical mechanism for fullerene-to-fullerene conversion by ingestion of two carbon atoms at a time (Fig. 7.1). As they point out, this is just the reverse of the C_2 extrusion process proposed earlier by the Rice group[10] in connection with experiments on laser-induced fragmentation of

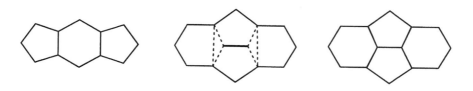

FIG. 7.1. A hypothetical mechanism for expansion of a fullerene by two carbon atoms. Insertion of a C_2 fragment across a hexagon with *para* pentagonal neighbours leads to a fullerene with one extra hexagonal face.

carbon clusters. The figure shows the topological requirement for C_2 inclusion by this mechanism: the parent fullerene, C_n should have somewhere on its surface a patch consisting of two pentagons linked to opposite edges of a central hexagon. For C_2 extrusion, the initial fullerene C_{n+2} should have a pair of fused pentagons embedded in a four-ring motif that is the 'inverse' of a Stone–Wales patch (compare Fig. 6.1). As drawn in Fig. 7.1, the process involves a four-centre transition state and is therefore Woodward–Hoffman forbidden.[10]

Mechanisms of conversion between fullerene isomers with different atom counts and between isomers with a fixed number of atoms are intimately linked. For example, a trivial consequence of the introduction of a pentagon adjacency within the active patch is that no two isolated-pentagon fullerene isomers C_n and C_{n+2} can be connected directly by the insertion pathway shown in Fig. 7.1. If the appropriate set of three rings is present, an isolated-pentagon isomer of C_n can gain two atoms in this way, but the process yields what is presumably a high-energy form of C_{n+2} with fused pentagons. A further rearrangement (perhaps of the Stone–Wales type) is needed to give a final product that has isolated pentagons. The corollary for C_2 extrusion is that, before ejecting C_2, any isolated-pentagon fullerene must first undergo rearrangement to a more strained isomer that has a pentagon adjacency.[15]

If the scope of possible insertion/extrusion processes is widened to include more than two atoms entering or leaving the cage at a time, plausible insertion mechanisms for C_4 and C_6 (Fig. 7.2) can be found by reversing the appropriate extrusion[10] or one of its variants. Each process would have characteristic symmetry and topological requirements. The present chapter concentrates on the C_2 processes because they are likely to have lower activation energies as

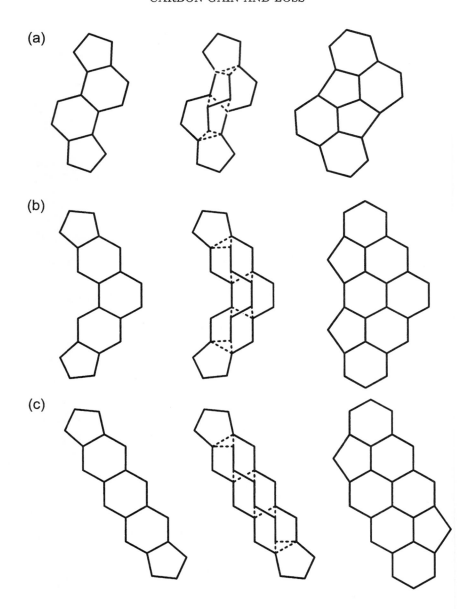

FIG. 7.2. Hypothetical mechanisms for insertion of more than two atoms at a time in a fullerene cage. Insertion of (a) a C_4 chain and (b),(c) C_6 chains in different conformations lead to fullerene products, but involve more bond-making and bond-breaking than the simpler C_2 insertion mechanism in Fig. 7.1.

fewer bonds are made and broken. From a topological point of view, any one of these processes is characterized by the molecular graphs of the reactant and product cages. Detailed consideration of the reaction path between these two cages and its energetics is beyond the qualitative model, and is likely to be difficult as most methods for the calculation of electronic structure are at their weakest when considering processes in which covalent links are made or broken. Accordingly, the present treatment is limited to an account of the restrictions imposed by symmetry and of the conversion maps that show the allowed C_n – C_{n+2} interconversions for small fullerenes. The same methods can be applied to other interconversion mechanisms if this becomes necessary.

7.2 Symmetry aspects of C_2 processes

Analogues of the selection rules discussed in connection with the Stone–Wales transformation (§ 6.2) can be derived for C_2 insertion and extrusion. Again the important site group is C_{2v}, the maximal site symmetry of an edge in a polyhedron, but more care is needed to define the symmetry that is conserved in the insertion/extrusion process.

Insertion of a C_2 fragment corresponds to decoration of a hexagonal face (maximal site symmetry C_{6v}) with a central C_2 pair to produce a pentagon–pentagon bond (maximal site symmetry C_{2v}). The symmetry that is preserved is the site group of the active patch (i.e., of the hexagon and its two pentagonal neighbours considered as a unit), which is equal to the site symmetry of the new edge in the product fullerene. As with a Stone–Wales rotation, each C_2 insertion/extrusion process can be assigned a site symmetry group C_{2v}, C_2, C_s, or C_1.

Both reactant and product patches can occur in multiple copies and all members of each equivalent set share a common site symmetry. Classification of product patches is straightforward: the possible equivalent sets in each fullerene point group are the edge orbits, exactly as listed in Table 6.1 for SW patches. Classification of reactant patches is complicated by the fact that up to three patches may be centred on a given hexagon, so that a symmetry operation can permute patches without shifting them to a new centre. The usual symmetry analysis in terms of orbits is not appropriate in this case. Instead, the allowed site symmetries of the reactant patches can be deduced by following the fate of an orbit of hexagons as the decorating patches are added.

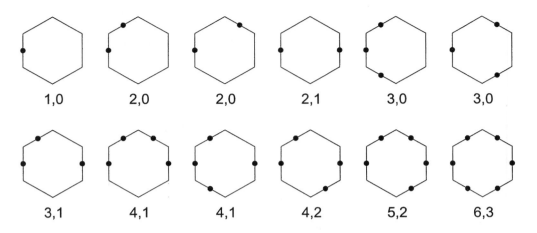

FIG. 7.3. Neighbour patterns for a hexagonal face of a fullerene. A given hexagon may have 0 to 6 pentagonal neighbours (represented by black dots) including from 1 to 3 *para* pairs and may therefore produce up to three distinct fullerene products on C_2 insertion. All distinct patterns involving at least one pentagonal neighbour are shown. The code (i, j) denotes a hexagon with i pentagonal neighbours amongst which are j *para* pairs.

A given hexagon may have 0, 1, 2, ... 6 pentagonal neighbours arranged in a pattern that has one of the symmetries C_{6v}, C_6, C_{3v}, C_3, C_{2v}, C_2, C_s, or C_1 (§ 5.1). The distinct patterns are indicated schematically in Fig 7.3, and the allowed orbits and site symmetries for hexagonal faces in the fullerene point groups are listed in Table 7.1. Reactant patch symmetries depend on the number, distribution, and symmetry of the neighbour pattern; if the hexagon has less than two pentagonal neighbours, then C_2 gain by the mechanism under consideration is impossible; if the hexagon has two or three pentagonal neighbours that include a *para* pair, then C_2 insertion can take place in just one way; with four pentagonal neighbours there is at least one and there may be two ways to insert C_2; with five pentagons there are two ways and with six there are three.

Table 7.2 lists the possible patch site groups for hexagons of each type. By combining the information given in Tables 6.1, 7.1, and 7.2, it is possible to deduce the possible symmetries of reactant patches for fullerenes in each allowed molecular point group. Some restrictions are thereby placed on the overall symmetries of two fullerenes C_n and C_{n+2} linked by a C_2 process because the two point groups must both have the patch site symmetry as a subgroup. For

Table 7.1. Sets of equivalent hexagonal faces in fullerenes. A hexagon site may have C_{6v}, C_6, C_{3v}, C_3, C_{2v}, C_2, C_s, or C_1 symmetry; the last four are also possible edge site symmetries and the orbits of hexagons in these cases are therefore identical with those listed for edges in Table 6.1. Only those fullerene groups admitting $C_{6(v)}$ and $C_{3(v)}$ sites are shown here. Orbits O_n are labelled by size (n).

		Site group		
G	C_{6v}	C_6	C_{3v}	C_3
I_h	-	-	O_{20}	-
I	-	-	-	O_{20}
T_h	-	-	-	O_8
T_d	-	-	O_4	-
T	-	-	-	O_4
D_{6h}	O_2	-	-	-
D_{6d}	O_2	-	-	-
D_6	-	O_2	-	-
D_{3h}	-	-	O_2	-
D_{3d}	-	-	O_2	-
D_3	-	-	-	O_2
S_6	-	-	-	O_2
C_{3h}	-	-	-	O_2
C_{3v}	-	-	O_1	-
C_3	-	-	-	O_1

example, a conversion $C_n(C_{2v}) \rightarrow C_{n+2}(T)$ may involve C_2 or C_1 patches but cannot proceed via C_{2v} or C_s reactant patches. These considerations show that chiral to chiral, chiral to/from achiral, and achiral to achiral conversions are all possible. Chiral insertion in an achiral hexagon produces enantiomeric products in equal amount.

Given the site symmetry, the number of patches in a set can be deduced. If the molecular point group is of order $|G|$ then, depending on the patch site symmetry, a fullerene has $|G|/4$ (C_{2v}), $|G|/2$ $(C_2$ or $C_s)$ or $|G|$ (C_1) equivalent patches by Eq. (5.1). Since the orders of the molecular point groups in initial and final isomers may be different, but the site group remains the same, it follows that the number of insertion patches leading from C_n to C_{n+2} is not necessarily equal to the number of extrusion patches in C_{n+2} leading back to C_n.

As each patch involves a pair of pentagons, there is an upper limit on the

Table 7.2. Products of C_2 insertion on a given hexagonal face of a fullerene. G is the site symmetry group of the 'active' hexagon in the reactant patch of the fullerene C_n and N is the number of pairs of *para* pentagons amongst its six neighbouring rings. In the body of the table are shown the site symmetry groups G' of the new edge in C_{n+2} that would be formed by addition of C_2 across each of these pairs. The notation $3 \times G'$ means that there are three copies of a single product, whereas G', G', G' means that there are three distinct products of the same symmetry. R and L denote enantiomeric products formed by asymmetric additions to an achiral reactant.

		N	
G	3	2	1
C_{6v}	$3 \times C_{2v}$	-	-
C_6	$3 \times C_2$	-	-
C_{3v}	$3 \times C_s$	-	-
C_3	$3 \times C_1$	-	-
C_{2v}	C_{2v}, C_2^L, C_2^R	C_2^L, C_2^R	C_{2v}
C_2	C_2, C_2, C_2	C_2, C_2	C_2
C_s	C_s, C_1^L, C_1^R	C_1^L, C_1^R	C_s
C_1	C_1, C_1, C_1	C_1, C_1	C_1

number of reactant and product patches, whether equivalent by symmetry or not, that can be simultaneously present on the surface of a fullerene. The maximum number of reactant patches for C_2 insertion in any fullerene is 30. The limit is realized uniquely in the icosahedral isolated-pentagon isomer C_{80}:7, where every hexagon belongs to a C_{2v} patch and every pentagon to five. This isomer of C_{80} has a special status as the only icosahedral (I or I_h) fullerene in which C_2 insertion by the mechanism of Fig. 7.1 is possible; C_{20} has no hexagonal faces, C_{60} has only *meta* pentagon pairs, and in C_{140} and beyond, the pentagons are all too far apart. A similar result holds for leapfrog fullerenes. A leapfrog fullerene contains neither fused pentagons nor *para* pairs of pentagons and so no leapfrog isomer can take part directly in C_2 insertion/extrusion of this type, though it may be able to rearrange to an isomer with the appropriate patches.

Insertion at a reactant site can interfere with other patches. In C_{80}, for example, attack of C_2 at any one site removes eight neighbouring potential reactant patches. A fullerene can contain at most 12 simultaneous 'product' patches (i.e., positions at which C_2 has been inserted or from which a C_2 fragment could be ejected). This maximum is realized, for example, in the 12 patches at C_s sites in the D_{6h} isomer C_{36}:15 and in an infinite series of larger carbon cylinders with the same caps. It is no coincidence that the limits of 30 and 12 are equal to those for the Stone–Wales patches of types **I** and **II**, respectively (§ 6.2).

Another piece of reasoning used earlier in connection with the Stone–Wales transformation can also be extended to the C_2 processes. If each member of a set of n_A equivalent patches in fullerene A leads to fullerene B on C_2 insertion, and if n_B equivalent patches in B lead back to A on C_2 extrusion, then the arguments of § 6.6 show that $n_A/n_B = |G_A|/|G_B|$ where $|G_X|$ is the order of the full point group of species X. However, an entropic model of the equilibrium distribution analogous to the one developed in § 6.6 for Stone–Wales isomerizations is implausible in this case because a substantial difference (of the order of twice the CC σ bond energy) is expected between the energies of the reactants and the product. For neutral C_{60} dissociating endothermically to $C_{58} + C_2$, this energy difference is estimated from experiment as 5.77 eV,[16] 7.64 ± 0.4 eV,[17] or 'well over 7.8 eV',[18] and calculations at various levels of sophistication[19–21] give figures in the range 11.8 to 12.0 eV.

7.3 Insertion/extrusion maps

Given a complete list of fullerenes C_n and $C_{n\pm2}$ represented by ring spirals, adjacency matrices or connection lists, it is straightforward to design a computer program to catalogue all C_2 insertion and extrusion processes for isomers of C_n. The molecular graph of each isomer is searched for the insertion patch of *para* pentagons straddling a hexagon or the extrusion patch of a pentagon–pentagon bond joining atoms in two hexagons (Fig. 7.1), and all possible product isomers are generated and identified by their eigenvalue spectra or otherwise. Results of direct calculations of this kind are presented in Table 7.3 for fullerenes C_{24} to C_{38}. The table is organized in terms of C_2 insertion reactions on these fullerenes, but can equally be read as a list of extrusions from C_{40} to C_{26}.

The table lists all products that can be formed by insertion of C_2 across an 'active' hexagon in each isomer n:m for n running from 24 to 38. $n = 24$ must be the lower limit for the process because no fullerene with 22 atoms can exist and C_{20}, having no hexagonal faces, has no sites for insertion or extrusion by this mechanism. For each conversion listed, the product isomer is given (without distinguishing enantiomers) along with n_A and n_B, the numbers of equivalent

Table 7.3. C_2 insertion pathways for small fullerenes C_{24} to C_{38}. For each fullerene isomer $n:m$, the table lists all possible insertions of C_2 across a pentagon–hexagon–pentagon triple. The notation $m'(n_A, n_B)$ for each product means that (in achiral-achiral or chiral-chiral reactions) n_A patches in m give isomer $n + 2 : m'$ on insertion and n_B patches in m' lead back to the isomer $n:m$ on extrusion. If m is achiral and m' chiral then $n_A/2$ patches lead to each enantiomer of m'; likewise if m is chiral and m' achiral, then $n_B/2$ patches lead to each enantiomer of m.

Reactant	Products of C_2 insertion
24:1	1(6,3)
26:1	1(6,2), 2(3,6)
28:1	2(4,4), 3(4,4)
28:2	3(12,2)
30:1	1(10,1)
30:2	1(4,2), 4(4,2), 6(2,3)
30:3	4(6,3), 5(1,3), 6(4,6)
32:1	1(2,2), 2(2,2), 3(2,2), 4(2,2), 5(2,2)
32:2	1(2,1), 2(4,2), 4(2,1)
32:3	2(6,1)
32:4	3(2,2), 4(2,2), 5(4,4), 6(2,6)
32:5	3(6,1)
32:6	3(6,2), 5(6,2)
34:1	1(2,2), 3(2,1), 5(1,2), 7(2,1), 10(1,1)
34:2	3(2,1), 4(1,1), 7(2,1), 9(2,4), 11(2,2), 12(2,2), 15(1,12)
34:3	5(2,4), 6(1,4), 7(2,1), 10(2,2), 11(2,2)
34:4	7(4,2), 8(2,2), 9(2,4), 10(1,1), 12(2,2), 14(1,4)
34:5	7(2,1), 10(2,2), 11(3,3), 12(3,3), 14(2,8)
34:6	8(3,1), 10(6,2)
36:1	1(2,2), 3(2,1), 4(2,1), 5(2,1)
36:2	4(4,1)
36:3	3(2,2), 4(1,1), 7(1,1), 8(1,1), 10(1,2), 11(1,1), 13(1,2), 15(1,4)
36:4	3(2,1), 6(2,2), 7(2,1), 8(2,1), 15(1,2)
36:5	7(4,1)
36:6	5(8,1)
36:7	5(1,1), 6(1,2), 7(2,2), 10(1,2), 11(1,1), 12(1,4), 14(2,2), 17(1,2)
36:8	5(2,1), 8(2,1), 11(2,1), 13(2,2), 14(2,1)
36:9	8(4,1), 14(4,1), 16(2,3), 17(4,2)
36:10	5(2,1), 11(4,2), 12(2,4), 14(2,1)
36:11	7(2,1), 10(2,2), 11(2,1), 14(2,1), 17(2,2)
36:12	11(2,1), 13(2,2), 14(4,2), 16(2,6), 17(2,2)
36:13	8(12,1)

Table 7.3 *(Continued)*

Reactant	Products of C_2 insertion
36:14	14(8,1), 17(8,2)
36:15	15(6,1), 17(12,1)
38:1	2(2,2), 4(2,1), 6(2,1), 9(1,1)
38:2	8(3,1)
38:3	4(2,2), 5(1,2), 6(1,1), 10(1,1), 14(1,2), 20(1,6), 21(1,2), 22(1,1)
38:4	6(2,2), 7(1,2), 10(1,1), 11(1,2), 12(1,1), 15(1,2), 16(1,2), 25(1,2)
38:5	9(1,2), 12(1,1), 13(1,2), 15(1,2), 22(2,2), 23(1,2), 24(1,2), 25(1,2), 29(1,2)
38:6	10(2,1), 11(1,1), 12(2,1), 13(2,2), 16(1,1), 26(2,1)
38:7	10(1,1), 11(1,2), 12(1,1), 19(1,2), 21(1,2), 34(2,2), 35(1,2)
38:8	14(1,2), 15(1,2), 17(1,1), 18(1,2), 26(2,2), 27(1,2), 28(1,2), 31(1,2), 34(2,2), 35(1,2)
38:9	15(3,1), 17(6,1), 18(3,1)
38:10	10(2,1), 21(2,2), 23(1,1), 24(2,2), 26(2,1), 29(1,1)
38:11	12(1,1), 22(2,2), 23(1,2), 24(1,2), 27(1,2), 30(1,3), 34(1,1), 35(1,2), 36(1,2)
38:12	13(2,1), 23(4,2), 37(1,1)
38:13	22(2,1), 25(2,2), 26(2,1), 28(2,2), 31(2,2), 35(1,1), 36(1,1)
38:14	24(1,2), 26(1,1), 27(1,2), 28(1,2), 29(1,2), 34(2,2), 35(1,2), 36(2,4), 37(1,4), 38(1,4)
38:15	20(2,3), 26(4,1), 32(2,2), 33(1,2), 34(4,1)
38:16	31(3,1), 36(6,2), 37(3,2)
38:17	26(2,1), 34(2,1), 35(2,2), 36(2,2), 37(2,4), 38(3,6), 39(1,10)

reactant patches in the reactant $A = C_n$ and extrusion patches in the product $B = C_{n+2}$.

When both isomers are achiral or both chiral, n_A and n_B straightforwardly describe the numbers of patches in one isomer that lead to the other. When one isomer (A, say) is achiral and the other (B) is chiral, then $n_A/2$ patches lead to each enantiomer of B. As noted in connection with the Stone–Wales transformation (§ 6.3 and Fig. 6.7), the $n_A/2$ patches that lead to a given enantiomer of B are exchanged by *proper* operations of the point group of isomer A. Patches exchanged by *improper* operations of this group yield opposite enantiomers of B. As an example, consider the conversion of the isomer $C_{32}:4$ (C_2) to $C_{34}:3$ (C_s) by addition of two carbon atoms. The entry 3(2,2) on the line 32:4 of Table 7.3 means that the smaller cage has two insertion patches (which

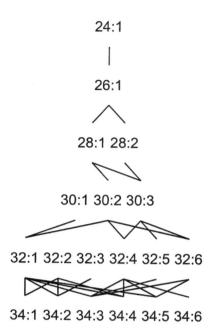

24:1

26:1

28:1 28:2

30:1 30:2 30:3

32:1 32:2 32:3 32:4 32:5 32:6

34:1 34:2 34:3 34:4 34:5 34:6

FIG. 7.4. Possible C_2 insertion/extrusion processes linking the smallest fullerenes.

must have site symmetry C_1) that lead to identical C_{34} cages, whereas two extrusion patches of C_{34}:3 (also with site symmetry C_1, and therefore related by a mirror plane) lead to opposite enantiomers of the C_{32} cage.

In many cases the site symmetry of the active patch can be deduced by checking n_A and n_B against the orders $|G_A|$ and $|G_B|$ of the molecular point groups. For example, Table 7.3 shows a transformation for C_{28}:2 (T_d) to C_{30}:3 (C_{2v}) for which $n_A = 12$ and $n_B = 2$. The order of the patch site group is $24/12 = 4/2 = 2$ and the choice is between C_2 and C_s; C_{2v} has no edge orbit of C_2 symmetry and so the group must be C_s. The 12 reactant patches of C_s symmetry in the tetrahedral isomer arise from four hexagons at C_{3v} sites, each completely surrounded by pentagons.

It would be useful to have a picture that conveyed the information summarized in Table 7.3 in more digestible form, by analogy with the Stone–Wales maps of Chapter 6. Figure 7.4 shows the first few steps in the cascade of C_2 insertions starting from a C_{24} fullerene precursor. The unique isomer C_{24}:1 yields the unique C_{26}:1, which has three equivalent equatorial hexagons surrounded by pentagons, each hexagon taking part in a single C_{2v} and a pair of C_2 reactant patches, leading to T_d and D_2 isomers of C_{28}, respectively. The number of conversions rapidly becomes too many for convenient visual representation; taking Fig. 7.4 as far as $C_{38} \rightarrow C_{40}$, for example, produces a blur of dozens of crisscrossing and overlapping lines. The picture becomes even more complicated when the 'vertical' C_2 processes are supplemented by the 'horizontal' Stone–Wales transformations. For example, in Fig 7.4 the isomer C_{30}:1 cannot be reached by insertion of C_2 into either isomer of C_{28}, but both C_{30}:2 and C_{30}:3 are accessible. As all three isomers of C_{30} make up a single Stone–Wales family (see Table 6.2 and Fig. 6.8), when Stone–Wales transformations are taken into account isomer C_{30}:1 can be reached by a two-step insertion/rearrangement pathway from C_{28}:1, or in three steps from C_{28}:2.

To show both C_2 and SW connections properly for larger fullerenes it would be necessary to branch out into the third dimension, with horizontal Stone–Wales maps for different values of n stacked in the vertical direction; this too would become unwieldy very quickly, as many of the maps cannot be drawn in the plane without crossings.

In Chapter 6 we saw that some fullerene isomers were islands on the Stone–Wales map, having no connections to other cages with equal numbers of atoms. In the more general conversion scheme an isomer may be linked to others by any or all of the three processes of insertion, extrusion or SW transformation. An 'island' isomer that has none of these links would be unreachable by a 'fullerene road' mechanism if C_2 fragments alone were involved and all annealing isomerizations were of the Stone–Wales type. Examples of such unconnected isomers can easily be devised. They include all icosahedral fullerenes that have more than 80 atoms, and all double-leapfrogs (leapfrogs of leapfrogs). Some specific results for the smaller fullerenes are:

(1) the smallest fullerene beyond C_{24} that *cannot* be produced directly by C_2 insertion is C_{30}:1, and the others with fewer than 42 atoms are C_{32}:2, C_{32}:3, C_{36}:2, C_{36}:13, C_{38}:2, C_{38}:9, C_{40}:1, C_{40}:3, C_{40}:40;

(2) the smallest fullerenes that can neither be produced by C_2 insertion nor by Stone–Wales transformation are C_{38}:2, C_{40}:1 and C_{40}:3;

(3) all fullerene isomers with fewer than 40 atoms have the pentagon–hexagon–pentagon patches requisite for C_2 insertion and the smallest isomer that *cannot* ingest C_2 by the Endo–Kroto mechanism is C_{40}:1;

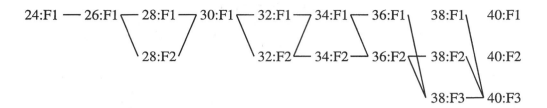

FIG. 7.5. A fullerene road linking fullerenes from C_{24} to C_{40}. The lines represent C_2 insertion/extrusion links between groups of isomers mutually connected by the Stone–Wales transformation. The families F1, F2, ... are listed in Table 6.2.

(4) C_{40}:1 is therefore the first 'island' off the C_2 fullerene road.

In Heath's fullerene road model for the formation of C_{60} and C_{70},[9] as rephrased by Curl,[1] the ring rearrangements that remove pentagon adjacencies are supposed to take place more rapidly than the growth steps. Calculations of activation energies[15] seem to support this hypothesis. This suggests a modification of the connection diagram (Fig. 7.4) in which the insertion/extrusion links are drawn not between individual isomers but between Stone–Wales families. Two families $\{C_n\}$ and $\{C_{n+2}\}$ would be considered to be linked if at least one member of $\{C_n\}$ gives a member of $\{C_{n+2}\}$ on insertion of C_2. Figure 7.5 shows this simplified diagram for C_{24} to C_{40}. Continuation to C_{60} and beyond would give an explicit realization of the fullerene road.

One final question is whether there is a route connecting the 'best' isomers of Table 4.1 and running all the way from C_{24} to C_{70} with only a small number of Stone–Wales 'detours'. Such detours must exist if the road is to arrive at C_{60} itself. The icosahedral isomer C_{60}:1812 (I_h) has neither insertion nor extrusion patches, but it is only one SW step away from an isomer (C_{60}:1809 (C_{2v}), Fig. 6.4b) that has both. C_2 extrusion from this C_{2v} isomer leads uniquely and directly to the fullerene C_{58}:1205 (C_{3v}) that is the most stable isomer of C_{58} according to the pentagon isolation criterion. Insertion of C_2 into the C_{2v} isomer of C_{60} again leads uniquely to the stable isomer C_{62}:2194 (C_1). The idea of a path that links stable isomers is therefore worth testing. Further research will certainly be done on this topic, as the prize could be an understanding of the enigma of fullerene formation.

References and notes

[1] R. F. Curl, Phil. Trans. Roy. Soc. (London) A 343 (1993) 19. Reprinted in H. W. Kroto and D. R. M. Walton, eds., The fullerenes: new horizons for the chemistry, physics and astrophysics of carbon (Cambridge University Press, Cambridge, 1993).

[2] H. Schwartz, Angew. Chemie Intl. Edn. 32 (1993) 1412.

[3] G. Meijer and D. S. Bethune, J. Chem. Phys. 93 (1990) 7800.

[4] H. W. Kroto and D. R. M. Walton in E. Osawa and O. Yonemitsu, eds., Chemistry of three dimensional polycyclic molecules (VCH, New York, 1992).

[5] T. Wakabayashi and Y. Achiba, Chem. Phys. Lett. 190 (1992) 465.

[6] R. Taylor, G. J. Langley, H. W. Kroto, and D. R. M. Walton, Nature 366 (1993) 728.

[7] J. R. Heath, S. C. O'Brien, R. F. Curl, H. W. Kroto, and R. E. Smalley, Comm. Cond. Mater. Phys. 13 (1987) 119.

[8] Q. L. Zhang, S. C. O'Brien, J. R. Heath, Y. Liu, R. F. Curl, H. W. Kroto, and R. E. Smalley, J. Phys. Chem. 90 (1986) 525.

[9] J. R. Heath in G. S. Hammond and V. J. Kuck, eds., Fullerenes: synthesis, properties and chemistry of large carbon clusters, ACS Symposium Series 481 (1991) 1.

[10] S. C. O'Brien, J. R. Heath, R. F. Curl, and R. E. Smalley, J. Chem. Phys. 88 (1988) 220.

[11] G. Ulmer, E. E. B. Campbell, R. Kuhnle, H-G. Busmann and I. V. Hertel, Chem. Phys. Lett. 182 (1991) 114.

[12] M. Endo and H. W. Kroto, J. Phys. Chem. 96 (1992) 6941.

[13] S. Iijima, Nature 354 (1991) 56.

[14] P. M. Ajayan and S. Iijima, Nature 361 (1993) 333.

[15] R. L. Murry, D. L. Strout, G. K. Odom, and G. E. Scuseria, Nature 366 (1993) 665. This paper considers an alternative proposal for C_2 loss from C_{60} in which the Stone–Wales transformation and C_2 extrusion steps are reversed, with the extrusion step occuring first through a dangling-bond intermediate. This alternative mechanism leads from C_{60} to a C_{58} cage that has a seven-membered ring and so in this case a generalized Stone–Wales transformation (Fig. 6.5) is needed to give the final fullerene product.

[16] P. Sandler, C. Lifshitz, and C. E. Klots, Chem. Phys. Lett. 200 (1992) 445.

[17] M. Foltin, M. Lezius, P. Scheier, and T. D. Märk, J. Chem. Phys. 98 (1993) 9624.

[18] R. L. Whetten and C. Yeretzian, Int. J. Mod. Phys. 6 (1992) 3801.

[19] R. E. Stanton, J. Phys. Chem. 96 (1992) 111.

[20] J. Yi and J. Bernholc, J. Chem. Phys. 96 (1992) 8634.

[21] W. C. Eckhoff and G. E. Scuseria, Chem. Phys. Lett. 216 (1993) 399.

APPENDIX

The spiral computer program

This appendix gives a FORTRAN 77 listing of the spiral computer program, along with sample output files for the general fullerene isomers of C_{36} and the isolated-pentagon fullerene isomers of C_{86}. The reason for including this listing here is that the program can be used to extend the main fullerene isomer tables, for example, by producing a similar catalogue of all 1812 general fullerene isomers of C_{60}.

The input to the program is very simple, consisting solely of the nuclearity of the required C_n cage and an integer flag which is 0 for general and 1 for isolated-pentagon isomers. For example, the C_{36} output file given below can be generated with the input 36,0 and the C_{86} output file with the input 86,1. The resulting output files consist of a one-line entry for each isomer giving the isomer label (for example, 36:6), its maximal point group (D2d), the positions of the 12 pentagons in its canonical ring spiral (1 2 3 4 7 10 12 15 17 18 19 20), and its idealized ^{13}C NMR signature (3 x 4, 3 x 8). This information corresponds to that given in the first four columns of each isomer entry in the main tables; we have chosen it as a compromise between a desire to convey as much useful information about each isomer as possible and a conflicting desire to keep both the program listing and the generated output as short as possible. In fact, the IR- and Raman-active vibrational mode counts of each isomer are also implicitly contained in the output, since they can be computed from the point group order and the NMR pattern using the method described in § 5.5 and the information in Table 5.4.

Other topological properties can also be obtained by appropriate modifications of the program, and this is how much of this book was prepared. The key thing that is required in order to obtain these other properties, which include the Hückel molecular orbital energy levels (§ 3.1) and the topological atomic coordinates (§ 5.2), is the adjacency matrix $\mathbf{A}$ of each fullerene isomer. This can be extracted from the fullerene dual adjacency matrix $\mathbf{D}$ that is used to represent the vertex connectivity of the fullerene dual (i.e., the face connectivity of the fullerene) in the spiral program.

The dual adjacency matrix $\mathbf{D}$ is defined by analogy with Eq. (3.1), with elements $D_{kl} = 1$ if vertices k and l of the dual (i.e., faces k and l of the fullerene) are adjacent and 0 otherwise. It has dimension $m \times m$, where $m = n/2+2$ is the number of faces in the fullerene, and it is generated within the spiral program by subroutine WINDUP as described below. Once $\mathbf{D}$ is available, the fullerene vertex adjacency matrix $\mathbf{A}$ can be obtained using the following self-contained subroutine, which is not used in the spiral program itself:

```
      SUBROUTINE DUAL (D,M,A,N,IER)
      IMPLICIT INTEGER (A-Z)
      PARAMETER (NMAX = 194)
      DIMENSION D(M,M),A(N,N)
      DIMENSION V(3,NMAX)
C         Given a fullerene dual adjacency matrix D, this subroutine
C         constructs the corresponding fullerene adjacency matrix A.
C         IER = 0 on return if the construction is successful.
      IF (N .GT. NMAX) STOP!    Increase NMAX
      I = 0
      DO 3 L = 1,M
         DO 2 K = 1,L
            IF (D(K,L) .EQ. 0) GO TO 2
            DO 1 J = 1,K
               IF (D(J,K).EQ.0 .OR. D(J,L).EQ.0) GO TO 1
               I = I+1
               IF (I .GT. N) GO TO 1
               V(1,I) = J!      Associate the three mutually adjacent
               V(2,I) = K!      dual vertices (fullerene faces) J,K,L
               V(3,I) = L!      with fullerene vertex I
1           CONTINUE
2        CONTINUE
3     CONTINUE
      IER = I-N
      IF (IER .NE. 0) RETURN!  D contains IER > 0 separating triangles
      DO 7 J = 1,N!            and is therefore NOT a fullerene dual
         DO 6 I = 1,J
            K = 0
            DO 5 JJ = 1,3
               DO 4 II = 1,3
                  IF (V(II,I) .EQ. V(JJ,J)) K = K+1
4              CONTINUE
5           CONTINUE
            IF (K .EQ. 2) THEN
               A(I,J) = 1!      Fullerene vertices I and J are adjacent
               A(J,I) = 1!      if they have 2 dual vertices in common
            ELSE
               A(I,J) = 0
               A(J,I) = 0
            ENDIF
6        CONTINUE
7     CONTINUE
      RETURN
      END
```

As stressed in § 2.2, this subroutine is exclusive to the fullerenes: it cannot be used reliably in its present form for other trivalent polyhedra containing square and/or triangular faces because their duals can have separating triangles. For example, if the subroutine is called with the dual adjacency matrix of a trigonal prism it returns unsuccessfully with IER = 1, indicating the presence of one separating triangle.

Now for the spiral program proper, which has three parts. The first part is a main program which does little more than implement the loop over spiral combinations in Eq. (2.11). A significantly more efficient (but less robust) version of this main program can be obtained by replacing the statement DO 1 J1 = 1,M-11*JPR with the statement DO 1 J1 = 1,1, which restricts the search over spiral combinations to those beginning on a pentagon:

```
      PROGRAM SPIRAL
      IMPLICIT INTEGER (A-Z)
      CHARACTER*3 GROUP
      PARAMETER (NMAX = 194)
      PARAMETER (MMAX = NMAX/2+2)
      DIMENSION D(MMAX*MMAX),S(MMAX)
      DIMENSION NMR(6)
C        This program catalogues fullerenes with a given number of
C        vertices using the spiral algorithm and a uniqueness test
C        based on equivalent spirals. The required input is N and IPR,
C        where N is the nuclearity of the fullerene and IPR = 0 for
C        general and 1 for isolated-pentagon isomers. The resulting
C        output is a catalogue of the isomers found containing their
C        idealized point groups, canonical spirals, and NMR patterns.
      READ (5,*) N,IPR
      IF (N .GT. NMAX) STOP!            Increase NMAX
      IF (2*(N/2) .NE. N) STOP!         N must be even
      IF (IPR.NE.0 .AND. IPR.NE.1) STOP!    IPR must be 0 or 1
      IF (IPR .EQ. 0) THEN
         IF (N .LT. 100) WRITE (6,601) N
601      FORMAT(1X,'GENERAL FULLERENE ISOMERS OF C',I2,':'/1X,77('-'))
         IF (N .GE. 100) WRITE (6,602) N
602      FORMAT(1X,'GENERAL FULLERENE ISOMERS OF C',I3,':'/1X,77('-'))
      ELSE
         IF (N .LT. 100) WRITE (6,603) N
603      FORMAT(1X,'ISOLATED-PENTAGON ISOMERS OF C',I2,':'/1X,77('-'))
         IF (N .GE. 100) WRITE (6,604) N
604      FORMAT(1X,'ISOLATED-PENTAGON ISOMERS OF C',I3,':'/1X,77('-'))
      ENDIF
      L = 0
      M = N/2+2
      JPR = IPR+1
      DO 1 J1 = 1,M-11*JPR!          Open loop over spiral
      DO 2 J2 = J1+JPR,M-10*JPR!     combinations
      DO 3 J3 = J2+JPR,M-9*JPR
      DO 4 J4 = J3+JPR,M-8*JPR
      DO 5 J5 = J4+JPR,M-7*JPR
      DO 6 J6 = J5+JPR,M-6*JPR
      DO 7 J7 = J6+JPR,M-5*JPR
      DO 8 J8 = J7+JPR,M-4*JPR
```

```
      DO 9 J9 = J8+JPR,M-3*JPR
      DO 10 J10 = J9+JPR,M-2*JPR
      DO 11 J11 = J10+JPR,M-JPR
      DO 12 J12 = J11+JPR,M
      DO 14 J = 1,M!                        Form spiral code in S
        S(J) = 6
 14   CONTINUE
      S(J1) = 5
      S(J2) = 5
      S(J3) = 5
      S(J4) = 5
      S(J5) = 5
      S(J6) = 5
      S(J7) = 5
      S(J8) = 5
      S(J9) = 5
      S(J10) = 5
      S(J11) = 5
      S(J12) = 5
      CALL WINDUP (S,M,D,IPR,IER)!           Wind up spiral into dual
      IF (IER .EQ. 12) GO TO 12!             and check for closure
      IF (IER .EQ. 11) GO TO 11
      IF (IER .EQ. 10) GO TO 10
      IF (IER .EQ. 9) GO TO 9
      IF (IER .EQ. 8) GO TO 8
      IF (IER .EQ. 7) GO TO 7
      IF (IER .EQ. 6) GO TO 6
      IF (IER .EQ. 5) GO TO 5
      IF (IER .EQ. 4) GO TO 4
      IF (IER .EQ. 3) GO TO 3
      IF (IER .EQ. 2) GO TO 2
      IF (IER .EQ. 1) GO TO 1
      CALL UNWIND (D,M,S,GROUP,NMR,IER)!     Unwind dual into spirals
      IF (IER .EQ. 13) GO TO 13!             and check for uniqueness
      K = 0
      L = L+1!                               Spiral S is canonical
      DO 15 J = 1,6
        IF (NMR(J) .EQ. 0) GO TO 16
        K = J
 15   CONTINUE
 16   WRITE (6,605) L,GROUP,
     +  J1,J2,J3,J4,J5,J6,J7,J8,J9,J10,J11,J12,(NMR(J),J=1,K)
605   FORMAT(1X,I8,2X,A3,1X,12I3,2X,3(I3,' x',I3,:,',,'))
 13   CONTINUE
 12   CONTINUE!                              Close loop over spiral
 11   CONTINUE!                              combinations
 10   CONTINUE
  9   CONTINUE
  8   CONTINUE
  7   CONTINUE
  6   CONTINUE
  5   CONTINUE
  4   CONTINUE
  3   CONTINUE
  2   CONTINUE
  1   CONTINUE
      WRITE (6,606)
606   FORMAT(1X,77('-'))
      STOP
      END
```

The second component of the spiral program is subroutine WINDUP, which attempts to wind up each tentative test spiral S in turn into a fullerene dual adjacency matrix D. The coding of this subroutine is a direct implementation of the discussion in §§ 2.5 and 2.6, so that, for example, each face k in the spiral after the second is connected to both the preceding face $i = k - 1$ and the first face j in the preceding spiral that still has an open edge (constraints (a) and (b) of the spiral conjecture, respectively). Notice also the statements beginning IF (IPR.EQ.1..., which detect pentagon adjacencies between successive coils of the winding spiral and allow an early exit (GO TO 10) if any such adjacencies are found during a search for isolated-pentagon fullerenes:

```
      SUBROUTINE WINDUP (S,M,D,IPR,IER)
      IMPLICIT INTEGER (A-Z)
      PARAMETER (NMAX = 194)
      PARAMETER (MMAX = NMAX/2+2)
      DIMENSION S(M),D(M,M)
      DIMENSION R(MMAX),C(MMAX,6)
C          This subroutine attempts to wind up an input spiral S into
C          a fullerene dual (face) adjacency matrix D. It returns with
C          IER = P if the spiral shorts or is discovered to be open-ended
C          after P pentagons have been added. Otherwise IER = 0 on return.
      IF (M .GT. MMAX) STOP!    Increase NMAX
      J = 1
      C(1,1) = 2
      C(2,1) = 1
      R(1) = 2
      R(2) = 2
      E = 1
      P = 6-S(1) + 6-S(2)
      DO 5 K = 3,M-1
         P = P + 6-S(K)
         R(K) = 1
         I = K-1
1        IF (IPR.EQ.1 .AND. S(I).EQ.5 .AND. S(K).EQ.5) GO TO 10
         IF (R(K) .GE. S(K)) GO TO 10
         C(I,R(I)) = K!          Connect face K to the last open face I
         C(K,R(K)) = I!          in the preceding spiral
         R(I) = R(I)+1
         R(K) = R(K)+1
         IF (R(I) .GT. S(I)) THEN
            L = I-1!             If this closes face I update I and go
            DO 2 I = L,J+1,-1!   back to connect face K to the new I
               IF (R(I) .LE. S(I)) GO TO 1
2           CONTINUE
            GO TO 10
         ENDIF
3        IF (IPR.EQ.1 .AND. S(J).EQ.5 .AND. S(K).EQ.5) GO TO 10
         IF (R(K) .GE. S(K)) GO TO 10
         C(J,R(J)) = K!          Connect face K to the first open face J
         C(K,R(K)) = J!          in the preceding spiral
         R(J) = R(J)+1
         R(K) = R(K)+1
         IF (R(J) .GT. S(J)) THEN
            L = J+1!             If this closes face J update J and go
```

```
        DO 4 J = L,I-1,+1! back to connect face K to the new J
           IF (R(J) .LE. S(J)) GO TO 3
4          CONTINUE
           GO TO 10
        ENDIF
        H = K-P
        E = E+R(K)-1!            Use Euler's theorem to streamline the
        V = 3+2*P+3*H-E!         search. F is a lower bound on the # of
        F = (V+1)/2+1!           additional faces required for closure
        IF (F .GT. M-K) GO TO 10
5   CONTINUE
        P = 12
        R(M) = 1
        DO 6 K = J,M-1
           IF (R(K) .LT. S(K)) GO TO 10
           IF (R(K) .GT. S(K)) GO TO 6
           IF (R(M) .GT. S(M)) GO TO 10
           IF (IPR.EQ.1 .AND. S(K).EQ.5 .AND. S(M).EQ.5) GO TO 10
           C(K,R(K)) = M!          Connect face M to all remaining
           C(M,R(M)) = K!          open faces (including face M-1)
           R(K) = R(K)+1
           R(M) = R(M)+1
6   CONTINUE
        IF (R(M) .LE. S(M)) GO TO 10
        P = 0!                   Successful spiral
        DO 9 J = 1,M!            Form dual adjacency matrix in D
           DO 7 I = 1,M
              D(I,J) = 0
7          CONTINUE
           DO 8 K = 1,S(J)
              I = C(J,K)
              D(I,J) = 1
8          CONTINUE
9   CONTINUE
10  IER = P
        RETURN
        END
```

The third and final component of the spiral program is subroutine UNWIND, which unwinds the surface of the dual **D** in as many of the $6n$ ways in Eq. (2.6) as are possible without shorting and checks to see whether any of the spirals so obtained is lexicographically smaller than the input spiral. The first part of this subroutine simply implements the spiral uniqueness test described in § 2.6, and needs no further comment here. However, if the test is passed, the input spiral **S** is canonical according to the definition in § 2.6 and the subroutine proceeds to calculate the idealized point group and ^{13}C NMR signature of the new fullerene. The algorithm used to do this is based on a complete set of symmetry-equivalent spiral unwindings of the fullerene surface that is generated automatically during the uniqueness test, and since this algorithm differs from the more intuitive topological coordinate method described in § 5.2 we shall now summarize its salient features.

Since each individual spiral has site symmetry C_1, as noted at the end of § 2.5, it has a total of $|G|$ symmetry-equivalent copies in the surface of the fullerene with one copy for each symmetry operation of the point group. The effect of each symmetry operation can therefore be catalogued as a permutation of the faces of the fullerene which retains the same spiral code of 5s and 6s, and this catalogue is stored in subroutine UNWIND in the array FP. Once the face permutations have been catalogued it is straightforward to use code similar to that in subroutine DUAL to construct the corresponding vertex (VP) and edge (EP) permutations for each symmetry operation, and hence to count the numbers of vertices, edges, and faces with each allowed site symmetry group order (§ 5.1). These counts are stored in the arrays MV, ME, and MF, respectively, and are added to give the complete special point orbit counts in MS. The idealized ^{13}C NMR pattern is then computed from the vertex orbit counts as described in § 5.4, and the special point orbit counts are combined with the point group order in a decision tree which assigns the point group:

```
      SUBROUTINE UNWIND (D,M,S,GROUP,NMR,IER)
      IMPLICIT INTEGER (A-Z)
      CHARACTER*3 GROUP
      PARAMETER (NMAX = 194)
      PARAMETER (LMAX = 3*NMAX/2)
      PARAMETER (MMAX = NMAX/2+2)
      DIMENSION D(M,M),S(M),NMR(6)
      DIMENSION P(MMAX),R(MMAX)
      DIMENSION V(3,NMAX),E(2,LMAX)
      DIMENSION VP(NMAX,120),EP(LMAX,120),FP(MMAX,120)
      DIMENSION MV(12),ME(12),MF(12),MS(12)
C        This subroutine unwinds a fullerene dual adjacency matrix D
C        into each of its constituent spirals and checks that none has
C        a lexicographically smaller code than the input spiral S. The
C        idealized point group and NMR signature of the fullerene are
C        also calculated if this test is passed, in which case the input
C        spiral is canonical and IER = 0 on return. Otherwise IER = 13.
      IF (M .GT. MMAX) STOP!          Increase NMAX
      SO = 0
      DO 10 I1 = 1,M!                 Begin multiple loop over all
         P(1) = I1!                   6*N possible spiral starts
         FLAG1 = 0!                   with initial faces I1,I2,I3
         IF (S(P(1)) .NE. S(1)) THEN
            IF (S(P(1)) .GT. S(1)) GO TO 10
            FLAG1 = 1
         ENDIF
         DO 9 I2 = 1,M
            IF (D(I1,I2) .EQ. 0) GO TO 9
            P(2) = I2
            FLAG2 = FLAG1
            IF (FLAG2.EQ.0 .AND. S(P(2)).NE.S(2)) THEN
               IF (S(P(2)) .GT. S(2)) GO TO 9
               FLAG2 = 2
            ENDIF
            DO 8 I3 = 1,M
```

```
         IF (D(I1,I3).EQ.0 .OR. D(I2,I3).EQ.0) GO TO 8
         IF (SO .EQ. 0) THEN
             SO = 1!                 Store a face permutation for
             DO 1 K = 1,M!           each symmetry operation in FP,
                FP(K,SO) = K!         with the identity operation
1            CONTINUE!               (here) in column 1
             GO TO 8
         ENDIF
         P(3) = I3
         FLAG3 = FLAG2
         IF (FLAG3.EQ.0 .AND. S(P(3)).NE.S(3)) THEN
             IF (S(P(3)) .GT. S(3)) GO TO 8
             FLAG3 = 3
         ENDIF
         DO 2 J = 1,M
             R(J) = 0
2        CONTINUE
         R(P(1)) = 2
         R(P(2)) = 2
         R(P(3)) = 2
         I = 1
         DO 6 J = 4,M
3            IF (R(P(I)) .EQ. S(P(I))) THEN
                 I = I+1
                 IF (I .EQ. J-1) GO TO 8
                 GO TO 3
             ENDIF
             IF = P(I)!              These are the first (IF)
             IL = P(J-1)!            and last (IL) open faces
             DO 5 IJ = 1,M!          in the preceding spiral
                 IF (D(IJ,IF).EQ.0 .OR. D(IJ,IL).EQ.0) GO TO 5
                 IF (R(IJ) .GT. 0) GO TO 5
                 P(J) = IJ
                 IF (FLAG3.EQ.0 .AND. S(P(J)).NE.S(J)) THEN
                     IF (S(P(J)) .GT. S(J)) GO TO 8
                     FLAG3 = J!       This spiral has a smaller
                 ENDIF!              code than S, but it may not
                 DO 4 K = 1,J-1!     close properly. Flag it
                     IF (D(P(J),P(K)) .EQ. 1) THEN
                         R(P(J)) = R(P(J))+1
                         R(P(K)) = R(P(K))+1
                     ENDIF
4                CONTINUE
                 GO TO 6
5            CONTINUE
             GO TO 8
6        CONTINUE
         IF (FLAG3 .EQ. 0) THEN
             SO = SO+1!              Arrive here once for each
             DO 7 K = 1,M!          spiral with the same code as
                 FP(K,SO) = P(K)!    S, which is once for each
7            CONTINUE!              symmetry operation SO
         ELSE
             IER = 13!               The flagged spiral has closed,
             RETURN!                 so call it a day
         ENDIF
8        CONTINUE
9     CONTINUE
10 CONTINUE
   IER = 0!                          Spiral S is canonical, and
   ORDER = SO!                       SO is the point group order.
```

```
            N = 0!                              Now calculate GROUP and NMR:
            L = 0
            DO 13 K = 2,M
               DO 12 J = 1,K-1
                  IF (D(J,K) .EQ. 0) GO TO 12
                  DO 11 I = 1,J-1
                     IF (D(I,J).EQ.0 .OR. D(I,K).EQ.0) GO TO 11
                     N = N+1
                     V(1,N) = I!                 Associate the three mutually
                     V(2,N) = J!                 adjacent faces I,J,K
                     V(3,N) = K!                 with vertex N
    11            CONTINUE
                  L = L+1
                  E(1,L) = J!                    And the two mutually adjacent
                  E(2,L) = K!                    faces J,K with edge L
    12         CONTINUE
    13      CONTINUE
            DO 18 SO = 1,ORDER
               DO 15 J = 1,N
                  J1 = FP(V(1,J),SO)
                  J2 = FP(V(2,J),SO)
                  J3 = FP(V(3,J),SO)
                  I1 = MIN(J1,J2,J3)
                  I3 = MAX(J1,J2,J3)
                  I2 = J1+J2+J3-I1-I3
                  DO 14 I = 1,N
                     IF (V(1,I).EQ.I1.AND.V(2,I).EQ.I2.AND.V(3,I).EQ.I3) THEN
                        VP(J,SO) = I!            Store a vertex permutation for
                        GO TO 15!                each symmetry operation in VP
                     ENDIF
    14            CONTINUE
    15         CONTINUE
               DO 17 J = 1,L
                  J1 = FP(E(1,J),SO)
                  J2 = FP(E(2,J),SO)
                  I1 = MIN(J1,J2)
                  I2 = J1+J2-I1
                  DO 16 I = 1,L
                     IF (E(1,I).EQ.I1 .AND. E(2,I).EQ.I2) THEN
                        EP(J,SO) = I!            And similarly an edge permutation
                        GO TO 17!                in EP
                     ENDIF
    16            CONTINUE
    17         CONTINUE
    18      CONTINUE
            DO 19 K = 1,12
               MV(K) = 0
               ME(K) = 0
               MF(K) = 0
    19      CONTINUE
            DO 21 J = 1,N
               IF (VP(J,1) .EQ. 0) GO TO 21
               VP(J,1) = 0
               K = 1
               DO 20 SO = 2,ORDER
                  I = VP(J,SO)
                  IF (VP(I,1) .EQ. 0) GO TO 20
                  VP(I,1) = 0
                  K = K+1
    20         CONTINUE
               K = ORDER/K!                      Count vertex orbits with
```

```
            MV(K) = MV(K)+1!           site group order K in MV(K)
21    CONTINUE
      DO 22 J = 1,N
         VP(J,1) = J
22    CONTINUE
      DO 24 J = 1,L
         IF (EP(J,1) .EQ. 0) GO TO 24
         EP(J,1) = 0
         K = 1
         DO 23 SO = 2,ORDER
            I = EP(J,SO)
            IF (EP(I,1) .EQ. 0) GO TO 23
            EP(I,1) = 0
            K = K+1
23       CONTINUE
         K = ORDER/K!                  And edge orbits with
         ME(K) = ME(K)+1!              site group order K in ME(K)
24    CONTINUE
      DO 25 J = 1,L
         EP(J,1) = J
25    CONTINUE
      DO 27 J = 1,M
         IF (FP(J,1) .EQ. 0) GO TO 27
         FP(J,1) = 0
         K = 1
         DO 26 SO = 2,ORDER
            I = FP(J,SO)
            IF (FP(I,1) .EQ. 0) GO TO 26
            FP(I,1) = 0
            K = K+1
26       CONTINUE
         K = ORDER/K!                  And face orbits with
         MF(K) = MF(K)+1!              site group order K in MF(K)
27    CONTINUE
      DO 28 J = 1,M
         FP(1,J) = J
28    CONTINUE
      DO 29 K = 1,12!                  And ALL special point orbits
         MS(K) = MV(K)+ME(K)+MF(K)!    with site group order K in MS(K)
29    CONTINUE
      DO 30 J = 1,6
         NMR(J) = 0
30    CONTINUE
      J = 0
      DO 31 K = 6,1,-1!                Use the vertex orbit counts
         IF (MV(K) .EQ. 0) GO TO 31!   to calculate the NMR pattern
         J = J+1
         NMR(J) = MV(K)
         J = J+1
         NMR(J) = ORDER/K
31    CONTINUE
      GROUP = '???'!                   And, finally, the full
      IF (ORDER .EQ. 1) THEN!          special point orbit counts
         GROUP = ' C1'!                (in conjunction with the
      ELSE IF (ORDER .EQ. 2) THEN!     point group order) to assign
         IF (MS(2) .EQ. 0) THEN!       the point group
            GROUP = ' Ci'
         ELSE IF (MS(2) .EQ. 2) THEN
            GROUP = ' C2'
         ELSE IF (MS(2) .GT. 2) THEN
            GROUP = ' Cs'
```

```
      ENDIF
ELSE IF (ORDER .EQ. 3) THEN
   GROUP = ' C3'
ELSE IF (ORDER .EQ. 4) THEN
   IF (MS(4) .EQ. 0) THEN
      IF (MS(2) .EQ. 1) THEN
         GROUP = ' S4'
      ELSE IF (MS(2) .EQ. 3) THEN
         GROUP = ' D2'
      ELSE IF (MS(2) .GT. 3) THEN
         GROUP = 'C2h'
      ENDIF
   ELSE IF (MS(4) .EQ. 2) THEN
      GROUP = 'C2v'
   ENDIF
ELSE IF (ORDER .EQ. 6) THEN
   IF (MS(6) .EQ. 0) THEN
      IF (MS(2) .EQ. 0) THEN
         GROUP = ' S6'
      ELSE IF (MS(2) .EQ. 2) THEN
         GROUP = ' D3'
      ELSE IF (MS(2) .GT. 2) THEN
         GROUP = 'C3h'
      ENDIF
   ELSE IF (MS(6) .EQ. 2) THEN
      GROUP = 'C3v'
   ENDIF
ELSE IF (ORDER .EQ. 8) THEN
   IF (MS(4) .EQ. 1) THEN
      GROUP = 'D2d'
   ELSE IF (MS(4) .EQ. 3) THEN
      GROUP = 'D2h'
   ENDIF
ELSE IF (ORDER .EQ. 10) THEN
   GROUP = ' D5'
ELSE IF (ORDER .EQ. 12) THEN
   IF (MS(6) .EQ. 0) THEN
      GROUP = '  T'
   ELSE IF (MS(6) .EQ. 1) THEN
      IF (MS(4) .EQ. 0) THEN
         IF (MS(2) .EQ. 2) THEN
            GROUP = ' D6'
         ELSE IF (MS(2) .GT. 2) THEN
            GROUP = 'D3d'
         ENDIF
      ELSE IF (MS(4) .EQ. 2) THEN
         GROUP = 'D3h'
      ENDIF
   ENDIF
ELSE IF (ORDER .EQ. 20) THEN
   IF (MS(4) .EQ. 0) THEN
      GROUP = 'D5d'
   ELSE IF (MS(4) .EQ. 2) THEN
      GROUP = 'D5h'
   ENDIF
ELSE IF (ORDER .EQ. 24) THEN
   IF (MS(12) .EQ. 0) THEN
      IF (MS(6) .EQ. 0) THEN
         GROUP = ' Th'
      ELSE IF (MS(6) .EQ. 2) THEN
         GROUP = ' Td'
```

```
            ENDIF
        ELSE IF (MS(12) .EQ. 1) THEN
            IF (MS(4) .EQ. 0) THEN
                GROUP = 'D6d'
            ELSE IF (MS(4) .EQ. 2) THEN
                GROUP = 'D6h'
            ENDIF
        ENDIF
    ELSE IF (ORDER .EQ. 60) THEN
        GROUP = ' I'
    ELSE IF (ORDER .EQ. 120) THEN
        GROUP = ' Ih'
    ENDIF
    RETURN
    END
```

This subroutine completes the spiral program, which generates the following example output files for the cases considered in Tables A.2 and A.13:

```
GENERAL FULLERENE ISOMERS OF C36:
---------------------------------------------------------------------------
     1    C2    1  2  3  4  5  7 14 16 17 18 19 20     18 x   2
     2    D2    1  2  3  4  5  8 13 16 17 18 19 20      9 x   4
     3    C1    1  2  3  4  5 12 13 14 16 17 19 20     36 x   1
     4    Cs    1  2  3  4  5 12 13 15 16 18 19 20      6 x   1, 15 x   2
     5    D2    1  2  3  4  7 10 11 14 17 18 19 20      9 x   4
     6    D2d   1  2  3  4  7 10 12 15 17 18 19 20      3 x   4,  3 x   8
     7    C1    1  2  3  4  7 11 12 14 16 17 19 20     36 x   1
     8    Cs    1  2  3  4  7 11 12 15 16 18 19 20      4 x   1, 16 x   2
     9    C2v   1  2  3  4 11 12 13 14 15 16 17 18      4 x   2,  7 x   4
    10    C2    1  2  3  5  7 10 11 14 16 18 19 20     18 x   2
    11    C2    1  2  3  5  7 10 12 15 16 18 19 20     18 x   2
    12    C2    1  2  3  5 10 11 13 14 15 16 17 19     18 x   2
    13    D3h   1  2  3  6 10 11 12 14 15 16 18 20      2 x   6,  2 x  12
    14    D2d   1  2  4  7  9 10 12 13 14 16 18 20      1 x   4,  4 x   8
    15    D6h   1  2  4  8  9 10 12 13 14 15 18 20      3 x  12
---------------------------------------------------------------------------


ISOLATED-PENTAGON ISOMERS OF C86:
---------------------------------------------------------------------------
     1    C1    1  7  9 11 13 24 28 31 33 37 39 43     86 x   1
     2    C2    1  7  9 11 13 24 28 32 37 39 41 43     43 x   2
     3    C2    1  7  9 11 13 25 29 31 34 38 43 45     43 x   2
     4    C2    1  7  9 11 14 22 28 31 34 37 39 43     43 x   2
     5    C1    1  7  9 11 14 22 28 31 35 37 39 42     86 x   1
     6    C2    1  7  9 11 14 23 28 31 33 37 39 43     43 x   2
     7    C1    1  7  9 11 22 24 26 29 32 38 40 43     86 x   1
     8    Cs    1  7  9 11 22 24 27 29 32 37 40 43      8 x   1, 39 x   2
     9    C2v   1  7  9 11 23 25 27 29 33 41 43 45      7 x   2, 18 x   4
    10    C2v   1  7  9 12 14 20 27 33 35 38 40 45      5 x   2, 19 x   4
    11    C1    1  7  9 12 14 20 28 32 35 37 39 42     86 x   1
    12    C1    1  7  9 12 14 20 28 33 35 37 39 41     86 x   1
    13    C1    1  7  9 12 14 20 28 33 35 37 40 45     86 x   1
    14    C2    1  7  9 12 14 21 28 31 35 37 39 42     43 x   2
    15    Cs    1  7  9 12 14 28 30 32 34 36 38 45      6 x   1, 40 x   2
    16    Cs    1  7  9 12 20 24 26 28 34 39 41 44      6 x   1, 40 x   2
    17    C2    1  7  9 12 20 24 26 29 33 38 40 43     43 x   2
    18    C3    1  7  9 12 20 24 27 30 34 37 39 42      2 x   1, 28 x   3
    19    D3    1  7 10 12 14 19 28 33 35 37 40 45      1 x   2, 14 x   6
---------------------------------------------------------------------------
```

ATLAS TABLES

Tables A.1 to A.22 summarize mathematical and chemical information on general isomers of fullerenes in the range C_{20} to C_{50} (Tables A.1 to A.9 and A.21) and isolated-pentagon isomers of fullerenes in the range C_{60} to C_{100} (Tables A.10 to A.20 and A.22). Each isomer is represented by a line of numerical data and a figure, with 24 isomers shown on each double page. The generation and use of this data is explained in the main text, but some brief notes are given below, with references to the sections where further explanations are to be found.

General isomers: C_{20} to C_{50}

Every isomer generated by the spiral algorithm (§ 2.6) is listed.

- The first field of each data line identifies the isomer by a label $n{:}m$, where n is the number of carbon atoms in the fullerene and m is its sequence number in the lexicographically ordered list of spirals (§ 2.6).

- The second field, 'Ring spiral', lists the canonical spiral of the isomer as a set of 12 integers which give the positions of the pentagons in the spiral code. All canonical spirals in this range of n start with a pentagon in position 1, but for larger fullerenes a more general start may be needed. The spiral can be 'wound up' to give adjacency information and used to estimate atomic coordinates (§ 5.2), assign symmetries (§ 5.3) and predict spectral signatures (§§ 5.4 and 5.5).

- The third field is the point group symbol for the isomer, the maximal symmetry group of rotations and reflections in three-dimensional space that is compatible with the molecular graph. Jahn–Teller distortion in a particular electronic state of the neutral or charged fullerene may lower the symmetry to a subgroup of this maximal group (§ 5.3).

- The fourth field is a list of distinct peaks and their intensities in an idealized (natural abundance) ^{13}C NMR spectrum of the isomer in maximal point group symmetry (§ 5.4). The format $i \times j$ implies i separate peaks of intensity j. If distortion from maximal symmetry to a subgroup occurs, then further

177

splittings are to be expected. However, regardless of molecular symmetry, no single isomer can have more than three different peak intensities (§ 5.4).

- The fifth field, 'Vibrations', lists the numbers of IR-active, Raman-active, and IR-Raman concident fundamentals in maximal symmetry. The number of polarized Raman fundamentals can also be calculated from the NMR signature as described in the text (§ 5.5).

- The sixth field, 'Pentagon indices', is a list of integers $p_0, p_1, \ldots p_5$ counting the pentagons that have $0, 1, \ldots 5$ pentagonal neighbours, respectively. Their total $p_0 + p_1 + \ldots + p_5$ is 12 and the weighted sum $0p_0 + 1p_1 + 2p_2 + 3p_3 + 4p_4 + 5p_5$ is equal to twice the number of pentagon–pentagon edges N_p. As the total number of edges and the total number of pentagon edges are both fixed, the counts of pentagon–pentagon, pentagon–hexagon, and hexagon–hexagon edges can all be obtained from these indices. Isomers can be ranked qualitatively in order of increasing steric strain using the pentagon–pentagon edge count N_p (§ 4.3).

- The seventh field, 'Band gap', lists the frontier orbital energy level difference $|\epsilon_{\text{HOMO}} - \epsilon_{\text{LUMO}}|$ in units of $|\beta|$, where β is the Hückel resonance integral (§ 3.1). The orbital energies for HOMO and LUMO and the resonance energy per atom are tabulated separately in Table A.21, with ϵ_{HOMO} and ϵ_{LUMO} in units of β and the resonance energy in units of $|\beta|$. All the entries in this table are positive because all C_n fullerenes in the range $n = 20$ to 50 have more bonding than antibonding orbitals in the 'topological' approximation in which all atoms share a common α parameter and all bonds a common β (§ 3.2).

- The final field, 'Transformations', lists the products of Stone–Wales transformations of isomer $n{:}m$ (§ 6.1). Each entry $m'(j)$ implies that isomer $n{:}m'$ can reached by rotation of j equivalent SW bonds in isomer $n{:}m$. In cases where j would be 1 the number in brackets is omitted to save space. The list does not distinguish enantiomeric products and so, for example, if isomer m is achiral and isomer m' is chiral, $j/2$ of the patches in isomer m produce the left- and $j/2$ the right-handed enantiomer of isomer m'. Neutral transformations from m to itself or from m to its enantiomer are not distinguished. The total number of patches, and in most cases their site symmetries, can be recovered from the information in this list (§ 6.4).

- The figure for each isomer is drawn from the 'topological' coordinates extracted from the eigenvectors of the adjacency matrix (§ 5.2). A view down one of the principal inertial axes is used in order to display the molecular symmetry, choosing a high-order axis of rotation or the normal to a reflection plane where possible. No perspective is applied, but front and back edges are distinguished as thick and thin lines.

Isolated-pentagon isomers: C_{60} to C_{100}

Every isolated-pentagon isomer generated by the spiral algorithm is listed, and all data fields are as in the tabulation of general isomers except as noted below.

- The sequence number m now refers to the position of the isomer in a list of lexicographically ordered spirals that describe isolated-pentagon isomers. Thus C_{60}:1 is the only isolated-pentagon isomer with 60 atoms, but the same fullerene cage would be labelled C_{60}:1812 in a list of general isomers.

- The sixth field, 'Hexagon indices', is a list of integers h_3, h_4, h_5, h_6 counting the hexagons that have 3, 4, 5, and 6 hexagonal neighbours, respectively. An isolated-pentagon fullerene has 60 pentagon–hexagon edges and $3n/2 - 60$ hexagon–hexagon edges. A discussion of the connection between the tabulated indices and steric strain is given in § 4.4.

- The frontier orbital energy levels ϵ_{HOMO} and ϵ_{LUMO} and resonance energies per atom of these isolated-pentagon isomers are collected together in Table A.22. In contrast to Table A.21, this table contains an occasional negative number in the LUMO column (marked by an overbar) denoting a properly closed-shell electronic configuration with an antibonding LUMO (§ 3.2). In the range of the table, these properly closed shells occur only for leapfrog fullerenes (§ 3.4); the remaining properly closed shells with non-bonding (0.0000) LUMOs are the first three carbon cylinders (§ 3.5).

- The final field, 'Transformations', now refers only to the isolate-to-isolate Stone–Wales transformations that occur via rotation of SW bonds embedded in the larger (type **II**) patches on the fullerene surface (§ 6.1).

Table A.1. Fullerene isomers of C_{20}, C_{24}, C_{26}, C_{28}, C_{30}, C_{32}, and C_{34}

Isomer	Ring spiral	Point group	NMR pattern	Vibrations	Pentagon indices	Band gap	Transformations
20:1	1 2 3 4 5 6 7 8 9 10 11 12	I_h	1×20	1, 4, 0	0 0 0 0 0 12	0.0000	
24:1	1 2 3 4 5 7 8 10 11 12 13 14	D_{6d}	2×12	8, 15, 0	0 0 0 0 12 0	0.0000	
26:1	1 2 3 4 5 7 9 11 12 13 14 15	D_{3h}	1×2, 2×6, 1×12	19, 32, 13	0 0 0 6 6 0	0.0669	
28:1	1 2 3 4 5 7 10 12 13 14 15 16	D_2	7×4	57, 78, 57	0 0 0 8 4 0	0.0934	1(2)
28:2	1 2 3 5 7 9 10 11 12 13 14 15	T_d	1×4, 2×12	11, 23, 11	0 0 0 12 0 0	0.0000	
30:1	1 2 3 4 5 6 12 13 14 15 16 17	D_{5h}	3×10	13, 23, 0	0 0 0 10 0 2	0.0000	2(5)
30:2	1 2 3 4 5 7 11 13 14 15 16 17	C_{2v}	3×2, 6×4	64, 84, 64	0 0 2 8 2 0	0.4045	1, 3(2)
30:3	1 2 3 4 7 10 11 12 13 14 15 16	C_{2v}	5×2, 5×4	65, 84, 65	0 0 2 10 0 0	0.0584	2(2)
32:1	1 2 3 4 5 7 12 14 15 16 17 18	C_2	16×2	90, 90, 90	0 0 4 6 2 0	0.0117	2(2), 4(2), 6
32:2	1 2 3 4 5 8 12 13 15 16 17 18	D_2	8×4	66, 90, 66	0 0 4 4 4 0	0.0000	1(4), 3(2)
32:3	1 2 3 4 5 9 12 13 14 16 17 18	D_{3d}	1×2, 1×6, 2×12	23, 24, 0	0 0 6 0 6 0	0.0000	2(6)
32:4	1 2 3 4 7 10 11 12 14 15 17 18	C_2	16×2	90, 90, 90	0 0 4 8 0 0	0.2527	1(2), 4(2)
32:5	1 2 3 4 7 10 11 13 14 16 17 18	D_{3h}	1×2, 3×6, 1×12	24, 40, 16	0 0 0 12 0 0	0.3943	
32:6	1 2 3 5 7 9 10 12 14 16 17 18	D_3	1×2, 5×6	44, 46, 30	0 0 6 6 0 0	0.3639	1(3)
34:1	1 2 3 4 5 7 13 15 16 17 18 19	C_2	17×2	96, 96, 96	0 0 4 6 2 0	0.0027	3(2)
34:2	1 2 3 4 5 12 13 14 15 16 17 18	C_s	6×1, 14×2	96, 96, 96	0 0 7 4 1 0	0.0397	2(2), 4(2), 5(2)
34:3	1 2 3 4 7 10 11 13 15 17 18 19	C_s	6×1, 14×2	96, 96, 96	0 1 4 7 0 0	0.1217	1(2), 6
34:4	1 2 3 4 7 11 12 13 15 16 17 19	C_2	17×2	96, 96, 96	0 0 6 6 0 0	0.1137	2(2), 4(2), 5(2)
34:5	1 2 3 5 7 10 11 13 15 16 18 19	C_2	17×2	96, 96, 96	0 0 8 4 0 0	0.0019	2(2), 4(2), 5
34:6	1 2 3 5 7 10 11 14 15 17 18 19	C_{3v}	1×1, 3×3, 4×6	50, 50, 50	0 0 6 6 0 0	0.0000	3(3)

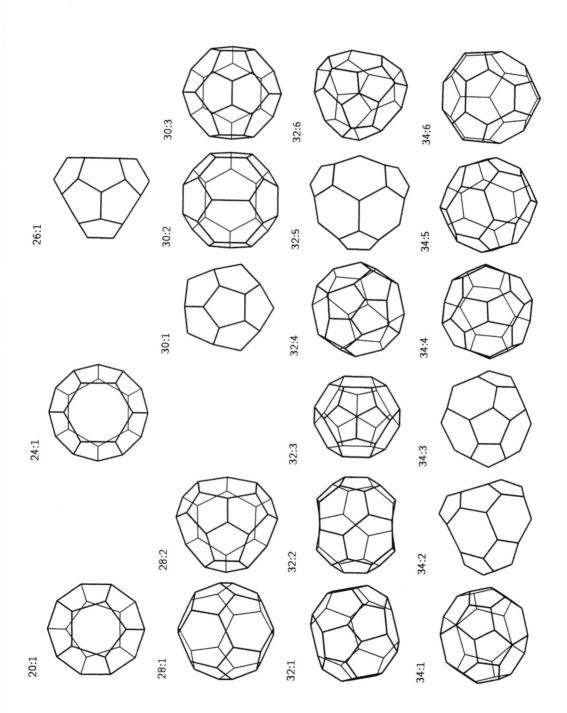

20:1

24:1

26:1

28:1

28:2

30:1

30:2

30:3

32:1

32:2

32:3

32:4

32:5

32:6

34:1

34:2

34:3

34:4

34:5

34:6

Table A.2. Fullerene isomers of C$_{36}$

Isomer	Ring spiral	Point group	NMR pattern	Vibrations	Pentagon indices	Band gap	Transformations
36:1	1 2 3 4 5 7 14 16 17 18 19 20	C_2	18×2	102, 102, 102	0 2 2 6 2 0	0.0051	2, 3(2), 6
36:2	1 2 3 4 5 8 13 16 17 18 19 20	D_2	9×4	75, 102, 75	0 0 4 4 4 0	0.1370	1(2)
36:3	1 2 3 4 5 12 13 14 16 17 19 20	C_1	36×1	102, 102, 102	0 1 5 5 1 0	0.0384	1, 3, 4, 7, 8, 11
36:4	1 2 3 4 5 12 13 15 16 18 19 20	C_s	6×1, 15×2	102, 102, 102	0 0 5 6 1 0	0.0157	3(2), 7(2)
36:5	1 2 3 4 7 10 11 14 17 18 19 20	D_2	9×4	75, 102, 75	0 0 4 8 0 0	0.1478	
36:6	1 2 3 4 7 10 12 15 17 18 19 20	D_{2d}	3×4, 3×8	39, 66, 39	0 4 0 8 0 0	0.1370	1(4), 8(4)
36:7	1 2 3 4 7 11 12 14 16 17 19 20	C_1	36×1	102, 102, 102	0 1 6 5 0 0	0.0579	3, 4, 7, 10, 11
36:8	1 2 3 4 7 11 12 15 16 18 19 20	C_s	4×1, 16×2	102, 102, 102	0 1 6 5 0 0	0.0960	3(2), 6, 11(2), 12(2), 13
36:9	1 2 3 4 11 12 13 14 15 16 17 18	C_{2v}	4×2, 7×4	78, 102, 78	0 0 10 2 0 0	0.1199	12(4), 13, 14(2)
36:10	1 2 3 5 7 10 11 14 16 18 19 20	C_2	18×2	102, 102, 102	0 0 8 4 0 0	0.0414	7(2)
36:11	1 2 3 5 7 10 12 15 16 18 19 20	C_2	18×2	102, 102, 102	0 2 6 4 0 0	0.0613	3(2), 7(2), 8(2), 12
36:12	1 2 3 5 10 11 13 14 15 16 17 19	C_2	18×2	102, 102, 102	0 0 10 2 0 0	0.0032	8(2), 9(2), 11, 12(2)
36:13	1 2 3 6 10 11 12 14 15 16 18 20	D_{3h}	2×6, 2×12	26, 44, 18	0 0 6 6 0 0	0.0000	8(6), 9(3)
36:14	1 2 4 7 9 10 12 13 14 16 18 20	D_{2d}	1×4, 4×8	38, 65, 38	0 0 12 0 0 0	0.3002	9(4), 15(2)
36:15	1 2 4 8 9 10 12 13 14 15 18 20	D_{6h}	3×12	13, 23, 0	0 0 12 0 0 0	0.0000	14(6)

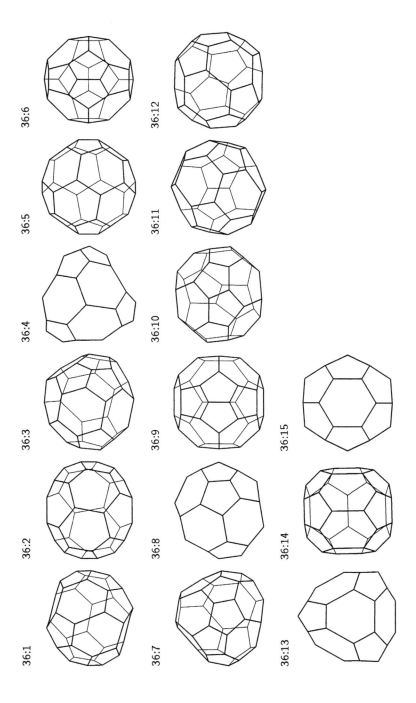

36:6

36:12

36:5

36:11

36:4

36:10

36:3

36:9

36:15

36:2

36:8

36:14

36:1

36:7

36:13

Table A.3. Fullerene isomers of C_{38}

Isomer	Ring spiral	Point group	NMR pattern	Vibrations	Pentagon indices	Band gap	Transformations
38:1	1 2 3 4 5 7 15 17 18 19 20 21	C_2	19×2	108, 108, 108	0 2 2 6 2 0	0.0312	3(2)
38:2	1 2 3 4 5 9 13 17 18 19 20 21	D_{3h}	1×2, 2×6, 2×12	28, 47, 19	0 0 6 0 6 0	0.0000	
38:3	1 2 3 4 5 12 13 15 17 19 20 21	C_1	38×1	108, 108, 108	0 3 3 5 1 0	0.0691	1, 3(2), 5, 10
38:4	1 2 3 4 5 13 14 15 17 18 19 21	C_1	38×1	108, 108, 108	0 1 5 5 1 0	0.1427	4, 5, 6, 7
38:5	1 2 3 4 7 11 12 15 17 19 20 21	C_1	38×1	108, 108, 108	0 3 4 5 0 0	0.1402	3, 4, 6, 8, 10, 11, 13
38:6	1 2 3 4 7 11 13 15 17 18 20 21	C_2	19×2	108, 108, 108	0 2 4 6 0 0	0.2069	4(2), 5(2), 10
38:7	1 2 3 4 7 11 13 16 17 19 20 21	C_1	38×1	108, 108, 108	0 1 6 5 0 0	0.0652	4, 7, 11
38:8	1 2 3 4 11 12 13 14 16 17 20 21	C_1	38×1	108, 108, 108	0 1 8 3 0 0	0.0982	5, 9, 11, 13, 14(2), 17
38:9	1 2 3 4 11 12 14 15 16 18 19 21	D_3	1×2, 6×6	53, 55, 36	0 0 6 6 0 0	0.1973	8(6)
38:10	1 2 3 5 7 10 12 15 17 19 20 21	C_2	19×2	108, 108, 108	0 4 4 4 0 0	0.0177	3(2), 5(2), 6, 10(2), 13(2)
38:11	1 2 3 5 7 10 14 16 17 18 19 20	C_1	38×1	108, 108, 108	0 1 8 3 0 0	0.1099	5, 7, 8, 14
38:12	1 2 3 5 7 11 14 16 17 18 20 21	C_{2v}	5×2, 7×4	83, 108, 83	0 0 8 4 0 0	0.0045	
38:13	1 2 3 5 10 11 13 15 16 19 20 21	C_2	19×2	108, 108, 108	0 2 8 2 0 0	0.1505	5(2), 8(2), 10(2), 13, 17(2)
38:14	1 2 3 5 10 12 14 15 16 17 19 20	C_1	38×1	108, 108, 108	0 1 10 1 0 0	0.1036	8(2), 11, 15, 16, 17
38:15	1 2 3 5 11 12 14 15 16 17 18 20	C_{2v}	7×2, 6×4	84, 108, 84	0 0 10 2 0 0	0.0311	14(4)
38:16	1 2 3 10 11 12 13 14 15 16 17 18	C_{3v}	2×1, 4×3, 4×6	57, 57, 57	0 0 12 0 0 0	0.0454	14(6)
38:17	1 2 4 7 9 10 12 13 15 18 20 21	C_2	19×2	108, 108, 108	0 2 10 0 0 0	0.3004	8(2), 13(2), 14(2), 17(2)

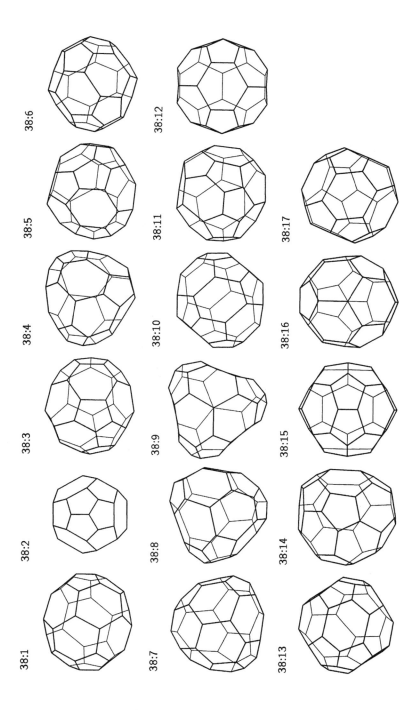

38:6

38:12

38:5

38:11

38:17

38:4

38:10

38:16

38:3

38:9

38:15

38:2

38:8

38:14

38:1

38:7

38:13

Table A.4. Fullerene isomers of C_{40}

Isomer	Ring spiral	Point group	NMR pattern	Vibrations	Pentagon indices	Band gap	Transformations
40:1	1 2 3 4 5 6 17 18 19 20 21 22	D_{5d}	4×10	18, 31, 0	0 0 0 10 0 2	0.0000	
40:2	1 2 3 4 5 7 16 18 19 20 21 22	C_2	20×2	114, 114, 114	0 2 2 6 2 0	0.1241	5(2)
40:3	1 2 3 4 5 8 17 18 19 20 21 22	D_2	10×4	84, 114, 84	0 0 4 4 4 0	0.0623	
40:4	1 2 3 4 5 12 13 16 19 20 21 22	C_1	40×1	114, 114, 114	0 2 3 6 1 0	0.1614	4, 5, 10
40:5	1 2 3 4 5 12 14 17 19 20 21 22	C_s	6×1, 17×2	114, 114, 114	1 2 2 6 1 0	0.0434	2(2), 4(2), 7, 9(2)
40:6	1 2 3 4 5 13 14 16 18 19 21 22	C_1	40×1	114, 114, 114	0 2 5 4 1 0	0.1469	8, 13, 21
40:7	1 2 3 4 5 13 14 17 18 20 21 22	C_s	4×1, 18×2	114, 114, 114	0 0 7 4 1 0	0.1886	5, 12(2)
40:8	1 2 3 4 5 13 15 16 18 19 20 22	C_{2v}	6×2, 7×4	88, 114, 88	0 2 4 4 2 0	0.1441	6(4)
40:9	1 2 3 4 7 11 12 16 19 20 21 22	C_2	20×2	114, 114, 114	0 4 2 6 0 0	0.0117	5(2), 10(2), 12(2), 14(2)
40:10	1 2 3 4 7 11 13 16 18 20 21 22	C_1	40×1	114, 114, 114	0 3 4 5 0 0	0.1790	4, 9, 10, 21, 22
40:11	1 2 3 4 7 11 14 17 18 20 21 22	C_2	20×2	114, 114, 114	0 0 6 6 0 0	0.1148	23
40:12	1 2 3 4 7 11 15 17 18 19 20 21	C_1	40×1	114, 114, 114	0 2 6 4 0 0	0.1861	7, 9, 17, 24, 25
40:13	1 2 3 4 7 12 15 17 18 19 21 22	C_s	6×1, 17×2	114, 114, 114	0 3 4 5 0 0	0.2432	6(2), 22(2)
40:14	1 2 3 4 11 12 13 15 17 20 21 22	C_s	6×1, 17×2	114, 114, 114	0 4 4 4 0 0	0.0099	9(2), 17(2), 22(2), 24, 29(2)
40:15	1 2 3 4 11 12 14 15 16 18 20 22	C_2	20×2	114, 114, 114	0 2 8 2 0 0	0.0528	16(2), 26(2), 35(2)
40:16	1 2 3 4 11 12 14 15 17 18 20 21	C_2	20×2	114, 114, 114	0 2 6 4 0 0	0.0964	15(2), 17(2), 19(2), 29
40:17	1 2 3 4 11 12 14 15 17 19 21 22	C_1	40×1	114, 114, 114	0 1 8 3 0 0	0.1804	12, 14, 16, 26, 27, 30, 31
40:18	1 2 3 4 11 12 15 16 19 20 21 22	C_2	20×2	114, 114, 114	0 0 8 4 0 0	0.2056	28(2), 34(2)
40:19	1 2 3 4 11 13 14 15 17 18 19 21	C_2	20×2	114, 114, 114	0 2 6 4 0 0	0.0456	16(2), 19(3), 30(2)
40:20	1 2 3 5 7 9 16 17 18 19 20 21	C_{3v}	1×1, 5×3, 4×6	60, 60, 60	0 3 6 3 0 0	0.0827	24(3)
40:21	1 2 3 5 7 10 13 16 18 20 21 22	C_2	20×2	114, 114, 114	0 4 4 4 0 0	0.1407	6(2), 10(2), 22(2), 25
40:22	1 2 3 5 7 10 14 16 18 19 21 22	C_1	40×1	114, 114, 114	0 3 6 3 0 0	0.1206	10, 13, 14, 21, 26, 27
40:23	1 2 3 5 7 10 14 17 18 20 21 22	C_2	20×2	114, 114, 114	0 2 6 4 0 0	0.1644	11, 28(2)
40:24	1 2 3 5 7 10 15 17 18 19 20 21	C_s	4×1, 18×2	114, 114, 114	0 4 6 2 0 0	0.0748	12(2), 14, 20, 26(2), 31(2)

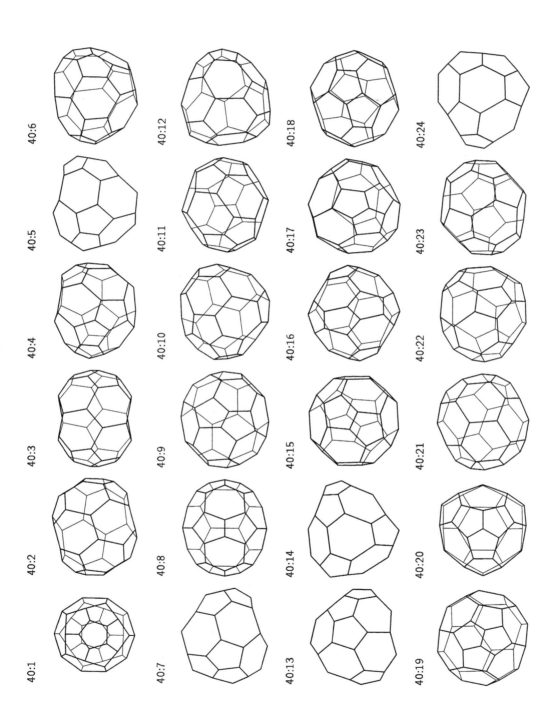

40:6
40:5
40:4
40:3
40:2
40:1

40:12
40:11
40:10
40:9
40:8
40:7

40:18
40:17
40:16
40:15
40:14
40:13

40:24
40:23
40:22
40:21
40:20
40:19

Table A.4. (Continued)

Isomer	Ring spiral	Point group	NMR pattern	Vibrations	Pentagon indices	Band gap	Transformations
40:25	1 2 3 5 10 11 14 15 16 19 20 22	C_2	20×2	114, 114, 114	0 2 8 2 0 0	0.1612	12(2), 21, 26(2), 34(2)
40:26	1 2 3 5 10 12 14 15 16 18 20 22	C_1	40×1	114, 114, 114	0 3 8 1 0 0	0.1026	15, 17, 22, 24, 25, 29, 36, 38
40:27	1 2 3 5 10 12 14 15 17 19 21 22	C_2	20×2	114, 114, 114	0 2 8 2 0 0	0.0814	17(2), 22(2), 32, 38
40:28	1 2 3 5 10 12 15 16 19 20 21 22	C_s	2×1, 19×2	114, 114, 114	1 0 9 2 0 0	0.1055	18(2), 23(2), 33, 37, 39
40:29	1 2 3 5 10 13 14 15 16 18 19 22	C_2	20×2	114, 114, 114	0 4 6 2 0 0	0.0864	14(2), 16, 26(2), 30(2), 31(4)
40:30	1 2 3 5 10 13 14 15 17 18 19 21	C_3	1×1, 13×3	74, 74, 74	0 3 6 3 0 0	0.0000	17(3), 19(3), 29(3)
40:31	1 2 3 5 10 13 15 16 18 19 20 22	C_s	6×1, 17×2	114, 114, 114	0 3 8 1 0 0	0.1613	17(2), 24(2), 29(4), 38(2), 40
40:32	1 2 3 5 11 12 14 15 17 18 21 22	D_2	10×4	84, 114, 84	0 0 8 4 0 0	0.0041	27(2)
40:33	1 2 3 5 11 12 15 16 18 20 21 22	D_{2h}	4×4, 3×8	44, 57, 0	0 0 8 4 0 0	0.0308	28(4)
40:34	1 2 3 5 11 14 15 16 17 19 20 21	C_1	40×1	114, 114, 114	0 1 10 1 0 0	0.1316	18, 25, 36, 37
40:35	1 2 4 7 9 10 12 14 18 20 21 22	C_2	20×2	114, 114, 114	0 2 10 0 0 0	0.0183	15(2), 35(2), 36
40:36	1 2 4 7 9 11 12 14 16 19 21 22	C_2	20×2	114, 114, 114	0 2 10 0 0 0	0.0066	26(2), 34(2), 35
40:37	1 2 4 7 9 11 12 15 17 19 20 21	C_{2v}	6×2, 7×4	88, 114, 88	0 2 10 0 0 0	0.2430	28(2), 34(4)
40:38	1 2 4 7 9 11 13 15 17 19 20 22	D_2	10×4	84, 114, 84	0 4 8 0 0 0	0.3135	26(4), 27(2), 31(4)
40:39	1 2 4 7 9 11 13 15 18 19 21 22	D_{5d}	2×10, 1×20	17, 30, 0	2 0 10 0 0 0	0.3731	28(10)
40:40	1 2 6 9 10 12 13 15 16 18 20 22	T_d	1×4, 1×12, 1×24	15, 31, 15	0 0 12 0 0 0	0.0000	31(12)

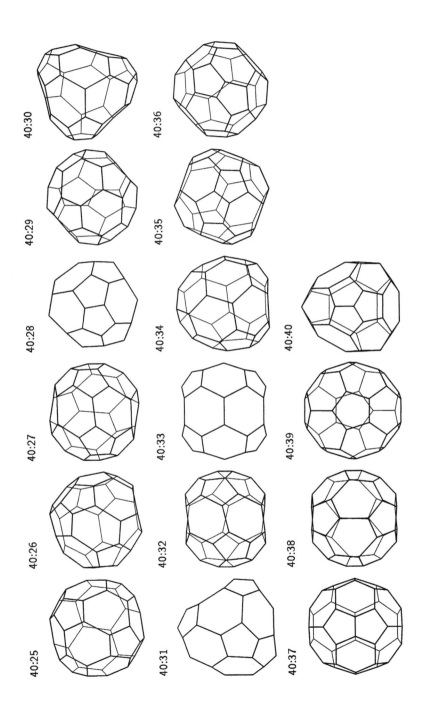

40:30

40:36

40:29

40:35

40:28

40:34

40:40

40:27

40:33

40:39

40:26

40:32

40:38

40:25

40:31

40:37

Table A.5. Fullerene isomers of C_{42}

Isomer	Ring spiral	Point group	NMR pattern	Vibrations	Pentagon indices	Band gap	Transformations
42:1	1 2 3 4 5 7 17 19 20 21 22 23	C_2	21×2	120, 120, 120	0 2 2 6 2 0	0.1083	
42:2	1 2 3 4 5 13 14 17 19 21 22 23	C_1	42×1	120, 120, 120	0 2 5 4 1 0	0.1481	3, 9, 10
42:3	1 2 3 4 5 13 15 17 19 20 22 23	C_1	42×1	120, 120, 120	1 1 4 5 1 0	0.0966	2, 5, 8, 13
42:4	1 2 3 4 5 13 15 18 19 21 22 23	C_1	42×1	120, 120, 120	0 2 5 4 1 0	0.1323	5, 27
42:5	1 2 3 4 5 13 16 17 19 20 21 23	C_2	21×2	120, 120, 120	0 2 4 4 2 0	0.1290	3(2), 4(2)
42:6	1 2 3 4 7 10 17 18 19 20 21 22	C_{2v}	7×2, 7×4	93, 120, 93	0 2 6 4 0 0	0.0556	12(2)
42:7	1 2 3 4 7 11 13 17 20 21 22 23	C_2	21×2	120, 120, 120	0 2 4 6 0 0	0.0908	8(2)
42:8	1 2 3 4 7 11 14 17 19 21 22 23	C_1	42×1	120, 120, 120	0 3 4 5 0 0	0.1286	3, 7, 9, 23
42:9	1 2 3 4 7 11 15 17 19 20 22 23	C_1	42×1	120, 120, 120	1 2 5 4 0 0	0.1488	2, 8, 11, 13, 14, 24, 28
42:10	1 2 3 4 7 11 15 18 19 21 22 23	C_1	42×1	120, 120, 120	0 3 4 5 0 0	0.2233	2, 24, 26
42:11	1 2 3 4 7 11 16 17 19 20 21 23	C_s	6×1, 18×2	120, 120, 120	0 3 4 5 0 0	0.2208	9(2), 12, 18(2), 19
42:12	1 2 3 4 7 11 16 18 19 20 21 22	C_s	4×1, 19×2	120, 120, 120	0 3 6 3 0 0	0.1725	6, 11, 21, 25(2)
42:13	1 2 3 4 7 12 15 17 18 20 22 23	C_{2v}	5×2, 8×4	92, 120, 92	2 2 2 6 0 0	0.0841	3(4), 9(4), 16
42:14	1 2 3 4 11 12 14 15 17 20 22 23	C_1	42×1	120, 120, 120	0 4 6 2 0 0	0.1383	9, 16, 19, 20, 29, 32, 35, 41
42:15	1 2 3 4 11 12 14 15 18 21 22 23	C_1	42×1	120, 120, 120	0 1 8 3 0 0	0.1077	30, 36, 37
42:16	1 2 3 4 11 12 14 16 17 20 21 23	C_{2v}	5×2, 8×4	92, 120, 92	0 4 4 4 0 0	0.0841	13, 14(4), 17(4)
42:17	1 2 3 4 11 12 14 16 18 20 21 22	C_1	42×1	120, 120, 120	0 3 6 3 0 0	0.1044	16, 18, 20, 29, 31, 32
42:18	1 2 3 4 11 12 14 16 19 21 22 23	C_1	42×1	120, 120, 120	0 2 6 4 0 0	0.1749	11, 17, 24, 25, 34
42:19	1 2 3 4 11 13 14 15 17 19 22 23	C_s	4×1, 19×2	120, 120, 120	0 4 6 2 0 0	0.1823	11, 14(2), 21, 32(2), 34(2)
42:20	1 2 3 4 11 13 14 16 18 19 21 22	C_1	42×1	120, 120, 120	0 2 8 2 0 0	0.0736	14, 17, 20, 33, 34, 39
42:21	1 2 3 4 11 13 15 17 19 20 22 23	C_{2v}	5×2, 8×4	92, 120, 92	0 4 6 2 0 0	0.2599	12(2), 19(2), 33(4)
42:22	1 2 3 5 7 9 16 17 19 20 22 23	C_s	8×1, 17×2	120, 120, 120	0 3 4 5 0 0	0.0304	24(2)
42:23	1 2 3 5 7 10 14 17 19 21 22 23	C_2	21×2	120, 120, 120	0 4 4 4 0 0	0.1290	8(2), 24(2)
42:24	1 2 3 5 7 10 15 17 19 20 22 23	C_1	42×1	120, 120, 120	1 3 5 3 0 0	0.0553	9, 10, 18, 22, 23, 30, 31, 35

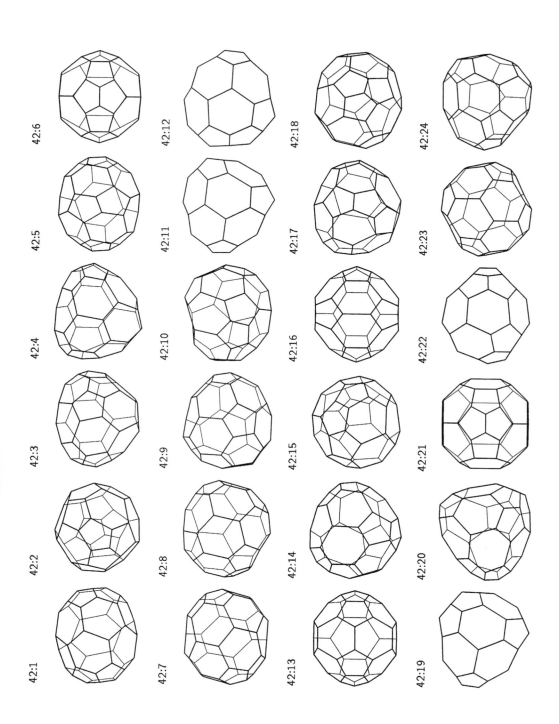

42:6
42:5
42:4
42:3
42:2
42:1

42:12
42:11
42:10
42:9
42:8
42:7

42:18
42:17
42:16
42:15
42:14
42:13

42:24
42:23
42:22
42:21
42:20
42:19

Table A.5. *(Continued)*

Isomer	Ring spiral	Point group	NMR pattern	Vibrations	Pentagon indices	Band gap	Transformations
42:25	1 2 3 5 7 10 16 18 19 20 21 22	C_1	42×1	120, 120, 120	0 4 6 2 0 0	0.0651	12, 18, 26, 29, 33, 36
42:26	1 2 3 5 7 11 16 17 19 20 21 22	C_1	42×1	120, 120, 120	0 3 6 3 0 0	0.1351	10, 25, 27, 31
42:27	1 2 3 5 7 12 16 17 18 20 21 22	C_2	21×2	120, 120, 120	0 4 4 4 0 0	0.1210	4(2), 26(2), 27
42:28	1 2 3 5 10 11 14 15 19 21 22 23	C_2	21×2	120, 120, 120	0 2 8 2 0 0	0.1035	9(2), 30(2), 41
42:29	1 2 3 5 10 12 14 15 17 20 22 23	C_1	42×1	120, 120, 120	0 3 8 1 0 0	0.0982	14, 17, 25, 32, 44
42:30	1 2 3 5 10 12 14 15 18 21 22 23	C_1	42×1	120, 120, 120	0 2 8 2 0 0	0.1876	15, 24, 28, 44
42:31	1 2 3 5 10 12 14 16 18 20 21 22	C_2	21×2	120, 120, 120	0 4 6 2 0 0	0.0954	17(2), 24(2), 26(2), 33(2)
42:32	1 2 3 5 10 13 14 15 17 19 22 23	C_1	42×1	120, 120, 120	0 5 6 1 0 0	0.1809	14, 17, 19, 29, 32, 33(2), 34, 39(2)
42:33	1 2 3 5 10 13 14 16 18 19 21 22	C_1	42×1	120, 120, 120	0 5 6 1 0 0	0.1745	20, 21, 25, 31, 32(2), 34, 35, 40, 45
42:34	1 2 3 5 10 13 14 16 18 20 22 23	C_1	42×1	120, 120, 120	0 3 8 1 0 0	0.1223	18, 19, 20, 32, 33, 34, 35, 39
42:35	1 2 3 5 10 13 15 16 18 20 21 23	C_s	8×1, 17×2	120, 120, 120	0 5 6 1 0 0	0.0496	14(2), 24(2), 33(2), 34(2), 44(2)
42:36	1 2 3 5 10 14 15 16 18 19 20 21	C_1	42×1	120, 120, 120	0 3 8 1 0 0	0.1608	15, 25, 36, 40, 43, 44
42:37	1 2 3 5 11 14 15 16 18 19 21 23	C_1	42×1	120, 120, 120	0 1 10 1 0 0	0.1549	15, 42, 43
42:38	1 2 3 5 11 15 16 19 20 21 22 23	C_2	21×2	120, 120, 120	0 0 10 2 0 0	0.2040	42(2)
42:39	1 2 3 10 11 12 13 15 16 17 21 23	C_1	42×1	120, 120, 120	0 4 8 0 0 0	0.1465	20, 32(2), 34, 39(2), 40, 41, 44, 45
42:40	1 2 3 10 11 12 14 16 17 20 21 23	C_2	21×2	120, 120, 120	0 2 10 0 0 0	0.0358	33(2), 36(2), 39(2), 44(2)
42:41	1 2 4 7 9 10 13 18 20 21 22 23	C_2	21×2	120, 120, 120	0 4 8 0 0 0	0.0446	14(2), 28, 39(2), 44(2)
42:42	1 2 4 7 9 11 12 15 19 21 22 23	C_s	6×1, 18×2	120, 120, 120	1 0 11 0 0 0	0.0148	37(2), 38(2)
42:43	1 2 4 7 9 11 12 16 19 20 21 22	C_2	21×2	120, 120, 120	0 2 10 0 0 0	0.1713	36(2), 37(2)
42:44	1 2 4 7 9 11 13 16 19 20 22 23	C_1	42×1	120, 120, 120	0 4 8 0 0 0	0.0646	29, 30, 35, 36, 39, 40, 41
42:45	1 2 4 7 9 12 13 16 18 19 21 22	D_3	7×6	59, 61, 40	0 6 6 0 0 0	0.2801	33(6), 39(6)

193

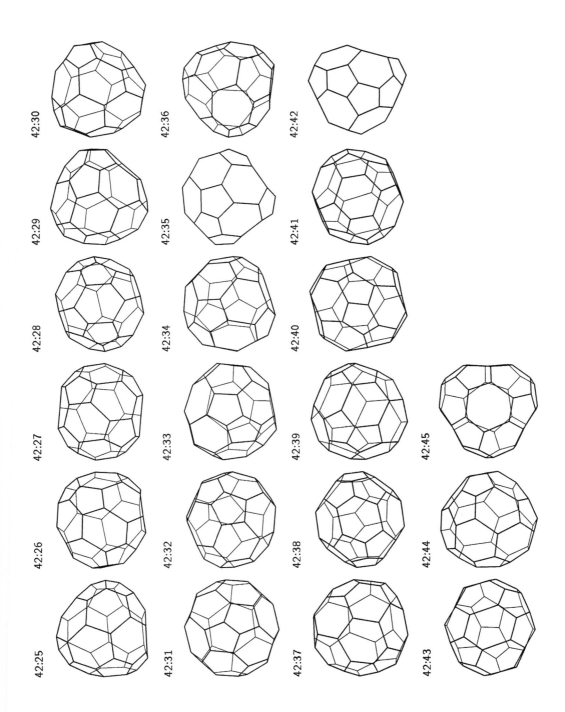

42:30

42:36

42:42

42:29

42:35

42:41

42:28

42:34

42:40

42:27

42:33

42:39

42:45

42:26

42:32

42:38

42:44

42:25

42:31

42:37

42:43

Table A.6. Fullerene isomers of C_{44}

Isomer	Ring spiral	Point group	NMR pattern	Vibrations	Pentagon indices	Band gap	Transformations
44:1	1 2 3 4 5 7 18 20 21 22 23 24	C_2	22×2	126, 126, 126	0 2 2 6 2 0	0.0979	5
44:2	1 2 3 4 5 8 18 19 21 22 23 24	D_2	11×4	93, 126, 93	0 0 4 4 4 0	0.0790	
44:3	1 2 3 4 5 9 18 19 20 22 23 24	D_{3d}	1×2, 1×6, 3×12	32, 33, 0	0 0 6 0 6 0	0.0000	
44:4	1 2 3 4 5 12 13 20 21 22 23 24	C_2	22×2	126, 126, 126	0 2 4 4 2 0	0.0584	4(2), 5(2)
44:5	1 2 3 4 5 12 14 19 21 22 23 24	C_2	22×2	126, 126, 126	2 0 4 4 2 0	0.0548	1, 4(2), 6(2), 7(2)
44:6	1 2 3 4 5 12 15 19 20 22 23 24	C_2	22×2	126, 126, 126	2 0 4 4 2 0	0.0384	5(2), 6(2), 8(2)
44:7	1 2 3 4 5 13 14 18 21 22 23 24	C_1	44×1	126, 126, 126	0 3 3 5 1 0	0.0531	5, 8, 10, 20
44:8	1 2 3 4 5 13 15 18 20 22 23 24	C_1	44×1	126, 126, 126	1 2 4 4 1 0	0.0979	6, 7, 8, 45
44:9	1 2 3 4 5 13 16 19 20 22 23 24	C_1	44×1	126, 126, 126	0 1 5 5 1 0	0.0586	21
44:10	1 2 3 4 5 13 17 19 20 21 22 23	C_1	44×1	126, 126, 126	0 1 7 3 1 0	0.1064	7, 19
44:11	1 2 3 4 5 14 17 19 20 21 23 24	C_s	6×1, 19×2	126, 126, 126	0 2 5 4 1 0	0.1061	15(2)
44:12	1 2 3 4 7 10 17 18 20 21 23 24	C_2	22×2	126, 126, 126	0 2 4 6 0 0	0.2517	16(2)
44:13	1 2 3 4 7 10 18 19 21 22 23 24	C_{2v}	8×2, 7×4	98, 126, 98	0 2 4 6 0 0	0.2838	
44:14	1 2 3 4 7 11 14 18 21 22 23 24	C_2	22×2	126, 126, 126	0 2 4 6 0 0	0.0371	15(2)
44:15	1 2 3 4 7 11 15 18 20 22 23 24	C_1	44×1	126, 126, 126	1 3 3 5 0 0	0.0855	11, 14, 16, 25, 39, 43
44:16	1 2 3 4 7 11 16 18 20 21 23 24	C_1	44×1	126, 126, 126	1 2 5 4 0 0	0.2708	12, 15, 17, 32, 40, 41
44:17	1 2 3 4 7 11 17 18 20 21 22 24	C_1	44×1	126, 126, 126	0 1 6 5 0 0	0.0275	16, 18, 26, 31
44:18	1 2 3 4 7 11 17 19 20 21 22 23	C_1	44×1	126, 126, 126	0 3 6 3 0 0	0.2174	17, 19, 30, 42
44:19	1 2 3 4 7 12 17 18 20 21 22 23	C_1	44×1	126, 126, 126	1 2 5 4 0 0	0.1519	10, 18, 20, 21, 23, 46
44:20	1 2 3 4 7 12 17 19 21 22 23 24	C_2	22×2	126, 126, 126	0 4 2 6 0 0	0.0500	7(2), 19(2)
44:21	1 2 3 4 7 13 17 18 19 21 22 23	C_1	44×1	126, 126, 126	0 3 4 5 0 0	0.0690	9, 19, 21
44:22	1 2 3 4 11 12 14 16 20 22 23 24	C_1	44×1	126, 126, 126	0 4 6 2 0 0	0.1932	24, 29, 36, 47, 74, 87
44:23	1 2 3 4 11 12 14 17 20 21 22 23	C_1	44×1	126, 126, 126	0 3 6 3 0 0	0.0750	19, 30, 48, 62, 65
44:24	1 2 3 4 11 12 15 16 20 21 23 24	D_2	11×4	93, 126, 93	0 4 4 4 0 0	0.0056	22(4), 25(4)

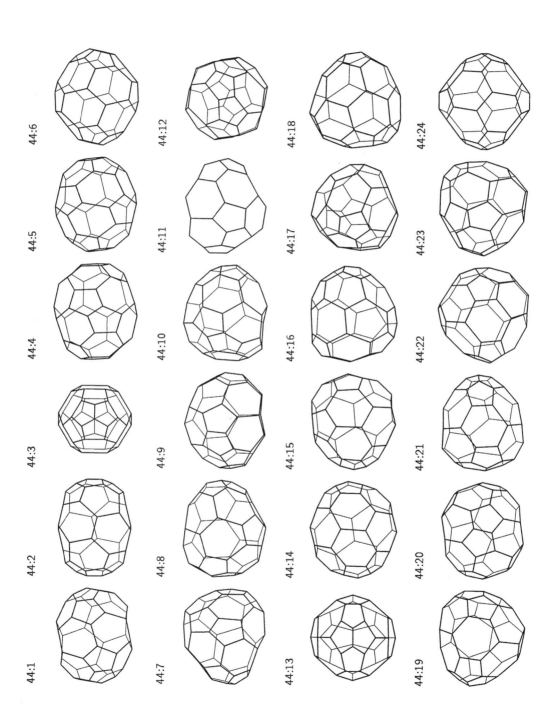

44:6

44:12

44:18

44:24

44:5

44:11

44:17

44:23

44:4

44:10

44:16

44:22

44:3

44:9

44:15

44:21

44:2

44:8

44:14

44:20

44:1

44:7

44:13

44:19

Table A.6. (*Continued*)

Isomer	Ring spiral	Point group	NMR pattern	Vibrations	Pentagon indices	Band gap	Transformations
44:25	1 2 3 4 11 12 15 17 20 21 22 24	C_1	44×1	126, 126, 126	1 3 5 3 0 0	0.1115	15, 24, 26, 29, 32, 47, 49, 50, 55
44:26	1 2 3 4 11 12 15 18 20 21 22 23	C_1	44×1	126, 126, 126	0 4 4 4 0 0	0.0408	17, 25, 31, 40, 42, 56
44:27	1 2 3 4 11 13 14 15 18 22 23 24	C_1	44×1	126, 126, 126	0 3 6 3 0 0	0.0954	28, 30, 50, 53, 58
44:28	1 2 3 4 11 13 14 16 18 21 23 24	C_s	6×1, 19×2	126, 126, 126	1 2 7 2 0 0	0.0210	27(2), 51(2), 84
44:29	1 2 3 4 11 13 14 16 19 22 23 24	C_1	44×1	126, 126, 126	0 4 6 2 0 0	0.1136	22, 25, 33, 52, 57, 78
44:30	1 2 3 4 11 13 14 17 19 21 22 23	C_1	44×1	126, 126, 126	0 4 6 2 0 0	0.1292	18, 23, 27, 31, 54, 57, 59
44:31	1 2 3 4 11 13 14 17 20 22 23 24	C_1	44×1	126, 126, 126	1 1 7 3 0 0	0.0312	17, 26, 30, 32, 34, 53, 56, 60
44:32	1 2 3 4 11 13 15 17 19 21 22 24	C_2	22×2	126, 126, 126	2 2 6 2 0 0	0.2415	16(2), 25(2), 31(2), 51(2), 52(2)
44:33	1 2 3 4 11 13 16 19 20 22 23 24	C_s	8×1, 18×2	126, 126, 126	0 3 6 3 0 0	0.0620	29(2), 56(2), 70
44:34	1 2 3 4 11 14 15 18 19 20 22 23	C_2	22×2	126, 126, 126	0 2 6 4 0 0	0.0917	31(2), 35, 61(2)
44:35	1 2 3 4 11 14 15 18 19 21 22 24	D_3	1×2, 7×6	62, 64, 42	0 0 6 6 0 0	0.1471	34(3)
44:36	1 2 3 4 12 14 16 18 19 20 22 23	C_2	22×2	126, 126, 126	0 4 6 2 0 0	0.0043	22(2), 36, 88(2)
44:37	1 2 3 5 7 9 16 18 20 22 23 24	D_{3h}	1×2, 3×6, 2×12	33, 55, 21	0 6 0 6 0 0	0.2067	39(3)
44:38	1 2 3 5 7 9 17 19 21 22 23 24	D_{3d}	1×2, 3×6, 2×12	33, 34, 0	0 6 0 6 0 0	0.1731	
44:39	1 2 3 5 7 10 15 18 20 22 23 24	C_{2v}	4×2, 9×4	96, 126, 96	2 4 2 4 0 0	0.2589	15(4), 37, 40(4), 55(2)
44:40	1 2 3 5 7 10 16 18 20 21 23 24	C_1	44×1	126, 126, 126	0 5 4 3 0 0	0.2129	16, 26, 39, 43, 49, 52, 56
44:41	1 2 3 5 7 10 16 19 20 22 23 24	C_1	44×1	126, 126, 126	0 3 6 3 0 0	0.1678	16, 48, 51
44:42	1 2 3 5 7 10 17 19 20 21 22 23	C_1	44×1	126, 126, 126	0 4 6 2 0 0	0.1177	18, 26, 47, 57
44:43	1 2 3 5 7 11 16 17 20 21 23 24	C_1	44×1	126, 126, 126	0 3 6 3 0 0	0.0549	15, 40, 47
44:44	1 2 3 5 7 11 16 19 21 22 23 24	C_2	22×2	126, 126, 126	0 4 4 4 0 0	0.0510	45
44:45	1 2 3 5 7 12 16 18 21 22 23 24	C_2	22×2	126, 126, 126	0 4 4 4 0 0	0.0383	8(2), 44
44:46	1 2 3 5 10 11 14 15 20 22 23 24	C_2	22×2	126, 126, 126	0 2 8 2 0 0	0.1093	19(2), 65(2)
44:47	1 2 3 5 10 12 14 16 20 22 23 24	C_1	44×1	126, 126, 126	1 3 7 1 0 0	0.1008	22, 25, 42, 43, 52, 64, 78
44:48	1 2 3 5 10 12 14 17 20 21 22 23	C_1	44×1	126, 126, 126	0 2 8 2 0 0	0.0953	23, 41, 54

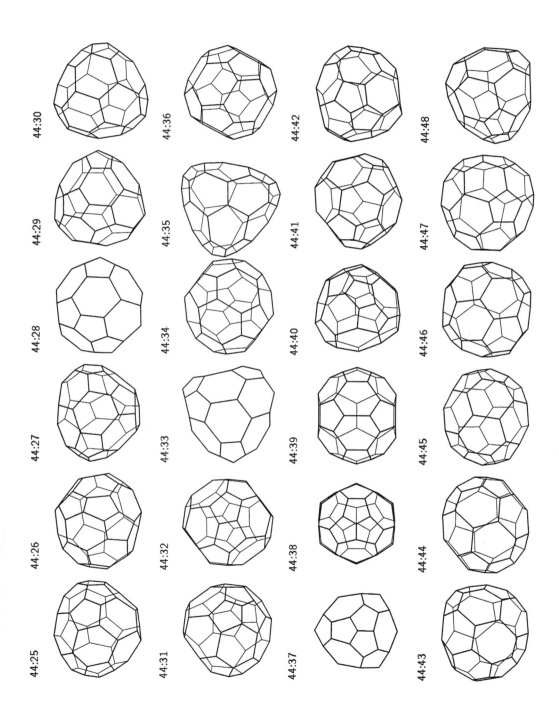

44:30

44:36

44:42

44:48

44:29

44:35

44:41

44:47

44:28

44:34

44:40

44:46

44:27

44:33

44:39

44:45

44:26

44:32

44:38

44:44

44:25

44:31

44:37

44:43

Table A.6. (*Continued*)

Isomer	Ring spiral	Point group	NMR pattern	Vibrations	Pentagon indices	Band gap	Transformations
44:49	1 2 3 5 10 12 15 17 20 21 22 24	C_2	22×2	126, 126, 126	0 4 6 2 0 0	0.0254	25(2), 40(2), 52(2)
44:50	1 2 3 5 10 13 14 15 18 22 23 24	C_1	44×1	126, 126, 126	0 4 6 2 0 0	0.1098	25, 27, 51, 53, 56, 57, 64
44:51	1 2 3 5 10 13 14 16 18 21 23 24	C_1	44×1	126, 126, 126	1 3 7 1 0 0	0.1533	28, 32, 41, 50, 53, 54, 57, 76
44:52	1 2 3 5 10 13 14 16 19 22 23 24	C_1	44×1	126, 126, 126	1 5 5 1 0 0	0.1754	29, 32, 40, 47, 49, 52, 55, 56, 57, 60, 69, 75
44:53	1 2 3 5 10 13 14 17 18 21 22 24	C_1	44×1	126, 126, 126	0 2 8 2 0 0	0.1438	27, 31, 50, 51, 53, 54
44:54	1 2 3 5 10 13 14 17 19 21 22 23	C_s	6×1, 19×2	126, 126, 126	1 3 7 1 0 0	0.1170	30(2), 48(2), 51(2), 53(2), 71
44:55	1 2 3 5 10 13 15 16 19 21 23 24	C_{2v}	6×2, 8×4	97, 126, 97	2 4 4 2 0 0	0.3107	25(4), 39(2), 52(4), 56(4), 72
44:56	1 2 3 5 10 13 15 18 19 21 22 23	C_1	44×1	126, 126, 126	1 4 5 2 0 0	0.1500	26, 31, 33, 40, 50, 52, 55, 57, 60, 61, 69
44:57	1 2 3 5 10 13 15 18 20 22 23 24	C_1	44×1	126, 126, 126	0 5 6 1 0 0	0.0461	29, 30, 42, 50, 51, 52, 56, 77
44:58	1 2 3 5 10 14 15 17 18 20 21 24	C_1	44×1	126, 126, 126	0 2 8 2 0 0	0.1538	27, 59, 62, 64
44:59	1 2 3 5 10 14 15 17 19 20 21 23	C_1	44×1	126, 126, 126	0 5 6 1 0 0	0.1421	30, 58, 59, 60, 62, 63, 67, 77
44:60	1 2 3 5 10 14 15 17 19 20 22 24	C_1	44×1	126, 126, 126	0 3 8 1 0 0	0.1486	31, 52, 56, 59, 60, 61, 64, 68, 69
44:61	1 2 3 5 10 14 15 18 19 20 22 23	C_2	22×2	126, 126, 126	0 4 6 2 0 0	0.0222	34(2), 56(2), 60(2), 80
44:62	1 2 3 5 10 14 16 17 19 20 21 22	C_1	44×1	126, 126, 126	0 4 6 2 0 0	0.1162	23, 58, 59, 62, 63(2)
44:63	1 2 3 5 11 14 15 17 18 20 21 23	C_1	44×1	126, 126, 126	0 3 8 1 0 0	0.0931	59, 62(2), 64, 65, 67
44:64	1 2 3 5 11 14 15 17 18 20 22 24	C_1	44×1	126, 126, 126	0 3 8 1 0 0	0.1294	47, 50, 58, 60, 63, 77
44:65	1 2 3 5 11 14 15 17 19 21 23 24	C_1	44×1	126, 126, 126	0 3 8 1 0 0	0.1897	23, 46, 63, 79
44:66	1 2 3 5 11 14 15 18 19 22 23 24	C_2	22×2	126, 126, 126	0 0 10 2 0 0	0.0622	81
44:67	1 2 3 10 11 12 14 15 17 20 22 23	C_1	44×1	126, 126, 126	0 4 8 0 0 0	0.0207	59, 63, 68, 77(2), 79, 88
44:68	1 2 3 10 11 12 14 15 17 21 23 24	C_2	22×2	126, 126, 126	0 4 8 0 0 0	0.1200	60(2), 67(2), 69(4), 73, 89
44:69	1 2 3 10 11 12 14 16 17 21 22 24	C_1	44×1	126, 126, 126	0 6 6 0 0 0	0.1468	52, 56, 60, 68(2), 70, 72, 75, 77(2), 80, 89
44:70	1 2 3 10 11 12 14 16 18 21 22 23	C_s	6×1, 19×2	126, 126, 126	0 4 8 0 0 0	0.0983	33, 69(2), 78(2), 88(2)
44:71	1 2 3 10 11 12 15 16 17 20 21 22	C_s	4×1, 20×2	126, 126, 126	0 2 10 0 0 0	0.0252	54, 76(2), 81(2)
44:72	1 2 3 10 11 13 14 16 17 20 22 24	D_{3h}	1×2, 3×6, 2×12	33, 55, 22	0 6 6 0 0 0	0.4241	55(3), 69(12)

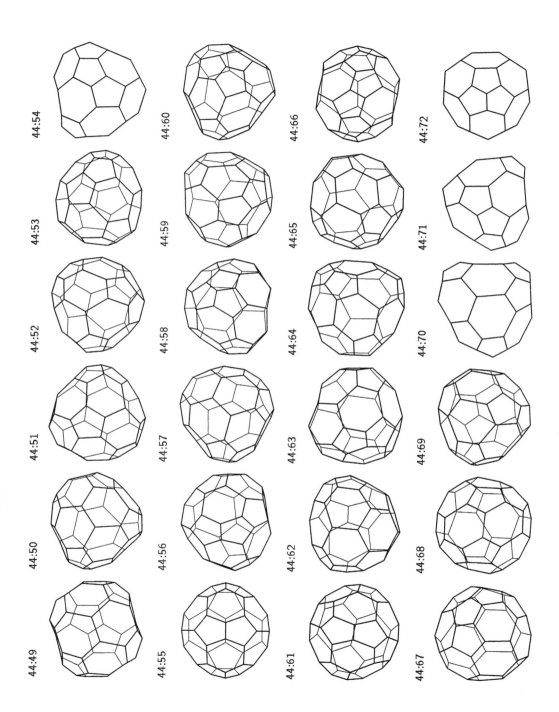

44:54

44:60

44:66

44:72

44:53

44:59

44:65

44:71

44:52

44:58

44:64

44:70

44:51

44:57

44:63

44:69

44:50

44:56

44:62

44:68

44:49

44:55

44:61

44:67

Table A.6. (*Continued*)

Isomer	Ring spiral	Point group	NMR pattern	Vibrations	Pentagon indices	Band gap	Transformations
44:73	1 2 3 10 12 13 15 16 18 19 21 23	T	2×4, 3×12	31, 53, 31	0 0 12 0 0 0	0.4059	68(6)
44:74	1 2 4 7 9 10 12 19 20 21 22 23	C_2	22×2	126, 126, 126	0 4 8 0 0 0	0.0572	22(2), 79, 88(2)
44:75	1 2 4 7 9 12 13 16 18 21 23 24	D_2	11×4	93, 126, 93	0 8 4 0 0 0	0.2460	52(4), 69(4), 78(4), 89(2)
44:76	1 2 4 7 9 12 13 16 19 22 23 24	C_2	22×2	126, 126, 126	0 4 8 0 0 0	0.3819	51(2), 71(2), 84(2)
44:77	1 2 4 7 9 12 13 17 19 21 22 23	C_1	44×1	126, 126, 126	0 6 6 0 0 0	0.1095	57, 59, 64, 67(2), 69(2), 78, 88
44:78	1 2 4 7 9 12 14 16 18 20 23 24	C_1	44×1	126, 126, 126	0 6 6 0 0 0	0.0709	29, 47, 70, 75, 77, 82, 87(2), 88
44:79	1 2 4 7 9 12 14 17 19 21 23 24	C_2	22×2	126, 126, 126	0 4 8 0 0 0	0.0948	65(2), 67(2), 74
44:80	1 2 4 7 9 13 17 18 19 21 22 23	D_3	1×2, 7×6	62, 64, 42	0 6 6 0 0 0	0.0151	61(3), 69(6)
44:81	1 2 4 7 9 14 17 18 19 20 21 22	C_2	22×2	126, 126, 126	0 2 10 0 0 0	0.1587	66, 71(2)
44:82	1 2 4 7 10 12 14 16 17 20 23 24	S_4	11×4	63, 95, 63	0 4 8 0 0 0	0.0000	78(4)
44:83	1 2 4 7 10 12 14 16 19 22 23 24	D_2	11×4	93, 126, 93	0 4 8 0 0 0	0.0296	84(4)
44:84	1 2 4 7 10 12 15 16 19 21 23 24	C_s	4×1, 20×2	126, 126, 126	0 4 8 0 0 0	0.2654	28, 76(2), 83(2)
44:85	1 2 4 7 10 14 17 18 20 21 22 24	D_2	11×4	93, 126, 93	0 0 12 0 0 0	0.2383	
44:86	1 2 4 7 10 14 18 19 20 22 23 24	D_{3d}	1×2, 3×6, 2×12	33, 34, 0	0 0 12 0 0 0	0.1848	
44:87	1 2 4 8 9 12 14 15 18 20 23 24	C_2	22×2	126, 126, 126	0 6 6 0 0 0	0.1081	22(2), 78(4), 87, 88(2)
44:88	1 2 4 8 9 12 14 18 19 21 22 23	C_1	44×1	126, 126, 126	0 6 6 0 0 0	0.1141	36, 67, 70, 74, 77, 78, 87, 88, 89
44:89	1 2 4 9 12 13 14 16 17 19 22 24	D_2	11×4	93, 126, 93	0 8 4 0 0 0	0.2543	68(2), 69(4), 75(2), 88(4), 89(2)

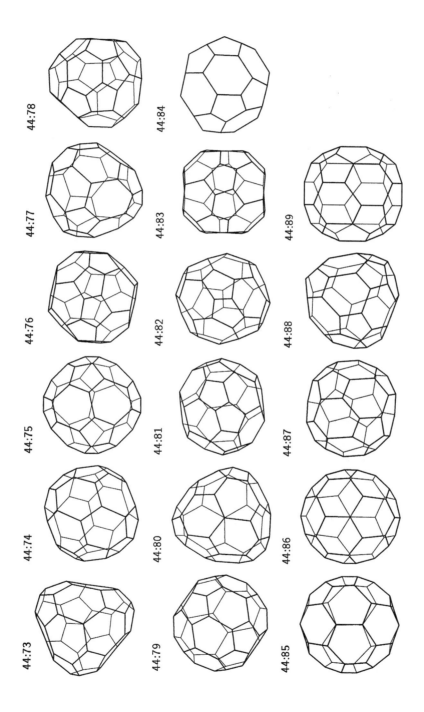

44:73

44:74

44:75

44:76

44:77

44:78

44:79

44:80

44:81

44:82

44:83

44:84

44:85

44:86

44:87

44:88

44:89

Table A.7. Fullerene isomers of C_{46}

Isomer	Ring spiral	Point group	NMR pattern	Vibrations	Pentagon indices	Band gap	Transformations
46:1	1 2 3 4 5 7 19 21 22 23 24 25	C_2	23×2	132, 132, 132	0 2 2 6 2 0	0.0199	
46:2	1 2 3 4 5 12 19 20 21 22 23 24	C_s	6×1, 20×2	132, 132, 132	1 0 8 2 1 0	0.0673	2(2), 8
46:3	1 2 3 4 5 13 15 19 22 23 24 25	C_1	46×1	132, 132, 132	1 1 4 5 1 0	0.1062	4, 21
46:4	1 2 3 4 5 13 16 19 21 23 24 25	C_1	46×1	132, 132, 132	1 2 4 4 1 0	0.0399	3, 5, 9, 49
46:5	1 2 3 4 5 13 17 19 21 22 24 25	C_1	46×1	132, 132, 132	1 1 6 3 1 0	0.0801	4, 7, 10, 18
46:6	1 2 3 4 5 13 17 20 21 23 24 25	C_1	46×1	132, 132, 132	0 2 5 4 1 0	0.0463	19
46:7	1 2 3 4 5 13 18 19 21 22 23 25	C_s	6×1, 20×2	132, 132, 132	0 2 5 4 1 0	0.1697	5(2), 8
46:8	1 2 3 4 5 13 18 20 21 22 23 24	C_s	4×1, 21×2	132, 132, 132	0 2 7 2 1 0	0.1397	2, 7
46:9	1 2 3 4 5 14 16 19 20 23 24 25	C_2	23×2	132, 132, 132	2 0 4 4 2 0	0.0665	4(2), 10(2)
46:10	1 2 3 4 5 14 17 19 20 22 24 25	C_s	6×1, 20×2	132, 132, 132	2 1 3 5 1 0	0.0593	5(2), 9(2), 12(2)
46:11	1 2 3 4 7 10 17 19 21 23 24 25	C_s	8×1, 19×2	132, 132, 132	0 5 0 7 0 0	0.2307	13
46:12	1 2 3 4 7 11 15 19 22 23 24 25	C_2	23×2	132, 132, 132	2 2 2 6 0 0	0.1767	10(2), 13(2), 18(2)
46:13	1 2 3 4 7 11 16 19 21 23 24 25	C_s	6×1, 20×2	132, 132, 132	2 3 2 5 0 0	0.2080	11, 12(2), 14(2), 43(2)
46:14	1 2 3 4 7 11 17 19 21 22 24 25	C_1	46×1	132, 132, 132	0 4 4 4 0 0	0.1664	13, 16, 18, 44
46:15	1 2 3 4 7 11 17 20 21 23 24 25	C_1	46×1	132, 132, 132	0 2 6 4 0 0	0.0681	27, 45
46:16	1 2 3 4 7 11 18 19 21 22 23 25	C_1	46×1	132, 132, 132	0 3 4 5 0 0	0.0842	14, 17, 24
46:17	1 2 3 4 7 11 18 20 21 22 23 24	C_1	46×1	132, 132, 132	0 3 6 3 0 0	0.0497	16, 29, 46
46:18	1 2 3 4 7 12 17 18 21 22 24 25	C_1	46×1	132, 132, 132	1 2 5 4 0 0	0.1284	5, 12, 14, 53
46:19	1 2 3 4 7 12 17 20 22 23 24 25	C_1	46×1	132, 132, 132	1 3 3 5 0 0	0.1463	6, 21, 24, 48
46:20	1 2 3 4 7 13 17 18 21 23 24 25	C_2	23×2	132, 132, 132	0 2 4 6 0 0	0.1297	20
46:21	1 2 3 4 7 13 17 19 22 23 24 25	C_1	46×1	132, 132, 132	0 3 4 5 0 0	0.0984	3, 19
46:22	1 2 3 4 11 12 14 15 22 23 24 25	C_2	23×2	132, 132, 132	0 4 4 4 0 0	0.0027	23(2), 25(2), 35(2)
46:23	1 2 3 4 11 12 14 16 21 23 24 25	C_1	46×1	132, 132, 132	1 3 5 3 0 0	0.0811	22, 24, 40, 54, 55, 73
46:24	1 2 3 4 11 12 14 17 21 22 24 25	C_1	46×1	132, 132, 132	1 4 3 4 0 0	0.1807	16, 19, 23, 35, 44, 46, 51

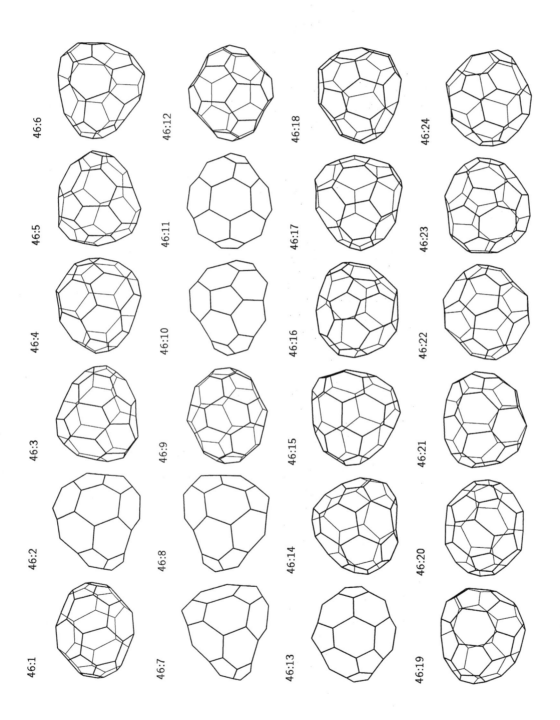

46:6 46:5 46:4 46:3 46:2 46:1

46:12 46:11 46:10 46:9 46:8 46:7

46:18 46:17 46:16 46:15 46:14 46:13

46:24 46:23 46:22 46:21 46:20 46:19

Table A.7. (*Continued*)

Isomer	Ring spiral	Point group	NMR pattern	Vibrations	Pentagon indices	Band gap	Transformations
46:25	1 2 3 4 11 12 15 20 22 23 24 25	C_1	46×1	132, 132, 132	0 4 6 2 0 0	0.1376	22, 29, 36, 55, 69, 97
46:26	1 2 3 4 11 12 16 20 21 22 23 25	C_1	46×1	132, 132, 132	0 3 6 3 0 0	0.1077	30, 41, 56, 76, 78
46:27	1 2 3 4 11 13 14 17 21 23 24 25	C_1	46×1	132, 132, 132	1 2 7 2 0 0	0.1056	15, 31, 57, 59, 61
46:28	1 2 3 4 11 13 14 18 21 22 23 24	C_s	8×1, 19×2	132, 132, 132	0 2 8 2 0 0	0.2160	58(2)
46:29	1 2 3 4 11 13 15 19 22 23 24 25	C_1	46×1	132, 132, 132	1 2 7 2 0 0	0.0662	17, 25, 58, 60, 70
46:30	1 2 3 4 11 13 16 19 21 22 23 25	C_1	46×1	132, 132, 132	1 2 7 2 0 0	0.0852	26, 37, 59, 60, 98
46:31	1 2 3 4 11 14 15 18 19 21 23 25	C_1	46×1	132, 132, 132	0 2 8 2 0 0	0.0017	27, 32, 33, 62, 64
46:32	1 2 3 4 11 14 15 18 20 21 23 24	C_2	23×2	132, 132, 132	0 4 6 2 0 0	0.1076	31(2), 34(2), 63(2)
46:33	1 2 3 4 11 14 16 18 19 21 22 25	C_s	6×1, 20×2	132, 132, 132	1 0 7 4 0 0	0.0224	31(2), 34(2)
46:34	1 2 3 4 11 14 16 18 20 21 22 24	C_1	46×1	132, 132, 132	1 3 5 3 0 0	0.0827	32, 33, 35, 38, 39, 62, 65
46:35	1 2 3 4 11 14 16 18 20 21 23 25	C_1	46×1	132, 132, 132	0 5 4 3 0 0	0.1381	22, 24, 34, 36, 40, 66, 69
46:36	1 2 3 4 11 14 16 19 20 21 23 24	C_1	46×1	132, 132, 132	0 4 6 2 0 0	0.1264	25, 35, 37, 67, 101
46:37	1 2 3 4 11 14 16 19 20 22 23 25	C_1	46×1	132, 132, 132	0 3 6 3 0 0	0.0403	30, 36, 68, 81
46:38	1 2 3 4 11 14 17 18 20 21 22 23	C_s	6×1, 20×2	132, 132, 132	0 4 4 4 0 0	0.0798	34(2), 52(2)
46:39	1 2 3 4 12 14 16 18 19 21 22 24	C_{2v}	5×2, 9×4	101, 132, 101	2 0 8 2 0 0	0.0962	34(4), 40(2), 89(2)
46:40	1 2 3 4 12 14 16 18 19 21 23 25	C_s	6×1, 20×2	132, 132, 132	1 4 5 2 0 0	0.1464	23(2), 35(2), 39, 99, 106
46:41	1 2 3 4 12 14 16 19 20 23 24 25	C_s	4×1, 21×2	132, 132, 132	1 2 7 2 0 0	0.1225	26(2), 42, 100(2)
46:42	1 2 3 4 12 16 19 20 21 23 24 25	C_{2v}	5×2, 9×4	101, 132, 101	0 2 6 4 0 0	0.0793	41(2), 83(2)
46:43	1 2 3 5 7 10 16 19 21 23 24 25	C_2	23×2	132, 132, 132	0 6 2 4 0 0	0.0514	13(2), 44(2)
46:44	1 2 3 5 7 10 17 19 21 22 24 25	C_1	46×1	132, 132, 132	0 5 4 3 0 0	0.0670	14, 24, 43, 54
46:45	1 2 3 5 7 10 17 20 21 23 24 25	C_1	46×1	132, 132, 132	0 3 6 3 0 0	0.0817	15, 56, 59
46:46	1 2 3 5 7 10 18 20 21 22 23 24	C_1	46×1	132, 132, 132	1 4 5 2 0 0	0.0504	17, 24, 48, 50, 55, 60, 69
46:47	1 2 3 5 7 11 16 20 22 23 24 25	C_2	23×2	132, 132, 132	0 4 4 4 0 0	0.0304	54
46:48	1 2 3 5 7 11 17 20 21 22 23 24	C_1	46×1	132, 132, 132	0 3 6 3 0 0	0.0820	19, 46, 56

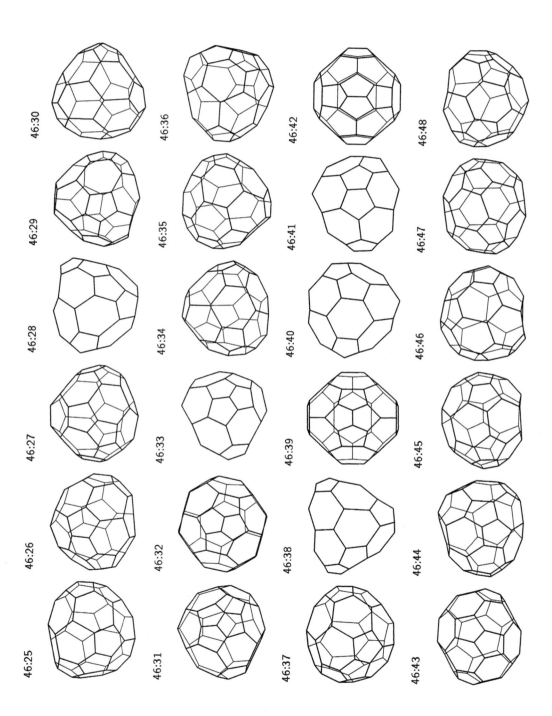

46:30
46:36
46:42
46:48
46:29
46:35
46:41
46:47
46:28
46:34
46:40
46:46
46:27
46:33
46:39
46:45
46:26
46:32
46:38
46:44
46:25
46:31
46:37
46:43

Table A.7. (*Continued*)

Isomer	Ring spiral	Point group	NMR pattern	Vibrations	Pentagon indices	Band gap	Transformations
46:49	1 2 3 5 7 12 16 17 21 23 24 25	C_2	23×2	132, 132, 132	0 4 4 4 0 0	0.0306	4(2)
46:50	1 2 3 5 7 16 17 18 19 21 22 25	C_1	46×1	132, 132, 132	0 4 6 2 0 0	0.1487	46, 50, 51, 67, 68
46:51	1 2 3 5 7 16 17 18 20 21 22 24	C_1	46×1	132, 132, 132	0 5 4 3 0 0	0.1004	24, 50, 52, 66
46:52	1 2 3 5 7 16 17 18 20 21 23 25	C_1	46×1	132, 132, 132	0 3 6 3 0 0	0.1331	38, 51, 65
46:53	1 2 3 5 10 11 14 20 21 22 23 24	C_2	23×2	132, 132, 132	0 2 8 2 0 0	0.0565	18(2), 77
46:54	1 2 3 5 10 12 14 16 21 23 24 25	C_2	23×2	132, 132, 132	2 2 6 2 0 0	0.1230	23(2), 44(2), 47, 72(2)
46:55	1 2 3 5 10 12 15 20 22 23 24 25	C_1	46×1	132, 132, 132	1 3 7 1 0 0	0.0858	23, 25, 46, 60, 75, 99
46:56	1 2 3 5 10 12 16 20 21 22 23 25	C_1	46×1	132, 132, 132	1 2 7 2 0 0	0.1312	26, 45, 48, 60, 79
46:57	1 2 3 5 10 13 14 17 21 23 24 25	C_s	8×1, 19×2	132, 132, 132	0 3 8 1 0 0	0.0838	27(2), 58(2)
46:58	1 2 3 5 10 13 14 18 21 22 23 24	C_1	46×1	132, 132, 132	0 3 8 1 0 0	0.1290	28, 29, 57, 59
46:59	1 2 3 5 10 13 15 18 21 22 24 25	C_1	46×1	132, 132, 132	1 3 7 1 0 0	0.2450	27, 30, 45, 58, 60, 64, 102
46:60	1 2 3 5 10 13 15 19 22 23 24 25	C_1	46×1	132, 132, 132	2 3 6 1 0 0	0.1772	29, 30, 46, 55, 56, 59, 60, 68, 88, 103
46:61	1 2 3 5 10 14 15 17 19 21 24 25	C_1	46×1	132, 132, 132	0 5 6 1 0 0	0.1151	27, 62, 70, 71, 85, 102
46:62	1 2 3 5 10 14 15 18 19 21 23 25	C_1	46×1	132, 132, 132	0 5 6 1 0 0	0.0622	31, 34, 61, 63, 69, 89, 111
46:63	1 2 3 5 10 14 15 18 20 21 23 24	C_1	46×1	132, 132, 132	0 5 6 1 0 0	0.1922	32, 62, 64, 65, 67, 112
46:64	1 2 3 5 10 14 15 18 20 22 24 25	C_1	46×1	132, 132, 132	0 3 8 1 0 0	0.1334	31, 59, 63, 68, 111
46:65	1 2 3 5 10 14 16 18 20 21 22 24	C_s	6×1, 20×2	132, 132, 132	1 4 5 2 0 0	0.0964	34(2), 52(2), 63(2), 66(2)
46:66	1 2 3 5 10 14 16 18 20 21 23 25	C_2	23×2	132, 132, 132	0 6 4 2 0 0	0.1159	35(2), 51(2), 65(2), 67(2)
46:67	1 2 3 5 10 14 16 19 20 21 23 24	C_1	46×1	132, 132, 132	1 5 5 1 0 0	0.2339	36, 50, 63, 66, 68, 69, 74, 82, 90, 109
46:68	1 2 3 5 10 14 16 19 20 22 23 25	C_1	46×1	132, 132, 132	1 3 7 1 0 0	0.0565	37, 50, 60, 64, 67, 75, 86, 90
46:69	1 2 3 5 10 15 16 19 20 21 22 24	C_1	46×1	132, 132, 132	1 5 5 1 0 0	0.0629	25, 35, 46, 62, 67, 70, 78, 81, 88, 99, 101
46:70	1 2 3 5 10 15 17 19 20 21 22 23	C_1	46×1	132, 132, 132	0 5 6 1 0 0	0.0432	29, 61, 69, 79, 87, 103
46:71	1 2 3 5 11 14 15 17 18 21 24 25	C_1	46×1	132, 132, 132	0 3 8 1 0 0	0.1549	61, 105
46:72	1 2 3 5 11 14 15 17 20 23 24 25	C_1	46×1	132, 132, 132	0 5 6 1 0 0	0.1823	54, 73(2), 80, 104

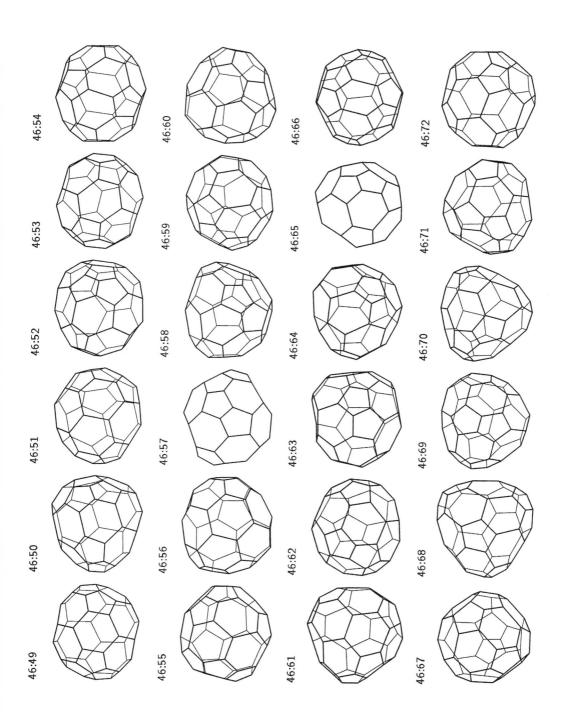

46:54

46:53

46:52

46:51

46:50

46:49

46:60

46:59

46:58

46:57

46:56

46:55

46:66

46:65

46:64

46:63

46:62

46:61

46:72

46:71

46:70

46:69

46:68

46:67

Table A.7. *(Continued)*

Isomer	Ring spiral	Point group	NMR pattern	Vibrations	Pentagon indices	Band gap	Transformations
46:73	1 2 3 5 11 14 16 17 20 22 24 25	C_1	46×1	132, 132, 132	0 5 6 1 0 0	0.0444	23, 72(2), 75, 106
46:74	1 2 3 5 11 14 16 18 20 21 23 24	C_1	46×1	132, 132, 132	0 5 6 1 0 0	0.2263	67, 75, 76, 78, 79, 108
46:75	1 2 3 5 11 14 16 18 20 22 23 25	C_1	46×1	132, 132, 132	0 5 6 1 0 0	0.1850	55, 68, 73, 74, 80, 107
46:76	1 2 3 5 11 14 16 19 21 23 24 25	C_1	46×1	132, 132, 132	0 3 8 1 0 0	0.2166	26, 74, 100
46:77	1 2 3 5 11 14 17 19 21 22 23 24	C_2	23×2	132, 132, 132	0 2 8 2 0 0	0.2270	53
46:78	1 2 3 5 11 15 16 18 20 21 22 24	C_1	46×1	132, 132, 132	0 5 6 1 0 0	0.0205	26, 69, 74, 78, 79, 98, 100
46:79	1 2 3 5 11 15 17 18 20 21 22 23	C_1	46×1	132, 132, 132	0 5 6 1 0 0	0.1916	56, 70, 74, 78, 80, 103
46:80	1 2 3 5 12 14 16 18 19 22 23 25	C_1	46×1	132, 132, 132	0 3 8 1 0 0	0.0957	72, 75, 79, 85
46:81	1 2 3 5 14 15 16 18 19 20 21 24	C_1	46×1	132, 132, 132	0 5 6 1 0 0	0.0540	37, 69, 82, 83, 86, 98, 101
46:82	1 2 3 5 14 15 17 18 19 20 21 23	C_1	46×1	132, 132, 132	0 3 8 1 0 0	0.1047	67, 81, 93, 110
46:83	1 2 3 7 10 11 18 20 21 22 23 24	C_s	6×1, 20×2	132, 132, 132	0 3 8 1 0 0	0.0908	42, 81(2), 92, 100(2)
46:84	1 2 3 10 11 12 14 15 16 23 24 25	C_2	23×2	132, 132, 132	0 4 8 0 0 0	0.0003	85(2), 86(2), 91(2), 96(2)
46:85	1 2 3 10 11 12 14 15 17 22 24 25	C_1	46×1	132, 132, 132	1 4 7 0 0 0	0.0829	61, 80, 84, 89, 93, 103, 104, 107, 111
46:86	1 2 3 10 11 12 14 16 21 23 24 25	C_1	46×1	132, 132, 132	0 6 6 0 0 0	0.1760	68, 81, 84, 88, 92, 93, 103, 107, 109, 116
46:87	1 2 3 10 11 12 14 17 21 22 23 24	C_1	46×1	132, 132, 132	0 4 8 0 0 0	0.0798	70, 88, 102, 105, 108, 110
46:88	1 2 3 10 11 13 14 16 20 23 24 25	C_1	46×1	132, 132, 132	0 6 6 0 0 0	0.2599	60, 69, 86, 87, 90, 102, 103, 110, 111, 114
46:89	1 2 3 10 11 13 15 16 19 22 24 25	C_s	2×1, 22×2	132, 132, 132	1 2 9 0 0 0	0.0179	39, 62(2), 85(2), 91, 99, 106
46:90	1 2 3 10 11 13 15 17 20 21 23 24	C_1	46×1	132, 132, 132	0 6 6 0 0 0	0.1166	67, 68, 88, 90, 93, 111, 112, 114, 116
46:91	1 2 3 10 11 14 15 16 19 21 24 25	C_{2v}	3×2, 10×4	100, 132, 100	0 4 8 0 0 0	0.0418	84(4), 89(2), 92, 107(2)
46:92	1 2 3 10 11 14 16 19 21 22 24 25	C_{2v}	3×2, 10×4	100, 132, 100	0 4 8 0 0 0	0.1540	83(2), 86(4), 91, 108(2)
46:93	1 2 3 10 12 13 15 16 18 19 22 25	C_1	46×1	132, 132, 132	0 4 8 0 0 0	0.1389	82, 85, 86, 90, 94, 96, 114
46:94	1 2 3 10 12 13 15 16 18 20 22 24	C_3	1×1, 15×3	86, 86, 86	0 6 6 0 0 0	0.0000	93(3), 95(3), 116(3)
46:95	1 2 3 10 12 13 15 17 18 20 22 23	C_2	23×2	132, 132, 132	0 4 8 0 0 0	0.0427	94(2), 95(3), 96(2)
46:96	1 2 3 10 12 13 15 17 18 21 23 25	C_2	23×2	132, 132, 132	0 4 8 0 0 0	0.0361	84(2), 93(2), 95(2), 116

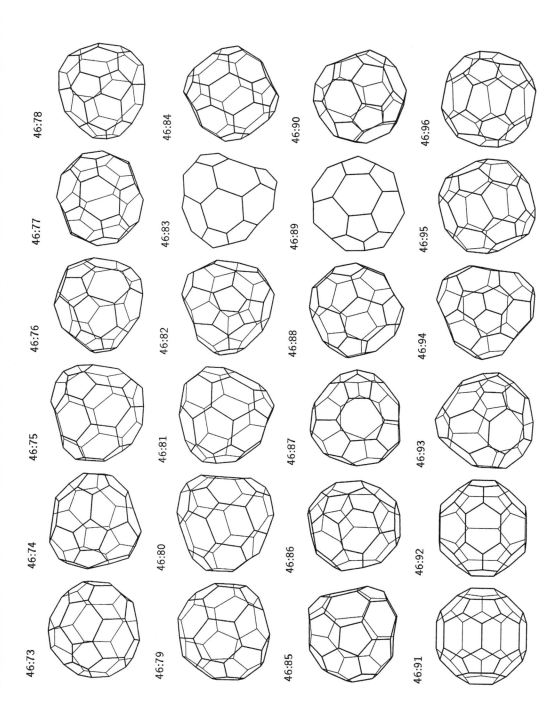

46:78

46:84

46:90

46:96

46:77

46:83

46:89

46:95

46:76

46:82

46:88

46:94

46:75

46:81

46:87

46:93

46:74

46:80

46:86

46:92

46:73

46:79

46:85

46:91

Table A.7. (*Continued*)

Isomer	Ring spiral	Point group	NMR pattern	Vibrations	Pentagon indices	Band gap	Transformations
46:97	1 2 4 7 9 10 18 19 20 21 22 23	C_2	23×2	132, 132, 132	0 4 8 0 0 0	0.2196	25(2), 101(2)
46:98	1 2 4 7 9 11 12 20 21 22 23 24	C_1	46×1	132, 132, 132	1 4 7 0 0 0	0.1097	30, 78, 81, 99, 100, 102, 103
46:99	1 2 4 7 9 11 13 19 21 22 23 24	C_s	4×1, 21×2	132, 132, 132	2 4 6 0 0 0	0.1285	40, 55(2), 69(2), 89, 98(2), 103(2), 107
46:100	1 2 4 7 9 11 15 19 20 21 23 24	C_1	46×1	132, 132, 132	1 4 7 0 0 0	0.1831	41, 76, 78, 83, 98, 101, 105, 108
46:101	1 2 4 7 9 11 16 19 20 21 22 24	C_1	46×1	132, 132, 132	0 6 6 0 0 0	0.2024	36, 69, 81, 97, 100, 101, 109, 110
46:102	1 2 4 7 9 12 13 17 21 23 24 25	C_1	46×1	132, 132, 132	0 6 6 0 0 0	0.3300	59, 61, 87, 88, 98, 105, 107, 111
46:103	1 2 4 7 9 12 13 18 21 22 23 24	C_1	46×1	132, 132, 132	1 6 5 0 0 0	0.1241	60, 70, 79, 85, 86, 88, 98, 99, 103, 108, 114
46:104	1 2 4 7 9 12 14 17 19 22 24 25	C_2	23×2	132, 132, 132	0 6 6 0 0 0	0.0832	72(2), 85(2), 106(2)
46:105	1 2 4 7 9 12 14 17 20 23 24 25	C_1	46×1	132, 132, 132	0 4 8 0 0 0	0.0175	71, 87, 100, 102
46:106	1 2 4 7 9 12 15 17 19 21 24 25	C_s	4×1, 21×2	132, 132, 132	0 6 6 0 0 0	0.1228	40, 73(2), 89, 104(2), 107
46:107	1 2 4 7 9 12 15 18 19 21 23 25	C_s	2×1, 22×2	132, 132, 132	1 6 5 0 0 0	0.3346	75(2), 85(2), 86(2), 91, 99, 102(2), 106, 108
46:108	1 2 4 7 9 12 15 18 20 21 23 24	C_s	4×1, 21×2	132, 132, 132	1 6 5 0 0 0	0.1941	74(2), 87(2), 92, 100(2), 103(2), 107, 109(2)
46:109	1 2 4 7 9 12 16 18 20 21 22 24	C_2	23×2	132, 132, 132	0 8 4 0 0 0	0.1648	67(2), 86(2), 101(2), 108(2), 110(2), 114(2)
46:110	1 2 4 7 9 12 17 18 20 21 22 23	C_1	46×1	132, 132, 132	0 6 6 0 0 0	0.1141	82, 87, 88, 101, 109, 114, 115
46:111	1 2 4 7 9 13 17 18 20 21 23 25	C_1	46×1	132, 132, 132	0 6 6 0 0 0	0.2486	62, 64, 85, 88, 90, 102, 112
46:112	1 2 4 7 9 13 17 19 20 21 23 24	C_2	23×2	132, 132, 132	0 6 6 0 0 0	0.0534	63(2), 90(2), 111(2)
46:113	1 2 4 7 10 12 15 19 22 23 24 25	C_2	23×2	132, 132, 132	0 2 10 0 0 0	0.2463	113(2)
46:114	1 2 4 8 12 13 15 17 18 19 22 25	C_1	46×1	132, 132, 132	0 8 4 0 0 0	0.1309	88, 90, 93, 103, 109, 110, 114(2), 115, 116(2
46:115	1 2 4 8 12 14 17 18 19 21 22 24	C_3	1×1, 15×3	86, 86, 86	0 6 6 0 0 0	0.0094	110(3), 114(3)
46:116	1 2 4 9 12 13 15 16 18 20 22 24	C_2	23×2	132, 132, 132	0 8 4 0 0 0	0.1064	86(2), 90(2), 94(2), 96, 114(4)

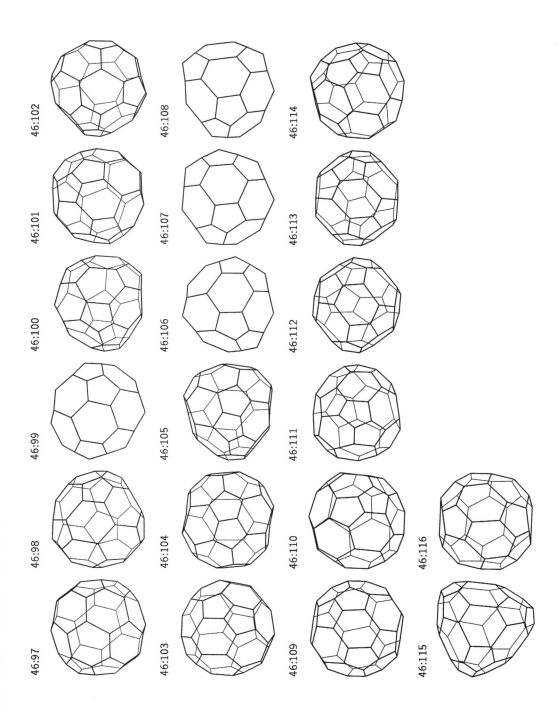

46:102

46:101

46:100

46:99

46:98

46:97

46:108

46:107

46:106

46:105

46:104

46:103

46:114

46:113

46:112

46:111

46:110

46:109

46:116

46:115

Table A.8. Fullerene isomers of C_{48}

Isomer	Ring spiral	Point group	NMR pattern	Vibrations	Pentagon indices	Band gap	Transformations
48:1	1 2 3 4 5 7 20 22 23 24 25 26	C_2	24×2	138, 138, 138	0 2 2 6 2 0	0.0717	
48:2	1 2 3 4 5 8 19 22 23 24 25 26	D_2	12×4	102, 138, 102	0 0 4 4 4 0	0.0135	
48:3	1 2 3 4 5 12 19 20 22 23 25 26	C_1	48×1	138, 138, 138	1 0 6 4 1 0	0.0761	3, 4, 8
48:4	1 2 3 4 5 12 20 21 23 24 25 26	C_s	8×1, 20×2	138, 138, 138	1 0 6 4 1 0	0.0095	3(2)
48:5	1 2 3 4 5 13 14 22 23 24 25 26	C_2	24×2	138, 138, 138	0 2 4 4 2 0	0.0948	12(2)
48:6	1 2 3 4 5 13 16 20 23 24 25 26	C_1	48×1	138, 138, 138	1 1 4 5 1 0	0.1218	7, 26
48:7	1 2 3 4 5 13 17 20 22 24 25 26	C_1	48×1	138, 138, 138	1 2 4 4 1 0	0.0850	6, 8, 23
48:8	1 2 3 4 5 13 18 20 22 23 25 26	C_1	48×1	138, 138, 138	1 1 6 3 1 0	0.1318	3, 7, 9
48:9	1 2 3 4 5 13 19 20 22 23 24 26	C_1	48×1	138, 138, 138	0 0 7 4 1 0	0.0995	8, 10
48:10	1 2 3 4 5 13 19 21 22 23 24 25	C_1	48×1	138, 138, 138	0 2 7 2 1 0	0.1566	9, 11
48:11	1 2 3 4 5 14 19 20 22 23 24 25	C_1	48×1	138, 138, 138	1 1 6 3 1 0	0.0621	10, 12, 13, 25
48:12	1 2 3 4 5 14 19 21 23 24 25 26	C_1	48×1	138, 138, 138	0 3 3 5 1 0	0.1635	5, 11, 24
48:13	1 2 3 4 5 15 19 20 21 23 24 25	C_1	48×1	138, 138, 138	1 2 4 4 1 0	0.0447	11, 13, 14, 70
48:14	1 2 3 4 5 15 19 20 22 24 25 26	C_2	24×2	138, 138, 138	0 2 4 4 2 0	0.0428	13(2)
48:15	1 2 3 4 7 10 17 20 23 24 25 26	D_{2h}	4×4, 4×8	53, 69, 0	0 4 0 8 0 0	0.0440	17(2)
48:16	1 2 3 4 7 10 18 21 23 24 25 26	D_2	12×4	102, 138, 102	0 4 0 8 0 0	0.0697	
48:17	1 2 3 4 7 11 16 20 23 24 25 26	C_{2v}	4×2, 10×4	105, 138, 105	2 2 2 6 0 0	0.0440	15, 18(4), 41
48:18	1 2 3 4 7 11 17 20 22 24 25 26	C_1	48×1	138, 138, 138	0 5 2 5 0 0	0.1836	17, 19, 38, 64
48:19	1 2 3 4 7 11 18 20 22 23 25 26	C_1	48×1	138, 138, 138	0 4 4 4 0 0	0.1961	18, 21, 42, 65
48:20	1 2 3 4 7 11 18 21 22 24 25 26	C_1	48×1	138, 138, 138	0 2 6 4 0 0	0.2179	66
48:21	1 2 3 4 7 11 19 20 22 23 24 26	C_1	48×1	138, 138, 138	1 3 3 5 0 0	0.0683	19, 22, 24, 29, 36, 43
48:22	1 2 3 4 7 11 19 21 22 23 24 25	C_1	48×1	138, 138, 138	1 3 5 3 0 0	0.1534	21, 25, 28, 67
48:23	1 2 3 4 7 12 17 21 23 24 25 26	C_1	48×1	138, 138, 138	1 3 3 5 0 0	0.1506	7, 69
48:24	1 2 3 4 7 12 18 20 22 23 24 26	C_2	24×2	138, 138, 138	2 2 2 6 0 0	0.0757	12(2), 21(2), 25(2), 27

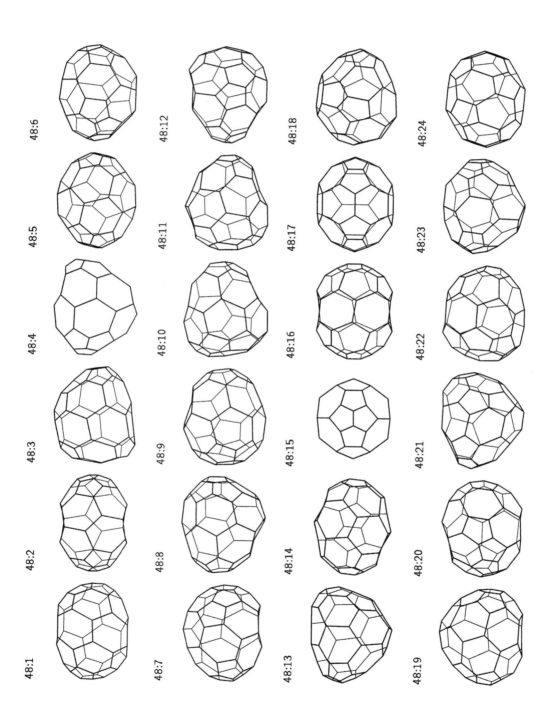

48:6

48:12

48:18

48:24

48:5

48:11

48:17

48:23

48:4

48:10

48:16

48:22

48:3

48:9

48:15

48:21

48:2

48:8

48:14

48:20

48:1

48:7

48:13

48:19

Table A.8. (*Continued*)

Isomer	Ring spiral	Point group	NMR pattern	Vibrations	Pentagon indices	Band gap	Transformations
48:25	1 2 3 4 7 12 18 21 22 23 24 25	C_1	48×1	138, 138, 138	1 2 5 4 0 0	0.0368	11, 22, 24, 76
48:26	1 2 3 4 7 13 17 18 22 24 25 26	C_1	48×1	138, 138, 138	0 3 4 5 0 0	0.1047	6
48:27	1 2 3 4 7 13 18 19 22 23 24 26	C_2	24×2	138, 138, 138	0 2 4 6 0 0	0.0757	24
48:28	1 2 3 4 7 17 18 19 20 22 23 26	C_1	48×1	138, 138, 138	0 3 6 3 0 0	0.0832	22, 28, 29
48:29	1 2 3 4 7 17 18 19 21 22 23 25	C_1	48×1	138, 138, 138	0 4 4 4 0 0	0.1342	21, 28, 30, 54
48:30	1 2 3 4 7 17 18 19 21 22 24 26	C_1	48×1	138, 138, 138	0 2 6 4 0 0	0.1038	29, 31, 55
48:31	1 2 3 4 7 17 18 20 21 22 24 25	C_s	6×1, 21×2	138, 138, 138	0 3 4 5 0 0	0.1082	30(2), 56
48:32	1 2 3 4 11 12 14 20 22 23 24 26	C_2	24×2	138, 138, 138	0 4 4 4 0 0	0.0671	33(2), 34(2), 40
48:33	1 2 3 4 11 12 14 21 22 23 24 25	C_1	48×1	138, 138, 138	0 4 6 2 0 0	0.0561	32, 62, 77, 157
48:34	1 2 3 4 11 12 15 20 21 23 24 26	C_1	48×1	138, 138, 138	1 3 5 3 0 0	0.1674	32, 36, 42, 77, 78, 100
48:35	1 2 3 4 11 12 15 21 22 24 25 26	C_1	48×1	138, 138, 138	0 3 6 3 0 0	0.0257	53, 79, 99, 120
48:36	1 2 3 4 11 12 16 20 21 22 24 26	C_1	48×1	138, 138, 138	1 4 3 4 0 0	0.0730	21, 34, 44, 65, 67, 72
48:37	1 2 3 4 11 13 14 16 23 24 25 26	C_2	24×2	138, 138, 138	2 2 4 4 0 0	0.1588	38(2), 44(2), 50(2)
48:38	1 2 3 4 11 13 14 17 22 24 25 26	C_1	48×1	138, 138, 138	2 3 4 3 0 0	0.1225	18, 37, 39, 41, 42, 45, 80, 82, 84
48:39	1 2 3 4 11 13 14 18 22 23 25 26	C_s	8×1, 20×2	138, 138, 138	0 5 4 3 0 0	0.0254	38(2), 40(2), 81(2)
48:40	1 2 3 4 11 13 14 19 22 23 24 26	C_2	24×2	138, 138, 138	0 4 4 4 0 0	0.0372	32, 39(2), 42(2)
48:41	1 2 3 4 11 13 15 17 21 24 25 26	D_{2h}	4×4, 4×8	53, 69, 0	4 0 4 4 0 0	0.0440	17(2), 38(8)
48:42	1 2 3 4 11 13 15 19 21 23 24 26	C_1	48×1	138, 138, 138	2 3 4 3 0 0	0.1323	19, 34, 38, 40, 43, 44, 47, 81, 83, 94
48:43	1 2 3 4 11 13 15 20 23 24 25 26	C_2	24×2	138, 138, 138	2 2 4 4 0 0	0.0259	21(2), 42(2), 44(2), 54(2)
48:44	1 2 3 4 11 13 16 19 21 22 24 26	C_1	48×1	138, 138, 138	2 3 4 3 0 0	0.0752	36, 37, 42, 43, 44, 51, 82, 83, 90
48:45	1 2 3 4 11 14 15 18 20 23 25 26	C_2	24×2	138, 138, 138	0 6 4 2 0 0	0.0252	38(2), 47(2), 50(2), 85(2)
48:46	1 2 3 4 11 14 15 18 21 24 25 26	C_2	24×2	138, 138, 138	0 2 8 2 0 0	0.0285	86(2)
48:47	1 2 3 4 11 14 15 19 20 23 24 26	C_1	48×1	138, 138, 138	0 4 6 2 0 0	0.1313	42, 45, 48, 51, 58, 87, 103
48:48	1 2 3 4 11 14 15 19 21 23 24 25	C_1	48×1	138, 138, 138	0 4 6 2 0 0	0.0504	47, 49, 52, 88, 102

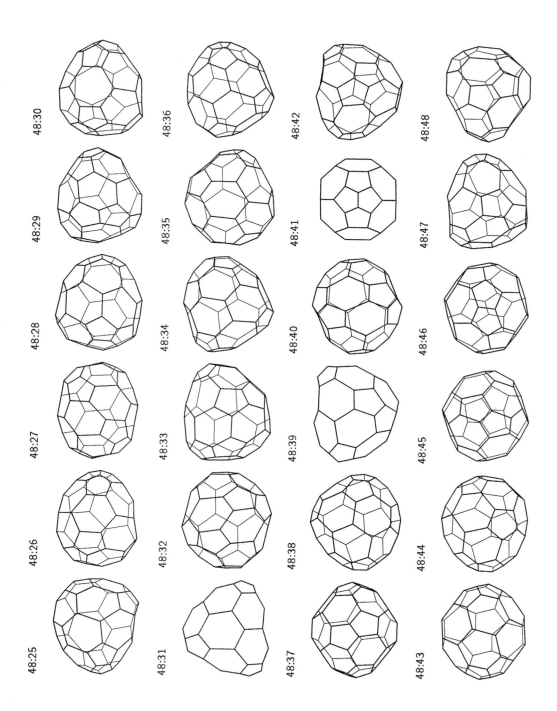

48:30

48:36

48:42

48:48

48:29

48:35

48:41

48:47

48:28

48:34

48:40

48:46

48:27

48:33

48:39

48:45

48:26

48:32

48:38

48:44

48:25

48:31

48:37

48:43

Table A.8. (*Continued*)

Isomer	Ring spiral	Point group	NMR pattern	Vibrations	Pentagon indices	Band gap	Transformations
48:49	1 2 3 4 11 14 15 19 22 24 25 26	C_1	48×1	138, 138, 138	0 1 8 3 0 0	0.0261	48, 89, 101
48:50	1 2 3 4 11 14 16 18 20 22 25 26	C_1	48×1	138, 138, 138	1 4 5 2 0 0	0.0721	37, 45, 51, 61, 90, 190
48:51	1 2 3 4 11 14 16 19 20 22 24 26	C_1	48×1	138, 138, 138	1 4 5 2 0 0	0.1038	44, 47, 50, 52, 54, 91, 164
48:52	1 2 3 4 11 14 16 19 21 22 24 25	C_1	48×1	138, 138, 138	1 2 7 2 0 0	0.0392	48, 51, 53, 55, 92, 132
48:53	1 2 3 4 11 14 16 19 21 23 25 26	C_1	48×1	138, 138, 138	0 4 6 2 0 0	0.1028	35, 52, 93, 172
48:54	1 2 3 4 11 14 17 19 20 22 23 26	C_1	48×1	138, 138, 138	1 3 5 3 0 0	0.0846	29, 43, 51, 55, 94, 97
48:55	1 2 3 4 11 14 17 19 21 22 23 25	C_1	48×1	138, 138, 138	1 3 5 3 0 0	0.1226	30, 52, 54, 56, 57, 95, 96
48:56	1 2 3 4 11 14 17 19 21 22 24 26	C_{2v}	6×2, 9×4	106, 138, 106	0 4 4 4 0 0	0.2227	31(2), 55(4)
48:57	1 2 3 4 11 14 18 19 21 22 23 24	C_1	48×1	138, 138, 138	0 1 8 3 0 0	0.0068	55, 98
48:58	1 2 3 4 11 15 19 20 21 23 24 25	C_2	24×2	138, 138, 138	0 4 6 2 0 0	0.1295	47(2), 59, 104(2)
48:59	1 2 3 4 11 15 19 20 22 23 24 26	C_2	24×2	138, 138, 138	0 2 6 4 0 0	0.0177	58, 105(2)
48:60	1 2 3 4 11 16 19 20 21 22 23 24	C_1	48×1	138, 138, 138	0 2 8 2 0 0	0.1414	106, 182
48:61	1 2 3 4 12 14 16 18 19 22 25 26	C_2	24×2	138, 138, 138	2 2 6 2 0 0	0.0741	50(2), 174(2)
48:62	1 2 3 4 12 14 18 20 22 23 24 25	C_s	8×1, 20×2	138, 138, 138	0 4 6 2 0 0	0.0508	33(2), 184
48:63	1 2 3 4 12 16 19 20 22 23 24 26	C_2	24×2	138, 138, 138	0 2 8 2 0 0	0.0476	181(2)
48:64	1 2 3 5 7 10 17 20 22 24 25 26	C_2	24×2	138, 138, 138	0 6 2 4 0 0	0.0155	18(2), 65(2), 82
48:65	1 2 3 5 7 10 18 20 22 23 25 26	C_1	48×1	138, 138, 138	2 3 4 3 0 0	0.0453	19, 36, 64, 68, 73, 78, 83, 99
48:66	1 2 3 5 7 10 18 21 22 24 25 26	C_1	48×1	138, 138, 138	1 3 5 3 0 0	0.1326	20, 69, 79, 100
48:67	1 2 3 5 7 10 19 21 22 23 24 25	C_1	48×1	138, 138, 138	1 4 5 2 0 0	0.1374	22, 36, 71, 77
48:68	1 2 3 5 7 11 17 20 22 23 25 26	C_2	24×2	138, 138, 138	0 4 4 4 0 0	0.0092	65(2)
48:69	1 2 3 5 7 11 17 21 22 24 25 26	C_1	48×1	138, 138, 138	0 3 6 3 0 0	0.0659	23, 66
48:70	1 2 3 5 7 12 17 21 22 23 24 25	C_2	24×2	138, 138, 138	0 4 4 4 0 0	0.0093	13(2)
48:71	1 2 3 5 7 16 17 18 20 22 25 26	C_1	48×1	138, 138, 138	0 6 4 2 0 0	0.2072	67, 71, 72
48:72	1 2 3 5 7 16 17 19 20 22 24 26	C_1	48×1	138, 138, 138	0 6 4 2 0 0	0.0740	36, 71, 73, 90

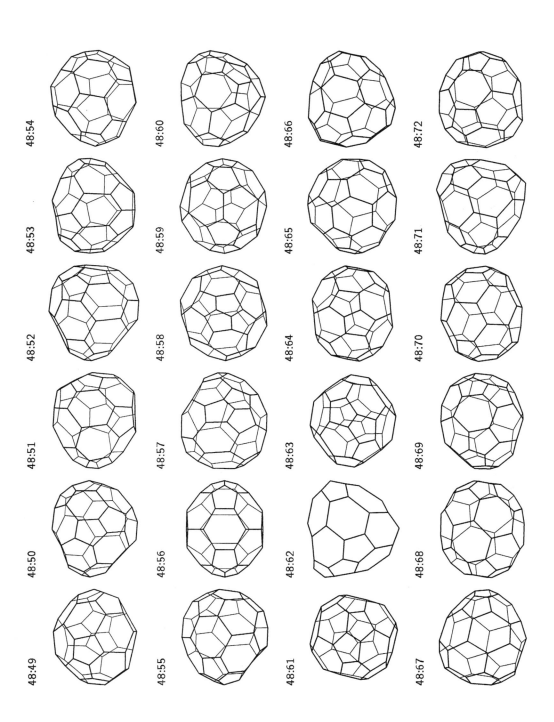

48:54

48:53

48:52

48:51

48:50

48:49

48:60

48:59

48:58

48:57

48:56

48:55

48:66

48:65

48:64

48:63

48:62

48:61

48:72

48:71

48:70

48:69

48:68

48:67

Table A.8. (*Continued*)

Isomer	Ring spiral	Point group	NMR pattern	Vibrations	Pentagon indices	Band gap	Transformations
48:73	1 2 3 5 7 16 17 19 21 22 24 25	C_1	48×1	138, 138, 138	0 6 4 2 0 0	0.1509	65, 72, 74, 91, 93
48:74	1 2 3 5 7 16 17 19 21 23 25 26	C_s	8×1, 20×2	138, 138, 138	0 4 6 2 0 0	0.0753	73(2), 92(2)
48:75	1 2 3 5 7 17 18 19 21 22 23 24	C_s	6×1, 21×2	138, 138, 138	0 4 6 2 0 0	0.0862	106(2)
48:76	1 2 3 5 10 11 14 20 22 23 25 26	C_2	24×2	138, 138, 138	0 2 8 2 0 0	0.1192	25(2)
48:77	1 2 3 5 10 12 14 21 22 23 24 25	C_1	48×1	138, 138, 138	1 3 7 1 0 0	0.1315	33, 34, 67, 113
48:78	1 2 3 5 10 12 15 20 21 23 24 26	C_2	24×2	138, 138, 138	2 2 6 2 0 0	0.0382	34(2), 65(2), 83, 108(2)
48:79	1 2 3 5 10 12 15 21 22 24 25 26	C_1	48×1	138, 138, 138	1 2 7 2 0 0	0.1167	35, 66, 108, 111
48:80	1 2 3 5 10 13 14 17 22 24 25 26	C_{2h}	4×2, 10×4	69, 69, 0	0 6 4 2 0 0	0.0420	38(4), 81(4)
48:81	1 2 3 5 10 13 14 18 22 23 25 26	C_2	24×2	138, 138, 138	0 6 4 2 0 0	0.1295	39(2), 42(2), 80(2), 82(2)
48:82	1 2 3 5 10 13 15 18 21 23 25 26	C_2	24×2	138, 138, 138	2 4 4 2 0 0	0.0604	38(2), 44(2), 64, 81(2), 83(2), 85(2)
48:83	1 2 3 5 10 13 15 19 21 23 24 26	C_2	24×2	138, 138, 138	4 2 4 2 0 0	0.0347	42(2), 44(2), 65(2), 78, 82(2), 83(2), 87(2), 91(2)
48:84	1 2 3 5 10 14 15 17 20 24 25 26	C_2	24×2	138, 138, 138	0 6 4 2 0 0	0.0608	38(2), 85(2), 94(2)
48:85	1 2 3 5 10 14 15 18 20 23 25 26	C_1	48×1	138, 138, 138	0 7 4 1 0 0	0.1151	45, 82, 84, 87, 90, 91, 103, 169
48:86	1 2 3 5 10 14 15 18 21 24 25 26	C_1	48×1	138, 138, 138	0 3 8 1 0 0	0.0592	46, 106, 170
48:87	1 2 3 5 10 14 15 19 20 23 24 26	C_1	48×1	138, 138, 138	1 5 5 1 0 0	0.0776	47, 83, 85, 87, 88, 91, 99, 104, 127, 137, 139
48:88	1 2 3 5 10 14 15 19 21 23 24 25	C_1	48×1	138, 138, 138	1 5 5 1 0 0	0.0744	48, 87, 89, 92, 101, 104, 128, 135, 168
48:89	1 2 3 5 10 14 15 19 22 24 25 26	C_s	4×1, 22×2	138, 138, 138	1 2 7 2 0 0	0.1207	49(2), 88(2), 105(2), 136
48:90	1 2 3 5 10 14 16 18 20 22 25 26	C_1	48×1	138, 138, 138	1 5 5 1 0 0	0.0523	44, 50, 72, 85, 90, 91, 164
48:91	1 2 3 5 10 14 16 19 20 22 24 26	C_1	48×1	138, 138, 138	2 5 4 1 0 0	0.1882	51, 73, 83, 85, 87, 90, 92, 94, 108, 127, 139, 163
48:92	1 2 3 5 10 14 16 19 21 22 24 25	C_1	48×1	138, 138, 138	2 3 6 1 0 0	0.0704	52, 74, 88, 91, 93, 95, 103, 109, 126, 133, 138
48:93	1 2 3 5 10 14 16 19 21 23 25 26	C_1	48×1	138, 138, 138	1 5 5 1 0 0	0.1326	53, 73, 92, 99, 102, 110, 139, 162
48:94	1 2 3 5 10 14 17 19 20 22 23 26	C_1	48×1	138, 138, 138	0 6 4 2 0 0	0.1215	42, 54, 84, 91, 95, 100, 103
48:95	1 2 3 5 10 14 17 19 21 22 23 25	C_2	24×2	138, 138, 138	0 6 4 2 0 0	0.0798	55(2), 92(2), 94(2), 98(2)
48:96	1 2 3 5 10 14 17 20 21 22 24 25	C_s	8×1, 20×2	138, 138, 138	0 4 6 2 0 0	0.1537	55(2), 97(2)

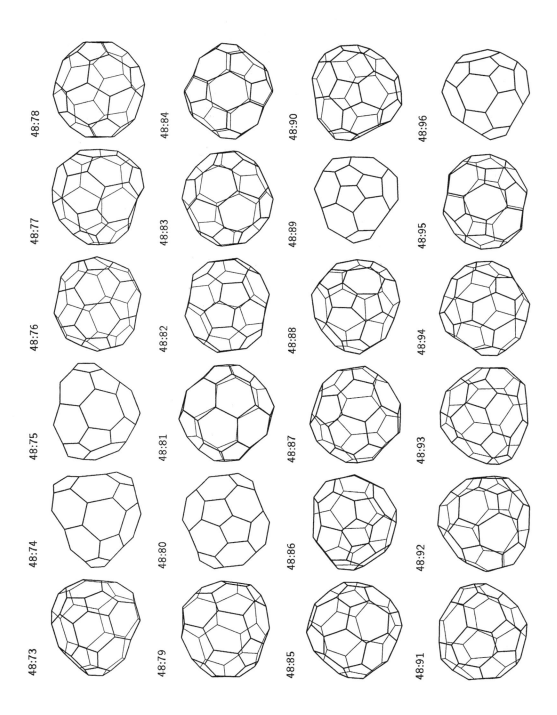

48:78

48:84

48:90

48:96

48:77

48:83

48:89

48:95

48:76

48:82

48:88

48:94

48:75

48:81

48:87

48:93

48:74

48:80

48:86

48:92

48:73

48:79

48:85

48:91

Table A.8. (*Continued*)

Isomer	Ring spiral	Point group	NMR pattern	Vibrations	Pentagon indices	Band gap	Transformations
48:97	1 2 3 5 10 14 17 20 21 23 24 26	C_2	24×2	138, 138, 138	0 4 6 2 0 0	0.1031	54(2), 96(2)
48:98	1 2 3 5 10 14 18 19 21 22 23 24	C_1	48×1	138, 138, 138	0 4 6 2 0 0	0.0966	57, 95, 100, 109
48:99	1 2 3 5 10 15 16 19 21 22 25 26	C_1	48×1	138, 138, 138	1 4 5 2 0 0	0.0453	35, 65, 87, 93, 101, 108, 114, 118
48:100	1 2 3 5 10 15 16 20 21 24 25 26	C_1	48×1	138, 138, 138	1 4 5 2 0 0	0.0695	34, 66, 94, 98, 108, 112, 113
48:101	1 2 3 5 10 15 18 19 21 22 23 26	C_1	48×1	138, 138, 138	0 4 6 2 0 0	0.0781	49, 88, 99, 102, 109
48:102	1 2 3 5 10 15 18 20 21 22 23 25	C_1	48×1	138, 138, 138	0 5 6 1 0 0	0.1707	48, 93, 101, 103, 168
48:103	1 2 3 5 10 15 18 20 21 23 24 26	C_1	48×1	138, 138, 138	0 5 6 1 0 0	0.1141	47, 85, 92, 94, 102, 104, 139
48:104	1 2 3 5 10 15 19 20 21 23 24 25	C_1	48×1	138, 138, 138	1 5 5 1 0 0	0.0365	58, 87, 88, 103, 105, 118, 129, 140, 171
48:105	1 2 3 5 10 15 19 20 22 23 24 26	C_1	48×1	138, 138, 138	1 2 7 2 0 0	0.0759	59, 89, 104, 119, 128, 152
48:106	1 2 3 5 10 16 19 20 21 22 23 24	C_1	48×1	138, 138, 138	1 3 7 1 0 0	0.1052	60, 75, 86, 121, 141, 173
48:107	1 2 3 5 11 14 16 17 21 24 25 26	C_2	24×2	138, 138, 138	0 4 6 2 0 0	0.0346	110(2)
48:108	1 2 3 5 11 14 16 18 20 22 24 26	C_1	48×1	138, 138, 138	2 3 6 1 0 0	0.1668	78, 79, 91, 99, 100, 109, 116, 161
48:109	1 2 3 5 11 14 16 18 21 22 24 25	C_1	48×1	138, 138, 138	1 1 9 1 0 0	0.0527	92, 98, 101, 108, 110, 143
48:110	1 2 3 5 11 14 16 18 21 23 25 26	C_1	48×1	138, 138, 138	0 5 6 1 0 0	0.0946	93, 107, 109, 114, 160
48:111	1 2 3 5 11 14 16 20 23 24 25 26	C_1	48×1	138, 138, 138	1 3 7 1 0 0	0.1698	79, 115, 117, 123, 161
48:112	1 2 3 5 11 14 17 20 21 23 24 25	C_1	48×1	138, 138, 138	0 4 6 2 0 0	0.0946	100, 113, 116
48:113	1 2 3 5 11 14 17 20 22 23 24 26	C_1	48×1	138, 138, 138	0 5 6 1 0 0	0.1201	77, 100, 112, 113
48:114	1 2 3 5 11 15 16 18 21 22 25 26	C_1	48×1	138, 138, 138	0 4 6 2 0 0	0.0830	99, 110, 116, 121
48:115	1 2 3 5 11 15 16 20 22 24 25 26	C_2	24×2	138, 138, 138	0 4 6 2 0 0	0.0697	111(2), 115, 117(2)
48:116	1 2 3 5 11 15 17 20 21 22 24 25	C_1	48×1	138, 138, 138	1 3 7 1 0 0	0.0759	108, 112, 114, 159
48:117	1 2 3 5 11 15 17 20 22 23 25 26	C_1	48×1	138, 138, 138	1 3 7 1 0 0	0.1171	111, 115, 119, 158
48:118	1 2 3 5 11 15 18 20 21 23 24 25	C_1	48×1	138, 138, 138	0 5 6 1 0 0	0.0903	99, 104, 119, 120, 121, 162
48:119	1 2 3 5 11 15 18 20 22 23 24 26	C_1	48×1	138, 138, 138	0 4 6 2 0 0	0.0879	105, 117, 118, 123, 131
48:120	1 2 3 5 11 15 19 21 23 24 25 26	C_1	48×1	138, 138, 138	0 3 8 1 0 0	0.0679	35, 118, 172

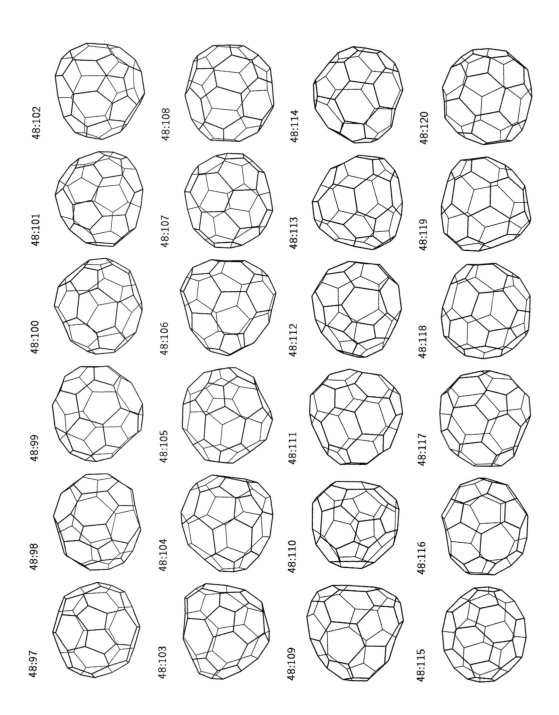

48:102

48:108

48:114

48:120

48:101

48:107

48:113

48:119

48:100

48:106

48:112

48:118

48:99

48:105

48:111

48:117

48:98

48:104

48:110

48:116

48:97

48:103

48:109

48:115

Table A.8. (*Continued*)

Isomer	Ring spiral	Point group	NMR pattern	Vibrations	Pentagon indices	Band gap	Transformations
48:121	1 2 3 5 11 16 18 20 21 22 23 24	C_1	48×1	138, 138, 138	1 3 7 1 0 0	0.0801	106, 114, 118, 122, 123, 160
48:122	1 2 3 5 11 16 19 21 22 23 24 26	C_2	24×2	138, 138, 138	0 2 8 2 0 0	0.1088	121(2)
48:123	1 2 3 5 12 15 18 19 22 23 24 26	C_1	48×1	138, 138, 138	0 2 8 2 0 0	0.0833	111, 119, 121, 143
48:124	1 2 3 5 14 15 17 18 19 20 22 26	C_1	48×1	138, 138, 138	0 5 6 1 0 0	0.0515	125, 128, 151, 166
48:125	1 2 3 5 14 15 17 18 19 21 22 25	C_2	24×2	138, 138, 138	0 4 6 2 0 0	0.2584	124(2)
48:126	1 2 3 5 14 15 17 18 20 21 24 26	C_1	48×1	138, 138, 138	0 3 8 1 0 0	0.2250	92, 127, 128, 146, 148
48:127	1 2 3 5 14 15 17 19 20 21 24 25	C_1	48×1	138, 138, 138	0 7 4 1 0 0	0.1861	87, 91, 126, 127, 128, 129, 149, 165
48:128	1 2 3 5 14 15 17 19 20 22 24 26	C_1	48×1	138, 138, 138	0 5 6 1 0 0	0.1718	88, 105, 124, 126, 127, 140, 150
48:129	1 2 3 5 14 16 18 19 20 21 23 24	C_1	48×1	138, 138, 138	0 5 6 1 0 0	0.0227	104, 127, 156, 167
48:130	1 2 3 10 11 12 14 15 22 23 24 25	C_1	48×1	138, 138, 138	0 4 8 0 0 0	0.1250	143, 153, 159, 173, 191
48:131	1 2 3 10 11 12 14 16 22 23 25 26	C_1	48×1	138, 138, 138	1 4 7 0 0 0	0.1194	119, 134, 143, 152, 158, 162, 191, 193
48:132	1 2 3 10 11 12 15 16 20 22 24 26	C_1	48×1	138, 138, 138	0 4 8 0 0 0	0.0521	52, 133, 164, 172, 179
48:133	1 2 3 10 11 12 15 16 21 22 24 25	C_1	48×1	138, 138, 138	0 6 6 0 0 0	0.0652	92, 132, 134, 135, 143, 146, 162, 163, 193
48:134	1 2 3 10 11 12 15 16 21 23 25 26	C_2	24×2	138, 138, 138	0 4 8 0 0 0	0.1787	131(2), 133(2), 136(2), 144, 198
48:135	1 2 3 10 11 13 15 16 20 22 24 25	C_1	48×1	138, 138, 138	1 6 5 0 0 0	0.0425	88, 133, 136, 137, 138, 142, 145, 150, 165, 171, 197, 198
48:136	1 2 3 10 11 13 15 16 20 23 25 26	C_s	6×1, 21×2	138, 138, 138	1 2 9 0 0 0	0.1618	89, 134(2), 135(2), 146(2), 152(2)
48:137	1 2 3 10 11 13 15 16 21 24 25 26	C_2	24×2	138, 138, 138	0 6 6 0 0 0	0.1181	87(2), 135(2), 139(2), 140(2), 149(2)
48:138	1 2 3 10 11 13 15 17 20 22 23 25	C_{2v}	6×2, 9×4	106, 138, 106	2 4 6 0 0 0	0.1211	92(4), 135(4), 139(4), 148(2)
48:139	1 2 3 10 11 13 15 17 20 22 24 26	C_1	48×1	138, 138, 138	1 6 5 0 0 0	0.2070	87, 91, 93, 103, 137, 138, 149, 165, 168, 169, 171
48:140	1 2 3 10 11 13 16 21 22 24 25 26	C_1	48×1	138, 138, 138	0 6 6 0 0 0	0.2422	104, 128, 137, 156, 165, 166, 168, 196
48:141	1 2 3 10 11 13 17 21 22 23 24 25	C_s	6×1, 21×2	138, 138, 138	0 4 8 0 0 0	0.2089	106(2), 170(2), 185(2)
48:142	1 2 3 10 11 14 15 16 20 21 24 25	C_s	8×1, 20×2	138, 138, 138	0 6 6 0 0 0	0.0172	135(2), 154(2), 155(2), 167(2)
48:143	1 2 3 10 11 14 15 19 21 23 24 25	C_1	48×1	138, 138, 138	1 2 9 0 0 0	0.0453	109, 123, 130, 131, 133, 144, 160, 161
48:144	1 2 3 10 11 14 15 19 22 24 25 26	D_2	12×4	102, 138, 102	0 0 12 0 0 0	0.1990	134(2), 143(4)

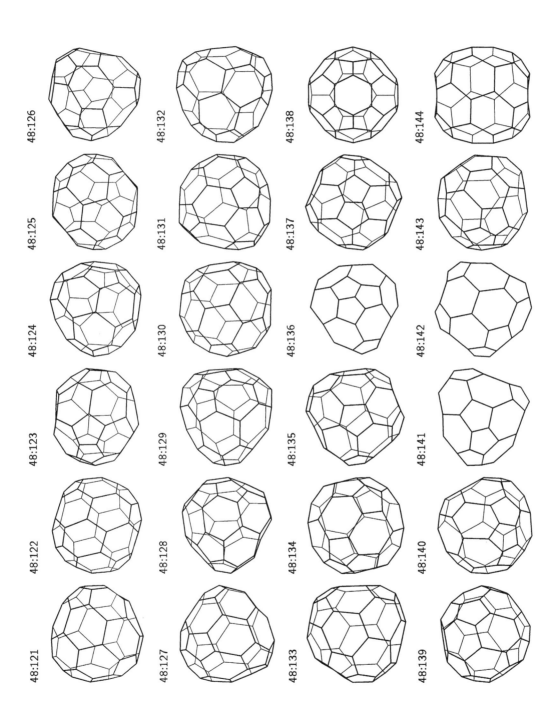

48:126

48:132

48:138

48:144

48:125

48:131

48:137

48:143

48:124

48:130

48:136

48:142

48:123

48:129

48:135

48:141

48:122

48:128

48:134

48:140

48:121

48:127

48:133

48:139

Table A.8. (Continued)

Isomer	Ring spiral	Point group	NMR pattern	Vibrations	Pentagon indices	Band gap	Transformations
48:145	1 2 3 10 12 13 15 16 19 22 24 25	C_1	48×1	138, 138, 138	0 6 6 0 0 0	0.0942	135, 146, 148, 155, 193, 195, 196
48:146	1 2 3 10 12 13 15 16 19 23 25 26	C_1	48×1	138, 138, 138	0 4 8 0 0 0	0.1211	126, 133, 136, 145, 150, 165
48:147	1 2 3 10 12 13 15 17 18 22 25 26	C_1	48×1	138, 138, 138	0 6 6 0 0 0	0.0833	147, 149, 154(2), 155, 195
48:148	1 2 3 10 12 13 15 17 19 22 23 25	C_s	6×1, 21×2	138, 138, 138	1 4 7 0 0 0	0.1131	126(2), 138, 145(2), 149(2), 150(2)
48:149	1 2 3 10 12 13 15 17 19 22 24 26	C_1	48×1	138, 138, 138	0 8 4 0 0 0	0.1427	127, 137, 139, 147, 148, 149, 150, 156, 196, 197, 199
48:150	1 2 3 10 12 13 15 17 20 22 24 25	C_1	48×1	138, 138, 138	1 6 5 0 0 0	0.0866	128, 135, 146, 148, 149, 151, 152, 155, 193, 196, 199
48:151	1 2 3 10 12 13 15 17 20 23 24 26	C_1	48×1	138, 138, 138	1 4 7 0 0 0	0.0422	124, 150, 154, 167, 179, 195
48:152	1 2 3 10 12 13 15 17 21 24 25 26	C_2	24×2	138, 138, 138	2 2 8 0 0 0	0.0473	105(2), 131(2), 136(2), 150(2), 171
48:153	1 2 3 10 12 13 15 18 21 23 24 25	C_2	24×2	138, 138, 138	0 2 10 0 0 0	0.0775	130(2), 185
48:154	1 2 3 10 12 14 15 16 19 20 24 26	C_1	48×1	138, 138, 138	0 6 6 0 0 0	0.1035	142, 147(2), 151, 155, 156, 197
48:155	1 2 3 10 12 14 15 16 19 21 24 25	C_1	48×1	138, 138, 138	0 6 6 0 0 0	0.0186	142, 145, 147, 150, 154, 155, 196
48:156	1 2 3 10 12 14 16 19 20 22 24 25	C_1	48×1	138, 138, 138	0 6 6 0 0 0	0.0121	129, 140, 149, 154, 167, 195, 196
48:157	1 2 4 7 9 10 19 20 21 22 24 25	C_2	24×2	138, 138, 138	0 4 8 0 0 0	0.2162	33(2), 184(2)
48:158	1 2 4 7 9 12 14 18 22 24 25 26	C_2	24×2	138, 138, 138	2 4 6 0 0 0	0.2105	117(2), 131(2), 161, 174(2), 194
48:159	1 2 4 7 9 12 14 19 22 23 24 25	C_1	48×1	138, 138, 138	1 4 7 0 0 0	0.1463	116, 130, 161, 174, 177
48:160	1 2 4 7 9 12 15 18 20 23 25 26	C_2	24×2	138, 138, 138	2 4 6 0 0 0	0.2810	110(2), 121(2), 143(2), 162(2), 173(2), 175
48:161	1 2 4 7 9 12 15 18 21 24 25 26	C_2	24×2	138, 138, 138	2 4 6 0 0 0	0.3750	108(2), 111(2), 143(2), 158, 159(2), 163
48:162	1 2 4 7 9 12 16 18 20 22 25 26	C_1	48×1	138, 138, 138	1 6 5 0 0 0	0.2100	93, 118, 131, 133, 160, 171, 172, 176, 180, 192
48:163	1 2 4 7 9 12 16 19 21 22 24 25	C_2	24×2	138, 138, 138	0 8 4 0 0 0	0.3843	91(2), 133(2), 161, 164(2), 165(2), 194
48:164	1 2 4 7 9 12 16 19 21 23 25 26	C_1	48×1	138, 138, 138	0 6 6 0 0 0	0.0727	51, 90, 132, 163, 178, 190
48:165	1 2 4 7 9 12 17 19 21 22 23 25	C_1	48×1	138, 138, 138	0 8 4 0 0 0	0.2492	127, 135, 139, 140, 146, 163, 167, 195, 199
48:166	1 2 4 7 9 12 17 20 21 22 24 25	C_2	24×2	138, 138, 138	0 6 6 0 0 0	0.3193	124(2), 140(2), 167(2)
48:167	1 2 4 7 9 12 18 19 21 22 23 24	C_1	48×1	138, 138, 138	1 6 5 0 0 0	0.0465	129, 142, 151, 156, 165, 166, 171, 176, 196
48:168	1 2 4 7 9 13 17 19 20 22 24 26	C_s	4×1, 22×2	138, 138, 138	1 6 5 0 0 0	0.2609	88(2), 102(2), 139(2), 140(2), 197

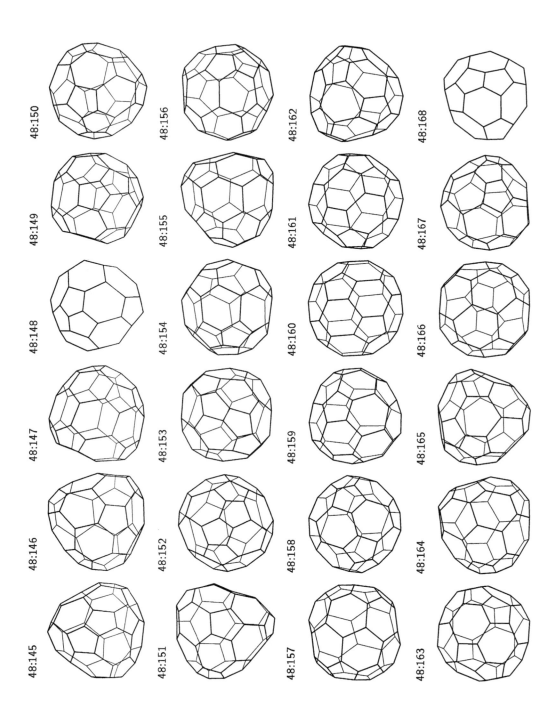

48:150

48:149

48:148

48:147

48:146

48:145

48:156

48:155

48:154

48:153

48:152

48:151

48:162

48:161

48:160

48:159

48:158

48:157

48:168

48:167

48:166

48:165

48:164

48:163

Table A.8. (*Continued*)

Isomer	Ring spiral	Point group	NMR pattern	Vibrations	Pentagon indices	Band gap	Transformations
48:169	1 2 4 7 9 13 17 19 21 23 25 26	D_2	12×4	102, 138, 102	0 8 4 0 0 0	0.1708	85(4), 139(4)
48:170	1 2 4 7 9 13 17 20 21 24 25 26	C_2	24×2	138, 138, 138	0 4 8 0 0 0	0.2246	86(2), 141(2)
48:171	1 2 4 7 9 13 18 19 20 22 23 24	C_2	24×2	138, 138, 138	2 6 4 0 0 0	0.1465	104(2), 135(2), 139(2), 152, 162(2), 167(2), 196(2)
48:172	1 2 4 7 9 14 18 19 20 22 24 26	C_1	48×1	138, 138, 138	1 4 7 0 0 0	0.0262	53, 120, 132, 162, 172, 176, 181
48:173	1 2 4 7 9 16 18 19 20 21 22 23	C_1	48×1	138, 138, 138	0 6 6 0 0 0	0.2730	106, 130, 160, 180, 182, 183, 185
48:174	1 2 4 7 10 12 14 17 22 24 25 26	C_1	48×1	138, 138, 138	1 4 7 0 0 0	0.0253	61, 158, 159, 190, 191
48:175	1 2 4 7 10 12 15 17 20 23 25 26	C_2	24×2	138, 138, 138	0 4 8 0 0 0	0.2288	160, 176(2)
48:176	1 2 4 7 10 12 16 17 20 22 25 26	C_1	48×1	138, 138, 138	0 6 6 0 0 0	0.1527	162, 167, 172, 175, 179, 192
48:177	1 2 4 7 10 12 16 19 21 23 24 26	C_2	24×2	138, 138, 138	0 6 6 0 0 0	0.0056	159(2), 178(2), 190
48:178	1 2 4 7 10 12 16 20 21 23 24 25	C_1	48×1	138, 138, 138	1 4 7 0 0 0	0.2028	164, 177, 179, 194
48:179	1 2 4 7 10 12 16 20 22 23 25 26	C_1	48×1	138, 138, 138	0 6 6 0 0 0	0.0491	132, 151, 176, 178, 193
48:180	1 2 4 7 10 14 17 19 20 21 23 24	C_2	24×2	138, 138, 138	0 6 6 0 0 0	0.2598	162(2), 173(2), 181(2), 191
48:181	1 2 4 7 10 14 17 19 20 22 24 26	C_1	48×1	138, 138, 138	0 4 8 0 0 0	0.0493	63, 172, 180, 182
48:182	1 2 4 7 10 15 18 20 21 22 24 25	C_1	48×1	138, 138, 138	0 4 8 0 0 0	0.1263	60, 173, 181, 184
48:183	1 2 4 7 10 16 17 19 20 21 22 23	C_2	24×2	138, 138, 138	0 6 6 0 0 0	0.1712	173(2), 184(2), 187(2)
48:184	1 2 4 7 10 16 18 20 21 22 23 25	C_s	6×1, 21×2	138, 138, 138	1 4 7 0 0 0	0.0893	62, 157(2), 182(2), 183(2)
48:185	1 2 4 7 11 15 17 18 20 21 22 24	C_2	24×2	138, 138, 138	0 6 6 0 0 0	0.2427	141(2), 153, 173(2), 187(2)
48:186	1 2 4 7 11 15 19 20 23 24 25 26	D_{6d}	2×12, 1×24	17, 30, 0	0 0 12 0 0 0	0.2135	
48:187	1 2 4 7 11 16 17 18 20 21 22 23	C_s	6×1, 21×2	138, 138, 138	0 6 6 0 0 0	0.0827	183(2), 185(2), 188(2)
48:188	1 2 4 7 12 16 17 18 19 21 22 23	D_3	8×6	68, 70, 46	0 6 6 0 0 0	0.1495	187(6)
48:189	1 2 4 8 9 10 19 20 21 22 23 25	D_{6d}	4×12	18, 31, 0	0 0 12 0 0 0	0.0000	
48:190	1 2 4 8 9 12 13 21 23 24 25 26	C_2	24×2	138, 138, 138	0 6 6 0 0 0	0.1482	50(2), 164(2), 174(2), 177
48:191	1 2 4 8 11 12 15 17 21 23 25 26	C_2	24×2	138, 138, 138	0 6 6 0 0 0	0.0988	130(2), 131(2), 174(2), 180
48:192	1 2 4 8 12 13 15 17 19 22 24 25	C_2	24×2	138, 138, 138	0 8 4 0 0 0	0.1698	162(2), 176(2), 192, 193(2), 196(2)

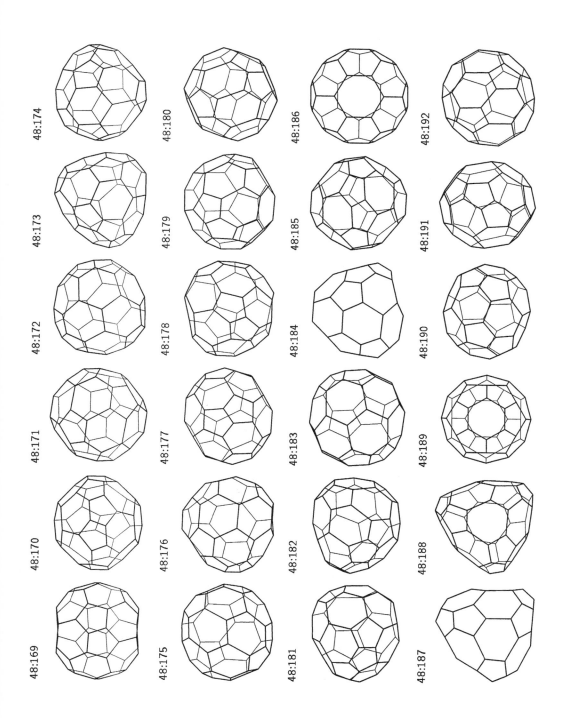

48:174

48:180

48:186

48:192

48:173

48:179

48:185

48:191

48:172

48:178

48:184

48:190

48:171

48:177

48:183

48:189

48:170

48:176

48:182

48:188

48:169

48:175

48:181

48:187

Table A.8. (*Continued*)

Isomer	Ring spiral	Point group	NMR pattern	Vibrations	Pentagon indices	Band gap	Transformations
48:193	1 2 4 8 12 13 15 17 19 23 25 26	C_1	48×1	138, 138, 138	0 8 4 0 0 0	0.1761	131, 133, 145, 150, 179, 192, 194, 197, 198
48:194	1 2 4 8 12 13 15 18 19 23 24 26	C_2	24×2	138, 138, 138	0 8 4 0 0 0	0.2283	158, 163, 178(2), 193(2), 195(2)
48:195	1 2 4 8 12 14 17 18 20 22 24 26	C_1	48×1	138, 138, 138	0 8 4 0 0 0	0.0713	145, 147, 151, 156, 165, 194, 197, 199
48:196	1 2 4 9 12 13 15 16 19 22 24 25	C_1	48×1	138, 138, 138	1 8 3 0 0 0	0.0986	140, 145, 149, 150, 155, 156, 167, 171, 192, 196, 197, 199(2)
48:197	1 2 4 9 12 13 15 17 19 22 24 26	C_s	6×1, 21×2	138, 138, 138	1 8 3 0 0 0	0.2468	135(2), 149(2), 154(2), 168, 193(2), 195(2), 196(2)
48:198	1 2 4 9 12 13 16 17 20 22 23 25	D_2	12×4	102, 138, 102	0 8 4 0 0 0	0.2453	134(2), 135(4), 193(4)
48:199	1 2 4 9 12 14 15 17 19 21 23 25	C_2	24×2	138, 138, 138	0 10 2 0 0 0	0.0833	149(2), 150(2), 165(2), 195(2), 196(4), 199

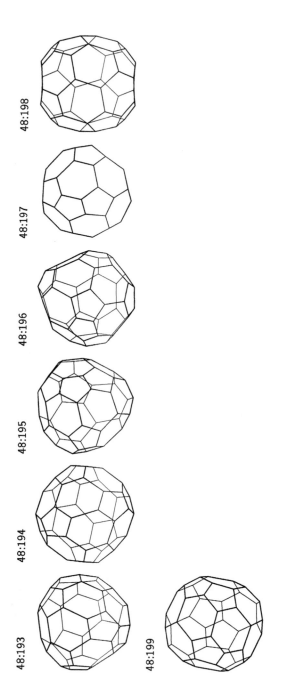

48:198

48:197

48:196

48:195

48:194

48:193

48:199

Table A.9. Fullerene isomers of C_{50}

Isomer	Ring spiral	Point group	NMR pattern	Vibrations	Pentagon indices	Band gap	Transformations
50:1	1 2 3 4 5 6 22 23 24 25 26 27	D_{5h}	5×10	23, 39, 0	0 0 0 10 0 2	0.0000	
50:2	1 2 3 4 5 7 21 23 24 25 26 27	C_2	25×2	144, 144, 144	0 2 2 6 2 0	0.0326	
50:3	1 2 3 4 5 9 19 23 24 25 26 27	D_{3h}	1×2, 2×6, 3×12	37, 62, 25	0 0 6 0 6 0	0.0000	
50:4	1 2 3 4 5 12 19 21 23 25 26 27	C_s	8×1, 21×2	144, 144, 144	1 3 2 5 1 0	0.1230	5(2), 8
50:5	1 2 3 4 5 12 20 22 24 25 26 27	C_s	8×1, 21×2	144, 144, 144	1 3 2 5 1 0	0.0777	4(2)
50:6	1 2 3 4 5 13 15 23 24 25 26 27	C_2	25×2	144, 144, 144	2 0 4 4 2 0	0.0690	17(2)
50:7	1 2 3 4 5 13 17 21 24 25 26 27	C_1	50×1	144, 144, 144	2 1 3 5 1 0	0.1137	8, 13, 14, 24
50:8	1 2 3 4 5 13 18 21 23 25 26 27	C_s	6×1, 22×2	144, 144, 144	2 2 3 4 1 0	0.1728	4, 7(2), 9(2)
50:9	1 2 3 4 5 13 19 21 23 24 26 27	C_1	50×1	144, 144, 144	0 3 5 3 1 0	0.1538	8, 11, 14
50:10	1 2 3 4 5 13 19 22 23 25 26 27	C_1	50×1	144, 144, 144	0 1 7 3 1 0	0.0639	
50:11	1 2 3 4 5 13 20 21 23 24 25 27	C_1	50×1	144, 144, 144	0 2 5 4 1 0	0.1243	9, 12
50:12	1 2 3 4 5 13 20 22 23 24 25 26	C_1	50×1	144, 144, 144	0 2 7 2 1 0	0.0861	11
50:13	1 2 3 4 5 14 17 20 24 25 26 27	C_{2v}	5×2, 10×4	110, 144, 110	2 0 4 4 2 0	0.1320	7(4)
50:14	1 2 3 4 5 14 19 20 23 24 26 27	C_1	50×1	144, 144, 144	1 1 6 3 1 0	0.0773	7, 9, 26
50:15	1 2 3 4 5 14 19 22 24 25 26 27	C_1	50×1	144, 144, 144	1 2 4 4 1 0	0.0229	17, 25
50:16	1 2 3 4 5 15 19 20 23 25 26 27	C_1	50×1	144, 144, 144	1 1 4 5 1 0	0.1231	16, 28
50:17	1 2 3 4 5 15 19 21 24 25 26 27	C_1	50×1	144, 144, 144	1 2 4 4 1 0	0.1317	6, 15, 80
50:18	1 2 3 4 7 11 17 21 24 25 26 27	C_2	25×2	144, 144, 144	0 4 2 6 0 0	0.0378	19(2)
50:19	1 2 3 4 7 11 18 21 23 25 26 27	C_1	50×1	144, 144, 144	0 5 2 5 0 0	0.2466	18, 20, 74
50:20	1 2 3 4 7 11 19 21 23 24 26 27	C_1	50×1	144, 144, 144	2 2 4 4 0 0	0.1750	19, 22, 25, 31, 75
50:21	1 2 3 4 7 11 19 22 23 25 26 27	C_1	50×1	144, 144, 144	1 2 5 4 0 0	0.1440	26, 76
50:22	1 2 3 4 7 11 20 21 23 24 25 27	C_1	50×1	144, 144, 144	1 3 3 5 0 0	0.1224	20, 23, 30, 36, 41
50:23	1 2 3 4 7 11 20 22 23 24 25 26	C_1	50×1	144, 144, 144	1 3 5 3 0 0	0.1068	22, 29, **40**, 77
50:24	1 2 3 4 7 12 17 22 24 25 26 27	C_2	25×2	144, 144, 144	2 2 2 6 0 0	0.0069	7(2), **26**(2)

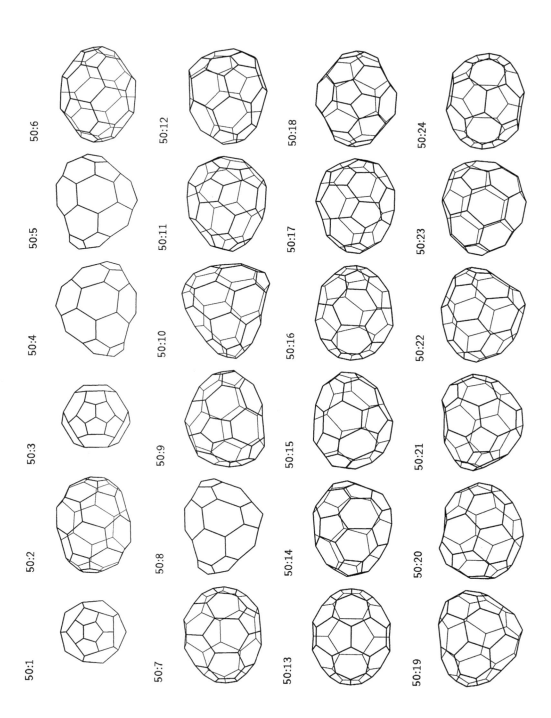

50:6

50:5

50:4

50:3

50:2

50:1

50:12

50:11

50:10

50:9

50:8

50:7

50:18

50:17

50:16

50:15

50:14

50:13

50:24

50:23

50:22

50:21

50:20

50:19

Table A.9. (*Continued*)

Isomer	Ring spiral	Point group	NMR pattern	Vibrations	Pentagon indices	Band gap	Transformations
50:25	1 2 3 4 7 12 18 21 23 24 26 27	C_1	50×1	144, 144, 144	1 3 3 5 0 0	0.1331	15, 20, 79
50:26	1 2 3 4 7 12 18 22 23 25 26 27	C_1	50×1	144, 144, 144	1 2 5 4 0 0	0.2377	14, 21, 24, 93
50:27	1 2 3 4 7 13 17 23 24 25 26 27	C_2	25×2	144, 144, 144	0 2 4 6 0 0	0.1565	
50:28	1 2 3 4 7 13 18 22 23 24 25 26	C_1	50×1	144, 144, 144	0 3 4 5 0 0	0.0817	16
50:29	1 2 3 4 7 17 18 19 21 23 26 27	C_1	50×1	144, 144, 144	0 5 4 3 0 0	0.0317	23, 29, 30, 55
50:30	1 2 3 4 7 17 18 20 21 23 25 27	C_1	50×1	144, 144, 144	0 5 4 3 0 0	0.1337	22, 29, 31, 54
50:31	1 2 3 4 7 17 18 20 22 23 25 26	C_1	50×1	144, 144, 144	0 5 4 3 0 0	0.0670	20, 30, 32
50:32	1 2 3 4 7 17 18 20 22 24 26 27	C_s	8×1, 21×2	144, 144, 144	0 3 6 3 0 0	0.1446	31(2)
50:33	1 2 3 4 7 18 19 20 22 23 24 25	C_s	6×1, 22×2	144, 144, 144	0 3 6 3 0 0	0.1082	
50:34	1 2 3 4 11 12 14 21 23 24 26 27	C_1	50×1	144, 144, 144	0 3 6 3 0 0	0.0420	94, 134
50:35	1 2 3 4 11 12 14 22 23 25 26 27	C_1	50×1	144, 144, 144	1 3 5 3 0 0	0.0911	36, 67, 72, 95, 96
50:36	1 2 3 4 11 12 15 21 23 25 26 27	C_1	50×1	144, 144, 144	1 4 3 4 0 0	0.0838	22, 35, 53, 75, 77, 87
50:37	1 2 3 4 11 12 20 21 22 23 24 25	C_1	50×1	144, 144, 144	0 4 6 2 0 0	0.1482	44, 67, 96, 210
50:38	1 2 3 4 11 13 14 21 23 24 25 27	C_1	50×1	144, 144, 144	1 2 5 4 0 0	0.0705	39, 59, 66, 73
50:39	1 2 3 4 11 13 14 22 23 24 25 26	C_1	50×1	144, 144, 144	1 4 5 2 0 0	0.1148	38, 40, 44, 58, 97, 99
50:40	1 2 3 4 11 13 15 21 23 24 25 26	C_1	50×1	144, 144, 144	3 2 5 2 0 0	0.0767	23, 39, 41, 42, 47, 55, 97, 98, 118
50:41	1 2 3 4 11 13 15 22 24 25 26 27	C_1	50×1	144, 144, 144	2 2 4 4 0 0	0.1415	22, 40, 53, 54, 68, 73
50:42	1 2 3 4 11 13 16 21 22 24 25 26	C_s	6×1, 22×2	144, 144, 144	3 2 5 2 0 0	0.1170	40(2), 43, 52(2), 98(2), 242
50:43	1 2 3 4 11 13 16 21 23 25 26 27	C_{2v}	7×2, 9×4	111, 144, 111	2 2 4 4 0 0	0.1044	42(2), 53(4)
50:44	1 2 3 4 11 13 19 21 22 23 24 25	C_1	50×1	144, 144, 144	1 4 5 2 0 0	0.0642	37, 39, 44, 45, 62, 99, 100
50:45	1 2 3 4 11 13 20 21 23 24 25 27	C_1	50×1	144, 144, 144	1 2 5 4 0 0	0.0815	44, 66, 67, 72
50:46	1 2 3 4 11 14 15 19 23 25 26 27	C_1	50×1	144, 144, 144	0 4 6 2 0 0	0.1062	50, 63, 101, 120
50:47	1 2 3 4 11 14 15 20 23 24 25 26	C_1	50×1	144, 144, 144	1 4 5 2 0 0	0.1521	40, 52, 57, 65, 102, 119
50:48	1 2 3 4 11 14 16 18 21 25 26 27	C_1	50×1	144, 144, 144	1 3 5 3 0 0	0.1230	49, 103, 126

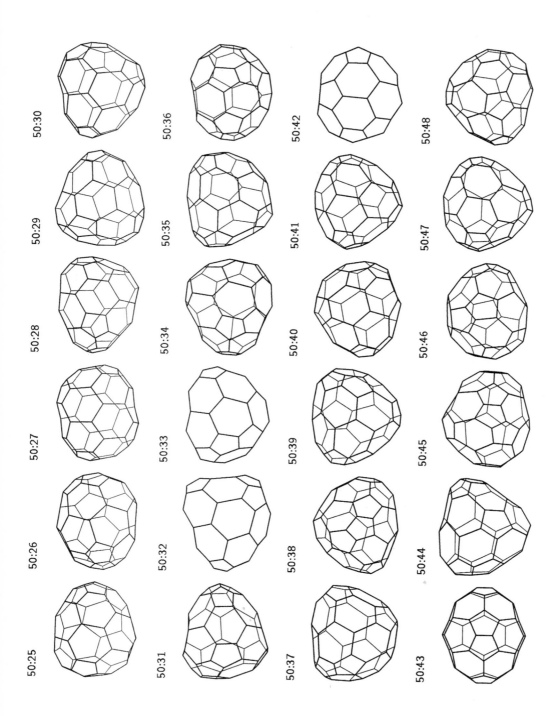

50:30

50:36

50:42

50:48

50:29

50:35

50:41

50:47

50:28

50:34

50:40

50:46

50:27

50:33

50:39

50:45

50:26

50:32

50:38

50:44

50:25

50:31

50:37

50:43

Table A.9. *(Continued)*

Isomer	Ring spiral	Point group	NMR pattern	Vibrations	Pentagon indices	Band gap	Transformations
50:49	1 2 3 4 11 14 16 19 21 24 26 27	C_1	50×1	144, 144, 144	1 4 5 2 0 0	0.0719	48, 51, 54, 104, 227
50:50	1 2 3 4 11 14 16 19 22 25 26 27	C_1	50×1	144, 144, 144	1 2 7 2 0 0	0.0578	46, 105, 235
50:51	1 2 3 4 11 14 16 20 21 24 25 27	C_1	50×1	144, 144, 144	2 2 6 2 0 0	0.0421	49, 52, 55, 68, 70, 106, 165
50:52	1 2 3 4 11 14 16 20 22 24 25 26	C_1	50×1	144, 144, 144	2 4 4 2 0 0	0.1047	42, 47, 51, 53, 55, 68, 69, 107, 211
50:53	1 2 3 4 11 14 16 20 23 25 26 27	C_1	50×1	144, 144, 144	2 3 4 3 0 0	0.0791	36, 41, 43, 52, 66, 72, 98, 108
50:54	1 2 3 4 11 14 17 19 21 23 26 27	C_1	50×1	144, 144, 144	1 4 5 2 0 0	0.0815	30, 41, 49, 55, 109, 112
50:55	1 2 3 4 11 14 17 20 21 23 25 27	C_1	50×1	144, 144, 144	2 4 4 2 0 0	0.0722	29, 40, 51, 52, 54, 55, 57, 58, 110, 111
50:56	1 2 3 4 11 14 18 19 22 23 26 27	C_1	50×1	144, 144, 144	0 2 6 4 0 0	0.0420	59, 89
50:57	1 2 3 4 11 14 18 20 21 23 24 27	C_1	50×1	144, 144, 144	0 4 6 2 0 0	0.0067	47, 55, 58, 69, 113
50:58	1 2 3 4 11 14 18 20 22 23 24 26	C_1	50×1	144, 144, 144	0 6 4 2 0 0	0.0605	39, 55, 57, 59, 62, 70, 114
50:59	1 2 3 4 11 14 18 20 22 23 25 27	C_1	50×1	144, 144, 144	1 3 5 3 0 0	0.0334	38, 56, 58, 60, 71, 115
50:60	1 2 3 4 11 14 18 21 22 23 25 26	C_1	50×1	144, 144, 144	0 3 6 3 0 0	0.0747	59, 61, 116
50:61	1 2 3 4 11 14 18 21 22 24 25 27	C_2	25×2	144, 144, 144	0 2 6 4 0 0	0.0479	60(2)
50:62	1 2 3 4 11 14 19 20 22 23 24 25	C_1	50×1	144, 144, 144	0 4 6 2 0 0	0.1052	44, 58, 117
50:63	1 2 3 4 11 15 19 20 22 23 25 27	C_1	50×1	144, 144, 144	0 4 6 2 0 0	0.1513	46, 64, 121, 123
50:64	1 2 3 4 11 15 19 21 22 23 25 26	C_2	25×2	144, 144, 144	0 4 6 2 0 0	0.1213	63(2), 122(2)
50:65	1 2 3 4 11 16 18 20 21 22 23 27	C_1	50×1	144, 144, 144	1 3 5 3 0 0	0.1010	47, 68, 124, 129
50:66	1 2 3 4 11 16 20 21 23 24 26 27	C_1	50×1	144, 144, 144	1 3 5 3 0 0	0.0656	38, 45, 53, 73, 99, 115
50:67	1 2 3 4 12 13 19 21 23 24 25 27	C_2	25×2	144, 144, 144	0 4 4 4 0 0	0.0355	35(2), 37(2), 45(2)
50:68	1 2 3 4 12 14 16 19 21 24 25 27	C_1	50×1	144, 144, 144	2 3 4 3 0 0	0.1041	41, 51, 52, 65, 68, 109, 118
50:69	1 2 3 4 12 14 18 19 21 23 24 27	C_s	6×1, 22×2	144, 144, 144	1 4 5 2 0 0	0.0331	52(2), 57(2), 70(2), 72, 168
50:70	1 2 3 4 12 14 18 19 22 23 24 26	C_1	50×1	144, 144, 144	1 4 5 2 0 0	0.0639	51, 58, 69, 71, 230
50:71	1 2 3 4 12 14 18 19 22 23 25 27	C_1	50×1	144, 144, 144	1 1 7 3 0 0	0.1000	59, 70, 148
50:72	1 2 3 4 12 14 18 20 23 24 26 27	C_s	8×1, 21×2	144, 144, 144	2 3 4 3 0 0	0.0734	35(2), 45(2), 53(2), 69, 100

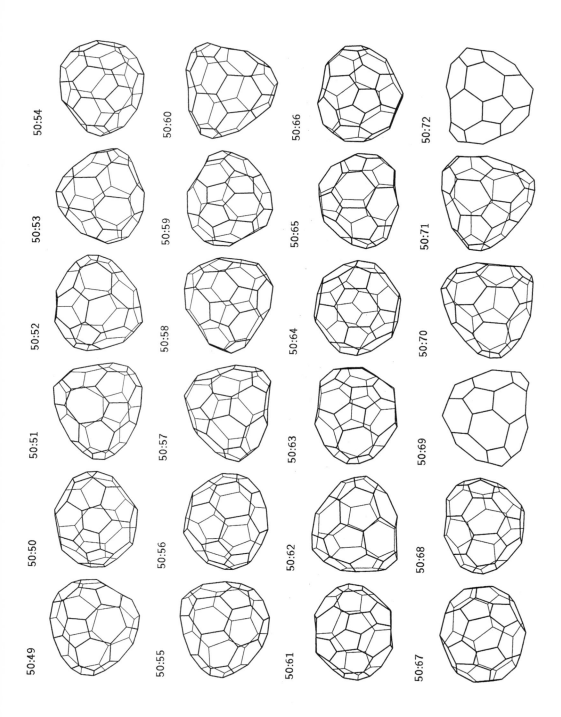

50:54

50:53

50:52

50:51

50:50

50:49

50:60

50:59

50:58

50:57

50:56

50:55

50:66

50:65

50:64

50:63

50:62

50:61

50:72

50:71

50:70

50:69

50:68

50:67

Table A.9. (*Continued*)

Isomer	Ring spiral	Point group	NMR pattern	Vibrations	Pentagon indices	Band gap	Transformations
50:73	1 2 3 4 12 16 19 20 22 23 25 27	C_1	50×1	144, 144, 144	1 3 5 3 0 0	0.1560	38, 41, 66, 73, 97, 112
50:74	1 2 3 5 7 10 18 21 23 25 26 27	C_2	25×2	144, 144, 144	2 4 2 4 0 0	0.0984	19(2), 75(2), 78, 87(2)
50:75	1 2 3 5 7 10 19 21 23 24 26 27	C_1	50×1	144, 144, 144	2 3 4 3 0 0	0.0336	20, 36, 74, 88, 95
50:76	1 2 3 5 7 10 19 22 23 25 26 27	C_1	50×1	144, 144, 144	1 3 5 3 0 0	0.0626	21, 85, 94
50:77	1 2 3 5 7 10 20 22 23 24 25 26	C_2	25×2	144, 144, 144	1 4 5 2 0 0	0.1321	23, 36, 83, 96, 98
50:78	1 2 3 5 7 11 17 21 23 25 26 27	C_2	25×2	144, 144, 144	0 4 4 4 0 0	0.0591	74
50:79	1 2 3 5 7 11 21 22 23 24 25 26	C_1	50×1	144, 144, 144	0 3 6 3 0 0	0.1244	25
50:80	1 2 3 5 7 12 17 22 23 25 26 27	C_2	25×2	144, 144, 144	0 4 4 4 0 0	0.0430	17(2)
50:81	1 2 3 5 7 16 17 18 21 25 26 27	C_1	50×1	144, 144, 144	0 5 4 3 0 0	0.1331	82, 83, 106
50:82	1 2 3 5 7 16 17 19 21 24 26 27	C_1	50×1	144, 144, 144	1 4 5 2 0 0	0.1652	81, 84, 104, 105
50:83	1 2 3 5 7 16 17 19 22 25 26 27	C_1	50×1	144, 144, 144	0 6 4 2 0 0	0.1751	77, 81, 87, 107
50:84	1 2 3 5 7 16 17 20 21 24 25 27	C_1	50×1	144, 144, 144	0 5 4 3 0 0	0.0568	82, 85, 103
50:85	1 2 3 5 7 16 17 20 22 24 25 26	C_1	50×1	144, 144, 144	1 4 5 2 0 0	0.2279	76, 84, 86
50:86	1 2 3 5 7 16 17 20 23 25 26 27	C_s	8×1, 21×2	144, 144, 144	0 3 6 3 0 0	0.1278	85(2)
50:87	1 2 3 5 7 16 18 19 22 24 26 27	C_1	50×1	144, 144, 144	1 5 3 3 0 0	0.0779	36, 74, 83, 88, 89, 108
50:88	1 2 3 5 7 16 18 20 22 24 25 27	C_s	6×1, 22×2	144, 144, 144	2 4 4 2 0 0	0.2767	75(2), 87(2), 92
50:89	1 2 3 5 7 17 18 19 22 23 26 27	C_1	50×1	144, 144, 144	0 5 4 3 0 0	0.0693	56, 87, 92, 115
50:90	1 2 3 5 7 17 18 20 21 23 24 27	C_1	50×1	144, 144, 144	0 5 4 3 0 0	0.0570	91, 124
50:91	1 2 3 5 7 17 18 20 22 23 24 26	C_s	6×1, 22×2	144, 144, 144	0 6 4 2 0 0	0.1724	90(2), 92
50:92	1 2 3 5 7 17 18 20 22 23 25 27	C_s	4×1, 23×2	144, 144, 144	1 2 7 2 0 0	0.0690	88, 89(2), 91
50:93	1 2 3 5 10 11 20 21 22 23 24 26	C_2	25×2	144, 144, 144	0 2 8 2 0 0	0.0266	26(2)
50:94	1 2 3 5 10 12 14 21 23 24 26 27	C_1	50×1	144, 144, 144	1 2 7 2 0 0	0.0970	34, 76, 130
50:95	1 2 3 5 10 12 14 22 23 25 26 27	C_2	25×2	144, 144, 144	2 2 6 2 0 0	0.0009	35(2), 75(2)
50:96	1 2 3 5 10 12 20 21 22 23 24 25	C_1	50×1	144, 144, 144	1 3 7 1 0 0	0.1159	35, 37, 77, 100, 133

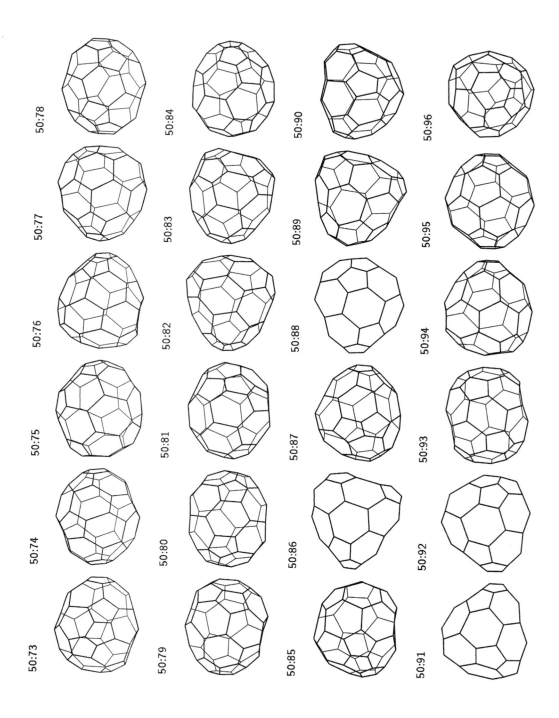

50:78

50:84

50:90

50:96

50:77

50:83

50:89

50:95

50:76

50:82

50:88

50:94

50:75

50:81

50:87

50:93

50:74

50:80

50:86

50:92

50:73

50:79

50:85

50:91

238

Table A.9. (*Continued*)

Isomer	Ring spiral	Point group	NMR pattern	Vibrations	Pentagon indices	Band gap	Transformations
50:97	1 2 3 5 10 13 14 22 23 24 25 26	C_1	50×1	144, 144, 144	1 5 5 1 0 0	0.0925	39, 40, 73, 97, 98, 99(2), 111
50:98	1 2 3 5 10 13 15 21 23 24 25 26	C_1	50×1	144, 144, 144	3 3 5 1 0 0	0.0689	40, 42, 53, 77, 97, 98, 99, 100, 102, 107, 211
50:99	1 2 3 5 10 13 18 21 22 23 24 26	C_1	50×1	144, 144, 144	1 5 5 1 0 0	0.0721	39, 44, 66, 97(2), 98, 100, 114
50:100	1 2 3 5 10 13 19 21 22 23 24 25	C_s	6×1, 22×2	144, 144, 144	3 3 5 1 0 0	0.0529	44(2), 72, 96(2), 98(2), 99(2), 117(2), 168
50:101	1 2 3 5 10 14 15 19 23 25 26 27	C_1	50×1	144, 144, 144	1 5 5 1 0 0	0.0583	46, 105, 121, 123, 126, 169, 224
50:102	1 2 3 5 10 14 15 20 23 24 25 26	C_1	50×1	144, 144, 144	1 5 5 1 0 0	0.0923	47, 98, 107, 117, 146, 223
50:103	1 2 3 5 10 14 16 18 21 25 26 27	C_1	50×1	144, 144, 144	1 4 5 2 0 0	0.0292	48, 84, 104, 124, 137
50:104	1 2 3 5 10 14 16 19 21 24 26 27	C_1	50×1	144, 144, 144	3 3 5 1 0 0	0.0918	49, 82, 103, 106, 109, 119, 127, 144, 170, 216
50:105	1 2 3 5 10 14 16 19 22 25 26 27	C_1	50×1	144, 144, 144	2 3 6 1 0 0	0.1347	50, 82, 101, 120, 128, 170, 215
50:106	1 2 3 5 10 14 16 20 21 24 25 27	C_1	50×1	144, 144, 144	2 3 6 1 0 0	0.0719	51, 81, 104, 107, 110, 113, 143, 163
50:107	1 2 3 5 10 14 16 20 22 24 25 26	C_1	50×1	144, 144, 144	2 5 4 1 0 0	0.1999	52, 83, 98, 102, 106, 108, 111, 114, 146, 214
50:108	1 2 3 5 10 14 16 20 23 25 26 27	C_2	25×2	144, 144, 144	2 4 4 2 0 0	0.1378	53(2), 87(2), 107(2), 115(2)
50:109	1 2 3 5 10 14 17 19 21 23 26 27	C_1	50×1	144, 144, 144	1 5 5 1 0 0	0.1035	54, 68, 104, 110, 118, 129, 165
50:110	1 2 3 5 10 14 17 20 21 23 25 27	C_1	50×1	144, 144, 144	0 7 4 1 0 0	0.2000	55, 106, 109, 110, 111, 113, 118
50:111	1 2 3 5 10 14 17 20 22 23 25 26	C_1	50×1	144, 144, 144	0 7 4 1 0 0	0.2207	55, 97, 107, 110, 111, 112, 114
50:112	1 2 3 5 10 14 17 20 22 24 26 27	C_1	50×1	144, 144, 144	0 5 6 1 0 0	0.0529	54, 73, 111, 112
50:113	1 2 3 5 10 14 18 20 21 23 24 27	C_1	50×1	144, 144, 144	0 5 6 1 0 0	0.0453	57, 106, 110, 114, 119
50:114	1 2 3 5 10 14 18 20 22 23 24 26	C_1	50×1	144, 144, 144	0 7 4 1 0 0	0.2057	58, 99, 107, 111, 113, 115, 117
50:115	1 2 3 5 10 14 18 20 22 23 25 27	C_1	50×1	144, 144, 144	1 4 5 2 0 0	0.1128	59, 66, 89, 108, 114, 116
50:116	1 2 3 5 10 14 18 21 22 23 25 26	C_2	25×2	144, 144, 144	0 4 6 2 0 0	0.1599	60(2), 115(2)
50:117	1 2 3 5 10 14 19 20 22 23 24 25	C_1	50×1	144, 144, 144	1 5 5 1 0 0	0.0756	62, 100, 102, 114, 133, 158, 171
50:118	1 2 3 5 10 15 17 20 21 22 25 27	C_1	50×1	144, 144, 144	1 5 5 1 0 0	0.0485	40, 68, 109, 110, 119, 211
50:119	1 2 3 5 10 15 18 20 21 22 24 27	C_1	50×1	144, 144, 144	1 5 5 1 0 0	0.0207	47, 104, 113, 118, 124, 223
50:120	1 2 3 5 10 15 18 20 22 23 26 27	C_1	50×1	144, 144, 144	0 5 6 1 0 0	0.0998	46, 105, 121, 224

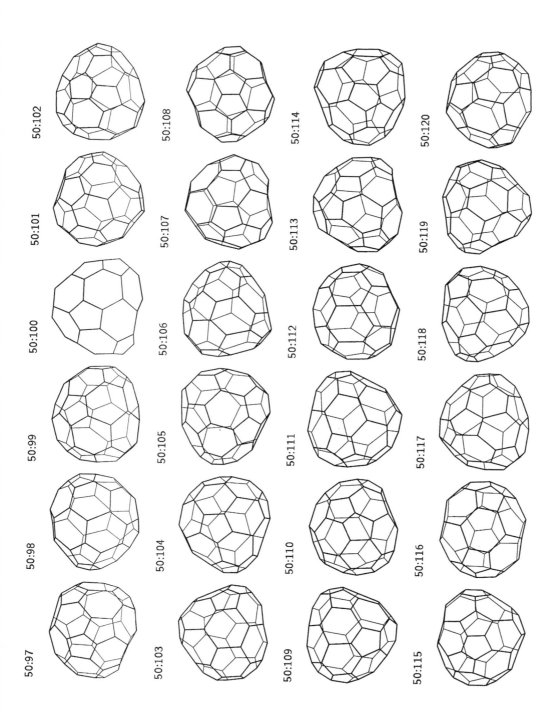

50:102

50:101

50:100

50:99

50:98

50:97

50:108

50:107

50:106

50:105

50:104

50:103

50:114

50:113

50:112

50:111

50:110

50:109

50:120

50:119

50:118

50:117

50:116

50:115

Table A.9. (*Continued*)

Isomer	Ring spiral	Point group	NMR pattern	Vibrations	Pentagon indices	Band gap	Transformations
50:121	1 2 3 5 10 15 19 20 22 23 25 27	C_1	50×1	144, 144, 144	1 5 5 1 0 0	0.0980	63, 101, 120, 122, 135, 153, 174, 225
50:122	1 2 3 5 10 15 19 21 22 23 25 26	C_1	50×1	144, 144, 144	1 5 5 1 0 0	0.2110	64, 121, 123(2), 136, 152, 172, 226
50:123	1 2 3 5 10 15 19 21 23 24 26 27	C_1	50×1	144, 144, 144	1 5 5 1 0 0	0.0642	63, 101, 122(2), 137, 173, 225
50:124	1 2 3 5 10 16 18 20 21 22 23 27	C_1	50×1	144, 144, 144	1 4 5 2 0 0	0.0654	65, 90, 103, 119, 132
50:125	1 2 3 5 11 12 21 22 23 24 25 26	C_{2v}	9×2, 8×4	112, 144, 112	0 0 10 2 0 0	0.0679	
50:126	1 2 3 5 11 14 15 18 23 25 26 27	C_1	50×1	144, 144, 144	0 5 6 1 0 0	0.0134	48, 101, 128, 137, 227
50:127	1 2 3 5 11 14 16 18 21 24 26 27	C_1	50×1	144, 144, 144	1 3 7 1 0 0	0.1078	104, 129, 213
50:128	1 2 3 5 11 14 16 18 22 25 26 27	C_1	50×1	144, 144, 144	1 3 7 1 0 0	0.0807	105, 126, 212
50:129	1 2 3 5 11 14 17 18 21 23 26 27	C_1	50×1	144, 144, 144	0 5 6 1 0 0	0.1701	65, 109, 127, 132
50:130	1 2 3 5 11 14 17 20 22 24 25 27	C_1	50×1	144, 144, 144	1 3 7 1 0 0	0.0848	94, 130, 131
50:131	1 2 3 5 11 14 17 21 22 24 25 26	C_1	50×1	144, 144, 144	0 3 8 1 0 0	0.0802	130, 132
50:132	1 2 3 5 11 14 17 21 23 24 26 27	C_1	50×1	144, 144, 144	0 5 6 1 0 0	0.1935	124, 129, 131
50:133	1 2 3 5 11 14 18 20 22 23 24 25	C_1	50×1	144, 144, 144	1 3 7 1 0 0	0.0387	96, 117, 134, 142
50:134	1 2 3 5 11 14 19 22 23 24 25 26	C_1	50×1	144, 144, 144	0 3 8 1 0 0	0.0460	34, 133
50:135	1 2 3 5 11 15 18 20 22 23 25 27	C_1	50×1	144, 144, 144	1 3 7 1 0 0	0.0515	121, 136, 138, 140, 229
50:136	1 2 3 5 11 15 18 21 22 23 25 26	C_1	50×1	144, 144, 144	0 3 8 1 0 0	0.1312	122, 135, 137, 166
50:137	1 2 3 5 11 15 18 21 23 24 26 27	C_1	50×1	144, 144, 144	0 5 6 1 0 0	0.0709	103, 123, 126, 136, 216
50:138	1 2 3 5 11 15 19 22 23 25 26 27	C_2	25×2	144, 144, 144	0 2 8 2 0 0	0.1010	135(2)
50:139	1 2 3 5 11 15 20 23 24 25 26 27	C_1	50×1	144, 144, 144	0 3 8 1 0 0	0.1423	141, 228
50:140	1 2 3 5 11 16 20 21 22 24 25 26	C_1	50×1	144, 144, 144	0 3 8 1 0 0	0.0538	135, 213
50:141	1 2 3 5 11 16 20 22 23 24 26 27	C_1	50×1	144, 144, 144	0 3 8 1 0 0	0.0677	139, 212
50:142	1 2 3 5 11 17 20 21 22 23 24 25	C_1	50×1	144, 144, 144	0 3 8 1 0 0	0.0397	133
50:143	1 2 3 5 14 15 17 18 19 21 26 27	C_1	50×1	144, 144, 144	0 7 4 1 0 0	0.1466	106, 144, 146, 151, 218
50:144	1 2 3 5 14 15 17 18 20 21 25 27	C_1	50×1	144, 144, 144	0 7 4 1 0 0	0.1215	104, 143, 145, 150, 183, 219

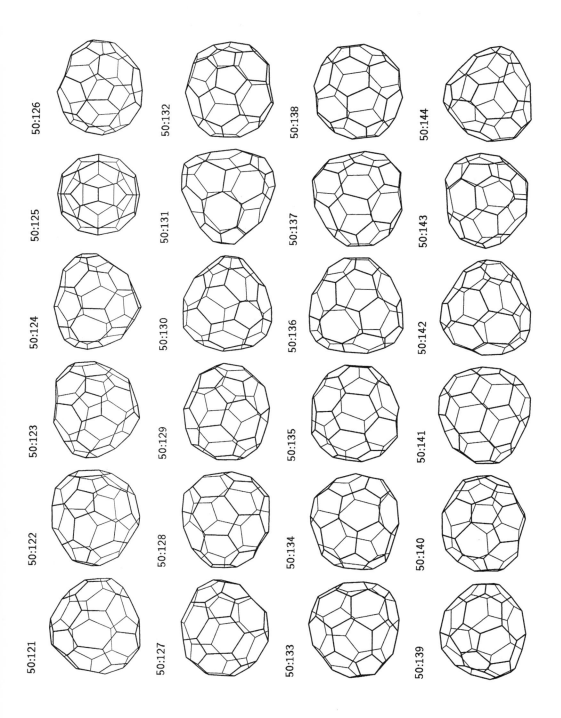

50:126

50:125

50:124

50:123

50:122

50:121

50:132

50:131

50:130

50:129

50:128

50:127

50:138

50:137

50:136

50:135

50:134

50:133

50:144

50:143

50:142

50:141

50:140

50:139

Table A.9. (*Continued*)

Isomer	Ring spiral	Point group	NMR pattern	Vibrations	Pentagon indices	Band gap	Transformations
50:145	1 2 3 5 14 15 17 18 20 22 25 26	C_1	50×1	144, 144, 144	0 5 6 1 0 0	0.1503	144, 184, 220
50:146	1 2 3 5 14 15 17 19 21 24 26 27	C_1	50×1	144, 144, 144	0 7 4 1 0 0	0.1015	102, 107, 143, 146, 217
50:147	1 2 3 5 14 15 18 19 20 22 24 25	C_1	50×1	144, 144, 144	0 5 6 1 0 0	0.1172	148, 149, 191, 231
50:148	1 2 3 5 14 15 18 19 20 23 25 27	C_1	50×1	144, 144, 144	0 3 8 1 0 0	0.0660	71, 147, 192, 230
50:149	1 2 3 5 14 16 18 19 20 22 23 25	C_1	50×1	144, 144, 144	1 3 7 1 0 0	0.0376	147, 150, 156, 203, 204
50:150	1 2 3 5 14 16 18 19 20 22 24 27	C_1	50×1	144, 144, 144	0 7 4 1 0 0	0.0031	144, 149, 151, 154, 159, 200, 221
50:151	1 2 3 5 14 16 18 19 21 22 24 26	C_1	50×1	144, 144, 144	0 5 6 1 0 0	0.0985	143, 150, 160, 184
50:152	1 2 3 5 14 16 18 20 21 24 25 27	C_1	50×1	144, 144, 144	1 3 7 1 0 0	0.0801	122, 153, 161, 189, 201
50:153	1 2 3 5 14 16 19 20 21 24 25 26	C_1	50×1	144, 144, 144	1 5 5 1 0 0	0.1721	121, 152, 154, 155, 162, 202, 222
50:154	1 2 3 5 14 16 19 20 22 24 25 27	C_1	50×1	144, 144, 144	0 5 6 1 0 0	0.1412	150, 153, 155, 161, 199, 202
50:155	1 2 3 5 14 16 19 20 23 24 26 27	C_1	50×1	144, 144, 144	0 4 6 2 0 0	0.1137	153, 154, 198
50:156	1 2 3 5 14 17 18 19 20 22 23 24	C_2	25×2	144, 144, 144	0 4 6 2 0 0	0.0325	149(2)
50:157	1 2 3 5 14 17 18 19 21 22 23 25	C_{3v}	2×1, 6×3, 5×6	76, 76, 76	0 3 6 3 0 0	0.2395	
50:158	1 2 3 5 14 17 19 20 21 23 24 25	C_s	6×1, 22×2	144, 144, 144	0 5 6 1 0 0	0.2121	117(2), 207
50:159	1 2 3 5 15 16 18 19 20 22 23 27	C_1	50×1	144, 144, 144	0 5 6 1 0 0	0.0047	150, 160, 161, 178, 194, 209
50:160	1 2 3 5 15 16 18 19 21 22 23 26	C_1	50×1	144, 144, 144	0 5 6 1 0 0	0.0332	151, 159, 160, 194, 206
50:161	1 2 3 5 15 16 18 20 21 23 25 27	C_1	50×1	144, 144, 144	0 5 6 1 0 0	0.0429	152, 154, 159, 162, 177, 193, 209
50:162	1 2 3 5 15 16 19 20 21 23 25 26	C_s	8×1, 21×2	144, 144, 144	0 5 6 1 0 0	0.0928	153(2), 161(2), 179, 193(2)
50:163	1 2 3 10 11 12 15 16 21 24 26 27	C_1	50×1	144, 144, 144	0 6 6 0 0 0	0.0099	106, 165, 167, 169, 214, 216, 218
50:164	1 2 3 10 11 12 15 16 22 25 26 27	C_1	50×1	144, 144, 144	0 4 8 0 0 0	0.0950	175, 192, 215, 228, 238
50:165	1 2 3 10 11 12 15 17 21 23 26 27	C_1	50×1	144, 144, 144	1 4 7 0 0 0	0.0923	51, 109, 163, 170, 211, 227, 230
50:166	1 2 3 10 11 12 16 21 22 23 24 26	C_1	50×1	144, 144, 144	1 4 7 0 0 0	0.2003	136, 167, 172, 180, 190, 216, 226, 229
50:167	1 2 3 10 11 12 16 21 22 24 25 27	C_2	25×2	144, 144, 144	0 4 8 0 0 0	0.1074	163(2), 166(2), 173(2)
50:168	1 2 3 10 11 13 14 22 23 24 25 26	C_s	4×1, 23×2	144, 144, 144	0 6 6 0 0 0	0.0462	69, 100, 171(2), 211(2), 230(2)

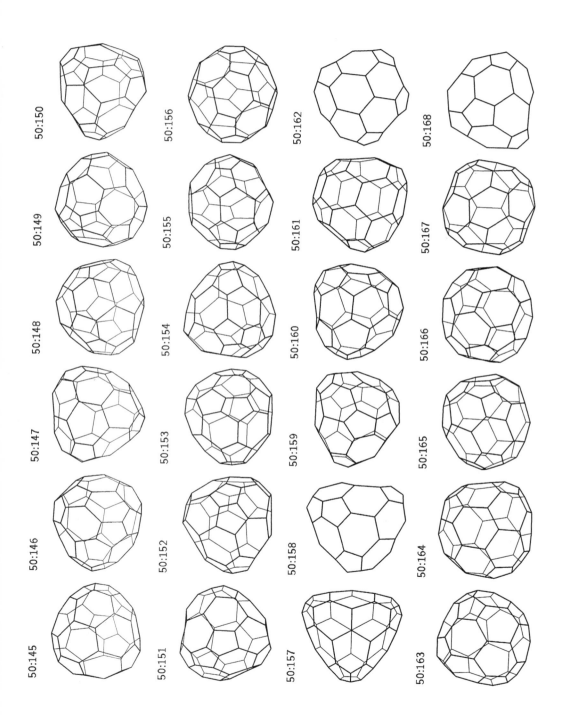

50:150

50:156

50:162

50:168

50:149

50:155

50:161

50:167

50:148

50:154

50:160

50:166

50:147

50:153

50:159

50:165

50:146

50:152

50:158

50:164

50:145

50:151

50:157

50:163

Table A.9. *(Continued)*

Isomer	Ring spiral	Point group	NMR pattern	Vibrations	Pentagon indices	Band gap	Transformations
50:169	1 2 3 10 11 13 15 16 20 24 26 27	C_1	50×1	144, 144, 144	1 6 5 0 0 0	0.0533	101, 163, 170, 173, 182, 217, 225, 244, 249
50:170	1 2 3 10 11 13 15 17 20 23 26 27	C_1	50×1	144, 144, 144	2 4 6 0 0 0	0.0847	104, 105, 165, 169, 183, 223, 224, 233, 239
50:171	1 2 3 10 11 13 15 21 23 24 25 26	C_1	50×1	144, 144, 144	0 6 6 0 0 0	0.0903	117, 168, 207, 223, 231, 239
50:172	1 2 3 10 11 13 16 20 22 23 24 26	C_1	50×1	144, 144, 144	1 6 5 0 0 0	0.1690	122, 166, 173, 174, 177, 187, 201, 219, 225, 244, 262
50:173	1 2 3 10 11 13 16 20 22 24 25 27	C_1	50×1	144, 144, 144	1 6 5 0 0 0	0.0625	123, 167, 169, 172, 178, 188, 218, 226, 260, 269
50:174	1 2 3 10 11 13 16 21 23 24 26 27	C_1	50×1	144, 144, 144	0 6 6 0 0 0	0.0405	121, 172, 202, 220, 224, 233, 248
50:175	1 2 3 10 11 14 15 19 23 25 26 27	C_2	25×2	144, 144, 144	0 2 10 0 0 0	0.1936	164(2), 212(2)
50:176	1 2 3 10 11 14 15 20 22 24 25 26	C_1	50×1	144, 144, 144	0 4 8 0 0 0	0.1704	177, 190, 198, 204, 205, 244
50:177	1 2 3 10 11 14 16 20 21 23 24 26	C_1	50×1	144, 144, 144	1 6 5 0 0 0	0.1276	161, 172, 176, 178, 179, 199, 201, 221, 222, 253, 261
50:178	1 2 3 10 11 14 16 20 21 24 25 27	C_1	50×1	144, 144, 144	1 4 7 0 0 0	0.1084	159, 173, 177, 182, 184, 186, 189, 200
50:179	1 2 3 10 11 14 16 20 22 24 26 27	C_{2v}	7×2, 9×4	111, 144, 111	0 6 6 0 0 0	0.0315	162(2), 177(4), 202(4)
50:180	1 2 3 10 11 14 19 22 23 24 25 26	C_2	25×2	144, 144, 144	0 2 10 0 0 0	0.2449	166(2), 213(2)
50:181	1 2 3 10 12 13 15 16 18 25 26 27	D_3	1×2, 8×6	71, 73, 48	0 6 6 0 0 0	0.1232	182(6)
50:182	1 2 3 10 12 13 15 16 19 24 26 27	C_1	50×1	144, 144, 144	0 8 4 0 0 0	0.0112	169, 178, 181, 183, 188(2), 260, 261, 265
50:183	1 2 3 10 12 13 15 17 19 23 26 27	C_1	50×1	144, 144, 144	1 6 5 0 0 0	0.0056	144, 170, 182, 184, 200, 246, 249, 251
50:184	1 2 3 10 12 13 15 17 20 23 25 27	C_1	50×1	144, 144, 144	2 4 6 0 0 0	0.0146	145, 151, 178, 183, 194, 195, 218, 221, 240
50:185	1 2 3 10 12 13 15 18 22 23 25 26	C_s	8×1, 21×2	144, 144, 144	0 4 8 0 0 0	0.1424	186(2), 196(2)
50:186	1 2 3 10 12 13 15 18 22 24 25 27	C_1	50×1	144, 144, 144	0 6 6 0 0 0	0.1206	178, 185, 195, 204, 241, 253
50:187	1 2 3 10 12 13 16 19 22 23 24 26	C_1	50×1	144, 144, 144	0 6 6 0 0 0	0.1870	172, 188, 199, 248, 259, 267
50:188	1 2 3 10 12 13 16 19 22 24 25 27	C_1	50×1	144, 144, 144	0 8 4 0 0 0	0.0265	173, 182(2), 187, 189, 200, 208, 262, 266, 268
50:189	1 2 3 10 12 13 16 20 22 24 25 26	C_1	50×1	144, 144, 144	2 4 6 0 0 0	0.0443	152, 178, 188, 190, 203, 204, 209, 222, 226, 264
50:190	1 2 3 10 12 13 16 21 24 25 26 27	C_1	50×1	144, 144, 144	1 2 9 0 0 0	0.1838	166, 176, 189, 201, 233
50:191	1 2 3 10 12 13 17 20 22 23 24 25	C_1	50×1	144, 144, 144	1 4 7 0 0 0	0.0608	147, 192, 203, 209, 237, 238, 241
50:192	1 2 3 10 12 13 17 21 23 24 25 27	C_1	50×1	144, 144, 144	1 2 9 0 0 0	0.0873	148, 164, 191, 196, 239

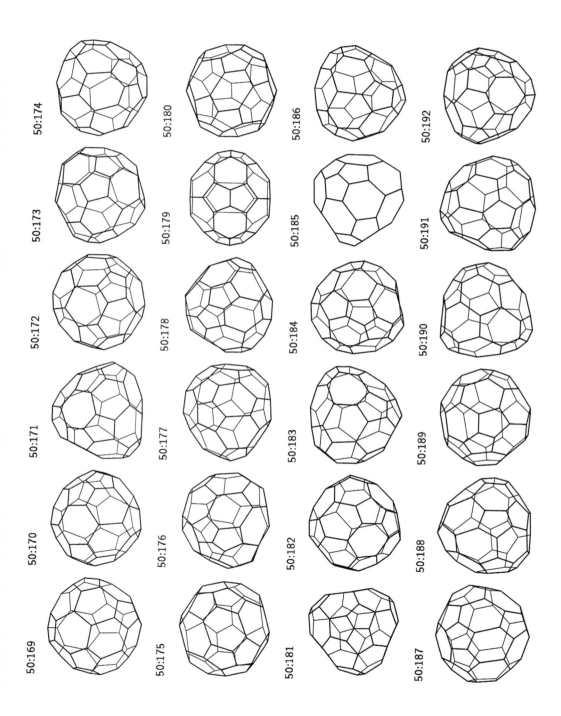

50:174

50:180

50:186

50:192

50:173

50:179

50:185

50:191

50:172

50:178

50:184

50:190

50:171

50:177

50:183

50:189

50:170

50:176

50:182

50:188

50:169

50:175

50:181

50:187

Table A.9. (*Continued*)

Isomer	Ring spiral	Point group	NMR pattern	Vibrations	Pentagon indices	Band gap	Transformations
50:193	1 2 3 10 12 14 15 16 19 25 26 27	C_1	50×1	144, 144, 144	1 4 7 0 0 0	0.1127	161, 162, 193, 202, 209, 222
50:194	1 2 3 10 12 14 15 16 20 24 25 26	C_1	50×1	144, 144, 144	1 4 7 0 0 0	0.0601	159, 160, 184, 206, 209, 237
50:195	1 2 3 10 12 14 15 18 20 23 24 26	C_1	50×1	144, 144, 144	0 6 6 0 0 0	0.0285	184, 186, 196, 197, 240, 252
50:196	1 2 3 10 12 14 15 18 20 23 25 27	C_1	50×1	144, 144, 144	0 6 6 0 0 0	0.0263	185, 192, 195, 196, 241, 251
50:197	1 2 3 10 12 14 15 19 20 23 24 25	C_1	50×1	144, 144, 144	0 4 8 0 0 0	0.1103	195, 237, 243
50:198	1 2 3 10 12 14 16 19 20 23 24 27	C_1	50×1	144, 144, 144	1 4 7 0 0 0	0.0660	155, 176, 199, 222, 243, 259
50:199	1 2 3 10 12 14 16 19 21 23 24 26	C_1	50×1	144, 144, 144	1 6 5 0 0 0	0.1936	154, 177, 187, 198, 200, 202, 205, 252, 263, 268
50:200	1 2 3 10 12 14 16 19 21 24 25 27	C_1	50×1	144, 144, 144	1 6 5 0 0 0	0.1056	150, 178, 183, 188, 199, 203, 241, 264, 265
50:201	1 2 3 10 12 14 16 19 22 23 25 26	C_1	50×1	144, 144, 144	2 4 6 0 0 0	0.0754	152, 172, 177, 190, 201, 202, 245, 246, 264
50:202	1 2 3 10 12 14 16 19 22 24 26 27	C_1	50×1	144, 144, 144	1 6 5 0 0 0	0.1802	153, 154, 174, 179, 193, 199, 201, 221, 222, 263
50:203	1 2 3 10 12 14 16 20 21 24 25 26	C_2	25×2	144, 144, 144	2 2 8 0 0 0	0.0286	149(2), 189(2), 191(2), 200(2)
50:204	1 2 3 10 12 15 16 19 20 22 23 25	C_1	50×1	144, 144, 144	1 4 7 0 0 0	0.0130	149, 176, 186, 189, 221, 245
50:205	1 2 3 10 12 15 16 19 21 22 24 26	C_2	25×2	144, 144, 144	0 4 8 0 0 0	0.0207	176(2), 199(2), 269
50:206	1 2 3 10 12 15 17 18 21 22 23 26	C_s	6×1, 22×2	144, 144, 144	1 4 7 0 0 0	0.1145	160(2), 194(2), 207
50:207	1 2 3 10 12 15 17 19 21 22 23 25	C_s	4×1, 23×2	144, 144, 144	0 6 6 0 0 0	0.2546	158, 171(2), 206, 237(2)
50:208	1 2 3 10 13 16 19 20 22 23 25 26	C_3	2×1, 16×3	92, 92, 92	0 6 6 0 0 0	0.2354	188(3), 267(3)
50:209	1 2 3 10 13 17 18 20 22 23 24 25	C_1	50×1	144, 144, 144	1 4 7 0 0 0	0.0127	159, 161, 189, 191, 193, 194, 221
50:210	1 2 4 7 9 10 19 20 21 23 25 27	C_2	25×2	144, 144, 144	0 4 8 0 0 0	0.0240	37(2)
50:211	1 2 4 7 9 12 13 22 23 24 25 26	C_1	50×1	144, 144, 144	1 6 5 0 0 0	0.0424	52, 98, 118, 165, 168, 214, 223, 242
50:212	1 2 4 7 9 12 15 19 23 25 26 27	C_1	50×1	144, 144, 144	1 4 7 0 0 0	0.1159	128, 141, 175, 215, 228(2)
50:213	1 2 4 7 9 12 15 20 23 24 25 26	C_1	50×1	144, 144, 144	1 4 7 0 0 0	0.1287	127, 140, 180, 216, 229, 234
50:214	1 2 4 7 9 12 16 19 21 24 26 27	C_2	25×2	144, 144, 144	0 8 4 0 0 0	0.2522	107(2), 163(2), 211(2), 217(2)
50:215	1 2 4 7 9 12 16 19 22 25 26 27	C_1	50×1	144, 144, 144	0 6 6 0 0 0	0.2168	105, 164, 212, 235, 236, 239
50:216	1 2 4 7 9 12 16 20 22 24 25 26	C_1	50×1	144, 144, 144	1 6 5 0 0 0	0.1030	104, 137, 163, 166, 213, 219, 225, 227, 233

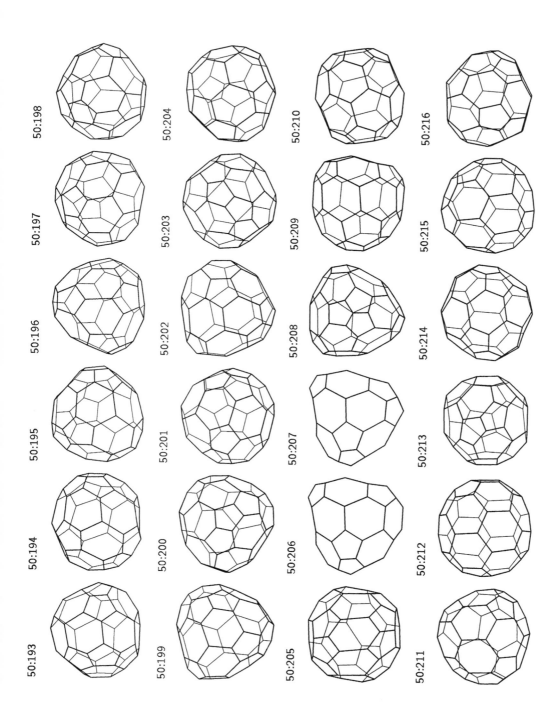

50:198

50:197

50:196

50:195

50:194

50:193

50:204

50:203

50:202

50:201

50:200

50:199

50:210

50:209

50:208

50:207

50:206

50:205

50:216

50:215

50:214

50:213

50:212

50:211

Table A.9. (*Continued*)

Isomer	Ring spiral	Point group	NMR pattern	Vibrations	Pentagon indices	Band gap	Transformations
50:217	1 2 4 7 9 12 17 19 21 23 26 27	C_1	50×1	144, 144, 144	0 8 4 0 0 0	0.1226	146, 169, 214, 218, 223, 250
50:218	1 2 4 7 9 12 17 20 21 23 25 27	C_1	50×1	144, 144, 144	0 8 4 0 0 0	0.1817	143, 163, 173, 184, 217, 219, 249
50:219	1 2 4 7 9 12 17 20 22 23 25 26	C_1	50×1	144, 144, 144	0 8 4 0 0 0	0.0597	144, 172, 216, 218, 220, 221, 246, 248
50:220	1 2 4 7 9 12 17 20 22 24 26 27	C_1	50×1	144, 144, 144	0 6 6 0 0 0	0.0701	145, 174, 219, 221, 247
50:221	1 2 4 7 9 12 18 20 22 23 24 26	C_1	50×1	144, 144, 144	2 6 4 0 0 0	0.0532	150, 177, 184, 202, 204, 209, 219, 220, 241, 246, 252, 264
50:222	1 2 4 7 9 12 18 21 22 23 25 26	C_1	50×1	144, 144, 144	2 6 4 0 0 0	0.1844	153, 177, 189, 193, 198, 202, 225, 233, 245, 263, 268
50:223	1 2 4 7 9 13 17 19 20 23 26 27	C_1	50×1	144, 144, 144	1 6 5 0 0 0	0.0778	102, 119, 170, 171, 211, 217
50:224	1 2 4 7 9 13 17 19 22 25 26 27	C_1	50×1	144, 144, 144	1 6 5 0 0 0	0.1100	101, 120, 170, 174, 225, 227, 249
50:225	1 2 4 7 9 13 18 19 22 24 26 27	C_1	50×1	144, 144, 144	2 6 4 0 0 0	0.1042	121, 123, 169, 172, 216, 222, 224, 226, 229, 248, 260
50:226	1 2 4 7 9 13 18 20 22 24 25 27	C_2	25×2	144, 144, 144	2 6 4 0 0 0	0.3275	122(2), 166(2), 173(2), 189(2), 225(2), 262(2)
50:227	1 2 4 7 9 14 17 19 21 25 26 27	C_1	50×1	144, 144, 144	0 6 6 0 0 0	0.0432	49, 126, 165, 216, 224
50:228	1 2 4 7 9 14 18 19 20 23 26 27	C_1	50×1	144, 144, 144	1 4 7 0 0 0	0.0476	139, 164, 212(2), 236
50:229	1 2 4 7 9 14 18 20 21 23 25 26	C_2	25×2	144, 144, 144	2 4 6 0 0 0	0.0714	135(2), 166(2), 213(2), 225(2)
50:230	1 2 4 7 9 17 18 20 21 22 23 26	C_1	50×1	144, 144, 144	0 6 6 0 0 0	0.0869	70, 148, 165, 168, 231, 239
50:231	1 2 4 7 9 17 19 20 21 22 23 25	C_1	50×1	144, 144, 144	0 6 6 0 0 0	0.0435	147, 171, 230, 237
50:232	1 2 4 7 10 12 16 20 23 24 26 27	C_1	50×1	144, 144, 144	0 4 8 0 0 0	0.0971	235, 236
50:233	1 2 4 7 10 12 17 21 22 23 25 26	C_1	50×1	144, 144, 144	1 6 5 0 0 0	0.1023	170, 174, 190, 216, 222, 243, 244, 246
50:234	1 2 4 7 10 15 17 20 21 22 25 26	C_2	25×2	144, 144, 144	0 4 8 0 0 0	0.1632	213(2)
50:235	1 2 4 7 11 12 15 17 23 25 26 27	C_1	50×1	144, 144, 144	0 4 8 0 0 0	0.0521	50, 215, 232
50:236	1 2 4 7 11 15 17 18 20 22 26 27	C_1	50×1	144, 144, 144	0 6 6 0 0 0	0.1092	215, 228, 232, 237, 238
50:237	1 2 4 7 11 16 18 20 21 22 24 25	C_1	50×1	144, 144, 144	1 6 5 0 0 0	0.1670	191, 194, 197, 207, 231, 236, 239, 240
50:238	1 2 4 7 11 16 19 21 23 24 26 27	C_2	25×2	144, 144, 144	0 6 6 0 0 0	0.2282	164(2), 191(2), 236(2)
50:239	1 2 4 7 12 14 17 18 19 23 26 27	C_1	50×1	144, 144, 144	1 6 5 0 0 0	0.2181	170, 171, 192, 215, 230, 237, 243, 251
50:240	1 2 4 7 12 16 17 18 20 22 23 26	C_1	50×1	144, 144, 144	0 8 4 0 0 0	0.1773	184, 195, 237, 240, 241, 251(2)

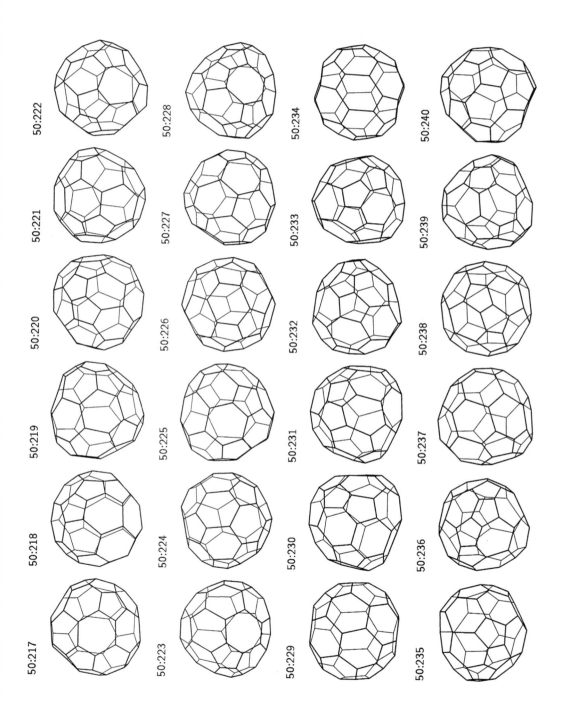

50:222

50:228

50:234

50:240

50:221

50:227

50:233

50:239

50:220

50:226

50:232

50:238

50:219

50:225

50:231

50:237

50:218

50:224

50:230

50:236

50:217

50:223

50:229

50:235

Table A.9. (*Continued*)

Isomer	Ring spiral	Point group	NMR pattern	Vibrations	Pentagon indices	Band gap	Transformations
50:241	1 2 4 7 12 16 17 19 20 22 23 25	C_1	50×1	144, 144, 144	0 8 4 0 0 0	0.0827	186, 191, 196, 200, 221, 240, 251, 252
50:242	1 2 4 8 9 13 15 19 20 24 26 27	C_{2v}	5×2, 10×4	110, 144, 110	2 4 6 0 0 0	0.1057	42(2), 211(4)
50:243	1 2 4 8 11 15 17 20 21 22 24 26	C_1	50×1	144, 144, 144	0 6 6 0 0 0	0.0927	197, 198, 233, 239, 252
50:244	1 2 4 8 12 13 17 18 22 24 26 27	C_1	50×1	144, 144, 144	0 8 4 0 0 0	0.0813	169, 172, 176, 233, 245, 259, 261, 269
50:245	1 2 4 8 12 14 17 18 21 24 26 27	C_1	50×1	144, 144, 144	1 6 5 0 0 0	0.0920	201, 204, 222, 244, 246, 253
50:246	1 2 4 8 12 14 17 19 21 23 24 26	C_1	50×1	144, 144, 144	1 8 3 0 0 0	0.0751	183, 201, 219, 221, 233, 245, 248, 252, 261, 263, 264
50:247	1 2 4 8 12 14 17 19 22 23 25 26	C_s	2×1, 24×2	144, 144, 144	1 6 5 0 0 0	0.0892	220(2), 248(2), 256, 264, 267
50:248	1 2 4 8 12 14 17 19 22 24 26 27	C_1	50×1	144, 144, 144	1 8 3 0 0 0	0.1285	174, 187, 219, 225, 246, 247, 249, 255, 262, 263, 266
50:249	1 2 4 8 12 14 18 19 22 24 25 27	C_1	50×1	144, 144, 144	1 8 3 0 0 0	0.0439	169, 183, 218, 224, 248, 250, 254, 259, 260, 265
50:250	1 2 4 8 12 14 18 20 22 24 25 26	C_2	25×2	144, 144, 144	0 8 4 0 0 0	0.0864	217(2), 249(2), 257
50:251	1 2 4 8 12 15 17 18 20 22 23 26	C_1	50×1	144, 144, 144	0 8 4 0 0 0	0.0909	183, 196, 239, 240(2), 241, 252
50:252	1 2 4 8 12 15 17 19 20 22 23 25	C_1	50×1	144, 144, 144	0 8 4 0 0 0	0.0806	195, 199, 221, 241, 243, 246, 251, 253
50:253	1 2 4 8 12 15 17 19 20 22 24 27	C_s	8×1, 21×2	144, 144, 144	0 8 4 0 0 0	0.0032	177(2), 186(2), 245(2), 252(2)
50:254	1 2 4 8 13 14 17 18 20 23 26 27	C_2	25×2	144, 144, 144	0 8 4 0 0 0	0.0815	249(2), 255(2), 257(2)
50:255	1 2 4 8 13 14 17 19 20 23 25 27	C_2	25×2	144, 144, 144	0 8 4 0 0 0	0.1432	248(2), 254(2), 256(2), 260
50:256	1 2 4 8 13 14 17 19 21 23 25 26	C_{2v}	5×2, 10×4	110, 144, 110	0 8 4 0 0 0	0.2396	247(2), 255(4), 262(2)
50:257	1 2 4 8 13 14 18 20 21 24 25 26	C_2	25×2	144, 144, 144	0 8 4 0 0 0	0.1521	250, 254(2), 257(2)
50:258	1 2 4 8 13 17 18 20 22 23 25 27	C_3	2×1, 16×3	92, 92, 92	0 6 6 0 0 0	0.0271	259(3)
50:259	1 2 4 8 13 17 18 20 22 23 25 27	C_1	50×1	144, 144, 144	0 8 4 0 0 0	0.0202	187, 198, 244, 249, 258, 268(2)
50:260	1 2 4 9 12 13 15 16 19 24 26 27	C_2	25×2	144, 144, 144	2 8 2 0 0 0	0.1401	173(2), 182(2), 225(2), 249(2), 255, 262(2), 266(2), 268(2)
50:261	1 2 4 9 12 13 16 18 22 24 26 27	C_2	25×2	144, 144, 144	0 10 2 0 0 0	0.2376	177(2), 182(2), 244(2), 246(2), 262(2), 268(2)
50:262	1 2 4 9 12 13 16 19 22 24 25 27	C_s	2×1, 24×2	144, 144, 144	2 8 2 0 0 0	0.2095	172(2), 188(2), 226(2), 248(2), 256, 260(2), 261(2), 264, 267, 271
50:263	1 2 4 9 12 14 16 19 21 23 24 26	C_2	25×2	144, 144, 144	2 8 2 0 0 0	0.3642	199(2), 202(2), 222(2), 246(2), 248(2), 263(2), 264(2), 270
50:264	1 2 4 9 12 14 16 19 21 24 25 27	C_s	4×1, 23×2	144, 144, 144	3 6 3 0 0 0	0.1591	189(2), 200(2), 201(2), 221(2), 246(2), 247, 262, 263(2), 266

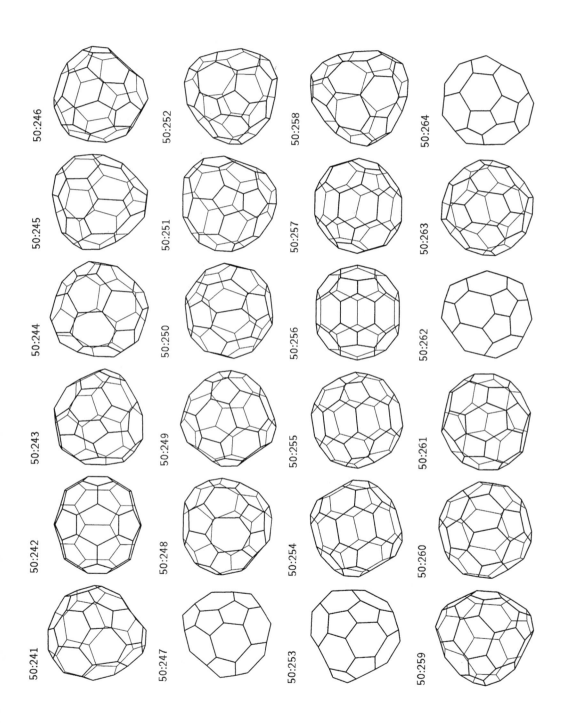

50:246

50:252

50:258

50:264

50:245

50:251

50:257

50:263

50:244

50:250

50:256

50:262

50:243

50:249

50:255

50:261

50:242

50:248

50:254

50:260

50:241

50:247

50:253

50:259

Table A.9. *(Continued)*

Isomer	Ring spiral	Point group	NMR pattern	Vibrations	Pentagon indices	Band gap	Transformations
50:265	1 2 4 9 13 15 16 18 20 22 26 27	C_2	25×2	144, 144, 144	0 10 2 0 0 0	0.1927	182(2), 200(2), 249(2), 266(2), 268(2)
50:266	1 2 4 9 13 15 16 19 20 22 25 27	C_s	6×1, 22×2	144, 144, 144	1 10 1 0 0 0	0.2481	188(2), 248(2), 260(2), 264, 265(2), 267(3), 270(2), 271
50:267	1 2 4 9 13 15 16 19 21 22 25 26	C_s	4×1, 23×2	144, 144, 144	1 8 3 0 0 0	0.2288	187(2), 208(2), 247, 262, 266(3), 268(2)
50:268	1 2 4 9 13 15 17 19 21 22 24 26	C_1	50×1	144, 144, 144	0 10 2 0 0 0	0.1969	188, 199, 222, 259(2), 260, 261, 265, 267, 269, 270
50:269	1 2 4 9 13 15 18 19 21 22 24 25	C_2	25×2	144, 144, 144	0 8 4 0 0 0	0.0159	173(2), 205, 244(2), 268(2)
50:270	1 2 9 10 12 13 15 17 20 23 25 27	D_3	1×2, 8×6	71, 73, 48	0 12 0 0 0 0	0.4679	263(3), 266(6), 268(6)
50:271	1 2 9 10 12 14 15 17 20 22 24 26	D_{5h}	3×10, 1×20	22, 38, 0	2 10 0 0 0 0	0.1031	262(10), 266(10)

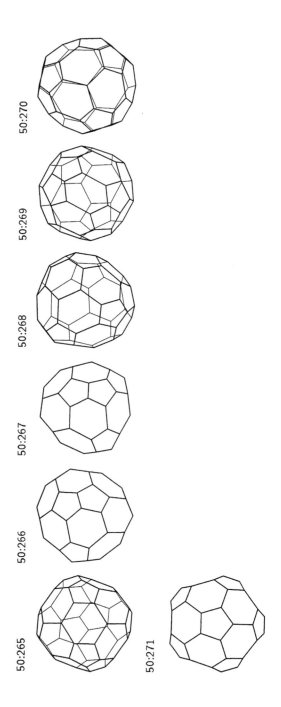

50:270

50:269

50:268

50:267

50:266

50:265

50:271

Table A.10. Isolated-pentagon fullerene isomers of C_{60}, C_{70}, C_{72}, C_{74}, C_{76}, C_{78}, and C_{80}

Isomer	Ring spiral	Point group	NMR pattern	Vibrations	Hexagon indices	Band gap	Transformations
60:1	1 7 9 11 13 15 18 20 22 24 26 32	I_h	1×60	4, 10, 0	20 0 0 0	0.7566	
70:1	1 7 9 11 13 15 27 29 31 33 35 37	D_{5h}	3×10, 2×20	31, 53, 0	10 15 0 0	0.5293	
72:1	1 7 9 11 13 18 22 24 27 34 36 38	D_{6d}	2×12, 2×24	26, 45, 0	12 12 0 2	0.7023	
74:1	1 7 9 11 14 23 26 28 30 32 35 38	D_{3h}	1×2, 4×6, 4×12	56, 93, 37	6 21 0 0	0.1031	
76:1	1 7 9 11 13 18 26 31 33 35 37 39	D_2	19×4	165, 222, 165	8 16 4 0	0.3436	
76:2	1 7 9 12 14 21 26 28 30 33 35 38	T_d	1×4, 2×12, 2×24	29, 59, 29	4 24 0 0	0.0000	
78:1	1 7 9 11 13 20 25 28 32 34 36 38	D_3	13×6	113, 115, 76	8 15 6 0	0.2532	
78:2	1 7 9 11 13 24 27 30 32 36 38 40	C_{2v}	3×2, 18×4	172, 228, 172	6 20 2 1	0.3481	3(2), 4
78:3	1 7 9 11 14 22 26 28 30 34 39 41	C_{2v}	5×2, 17×4	173, 228, 173	4 23 2 0	0.1802	2(2), 5
78:4	1 7 9 11 15 18 22 25 33 37 39 41	D_{3h}	3×6, 5×12	58, 97, 38	8 18 0 3	0.6333	2(3)
78:5	1 7 9 12 14 21 26 28 30 34 39 41	D_{3h}	3×6, 5×12	58, 97, 39	2 27 0 0	0.0730	3(3)
80:1	1 7 9 11 13 15 28 30 32 34 36 42	D_{5d}	2×10, 3×20	35, 60, 0	10 10 10 0	0.0728	
80:2	1 7 9 11 13 18 25 30 32 34 36 42	D_2	20×4	174, 234, 174	8 14 8 0	0.1749	
80:3	1 7 9 11 14 22 27 30 34 36 38 40	C_{2v}	6×2, 17×4	178, 234, 178	4 22 4 0	0.0338	5(2)
80:4	1 7 9 11 14 23 28 30 33 35 37 39	D_3	1×2, 13×6	116, 118, 78	6 18 6 0	0.1351	
80:5	1 7 9 12 14 20 26 28 32 34 39 42	C_{2v}	4×2, 18×4	177, 234, 177	2 26 2 0	0.0987	3(2), 6
80:6	1 7 10 12 14 19 26 28 32 34 39 42	D_{5h}	4×10, 2×20	36, 61, 0	0 30 0 0	0.0000	5(5)
80:7	1 8 10 12 14 16 28 30 32 34 36 42	I_h	1×20, 1×60	6, 14, 0	0 30 0 0	0.0000	

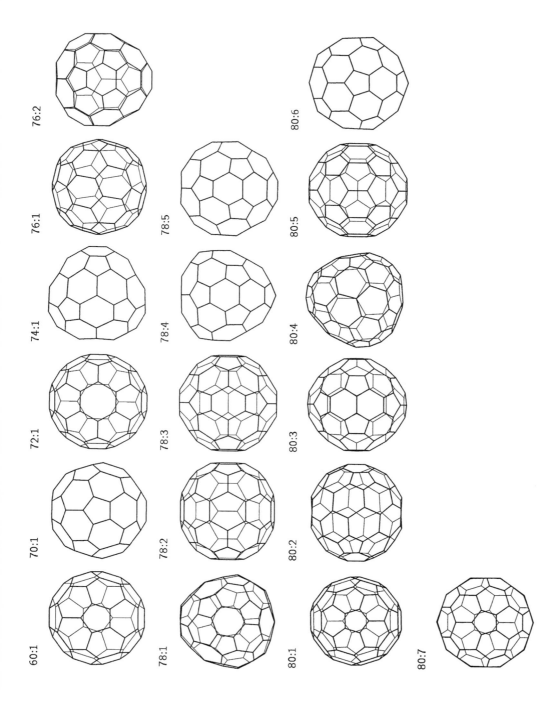

76:2

80:6

76:1

78:5

80:5

74:1

78:4

80:4

72:1

78:3

80:3

70:1

78:2

80:2

60:1

78:1

80:1

80:7

Table A.11. Isolated-pentagon fullerene isomers of C_{82}

Isomer	Ring spiral	Point group	NMR pattern	Vibrations	Hexagon indices	Band gap	Transformations
82:1	1 7 9 11 13 24 26 29 31 33 37 43	C_2	41×2	240, 240, 240	6 17 8 0	0.1495	2(2)
82:2	1 7 9 11 13 24 27 29 31 33 36 43	C_s	6×1, 38×2	240, 240, 240	5 20 5 1	0.3313	1(2), 3(2)
82:3	1 7 9 11 14 22 26 29 31 34 37 43	C_2	41×2	240, 240, 240	4 21 6 0	0.2568	2(2), 4(2)
82:4	1 7 9 11 14 22 27 29 31 34 36 43	C_s	6×1, 38×2	240, 240, 240	3 23 5 0	0.2450	3(2), 5(2)
82:5	1 7 9 12 14 20 26 29 32 34 37 43	C_2	41×2	240, 240, 240	2 25 4 0	0.1300	4(2), 6(2)
82:6	1 7 9 12 14 20 27 29 32 34 36 43	C_s	6×1, 38×2	240, 240, 240	1 27 3 0	0.0683	5(2), 7, 9(2)
82:7	1 7 9 12 14 20 27 32 34 36 38 40	C_{3v}	1×1, 3×3, 12×6	122, 122, 122	3 24 3 1	0.0000	6(3)
82:8	1 7 9 13 20 22 26 28 30 35 41 43	C_{3v}	1×1, 5×3, 11×6	123, 123, 123	1 27 3 0	0.0467	9(3)
82:9	1 7 10 12 14 18 26 30 32 34 37 43	C_{2v}	7×2, 17×4	183, 240, 183	0 29 2 0	0.0160	6(4), 8(2)

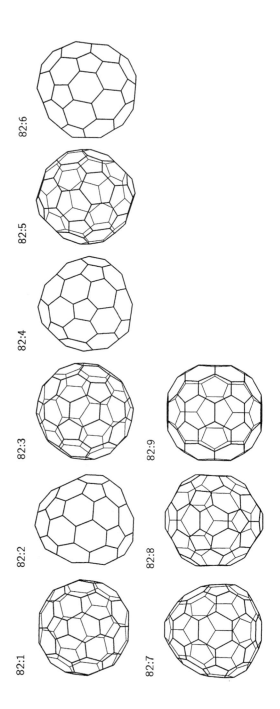

82:6

82:5

82:4

82:3

82:9

82:2

82:8

82:1

82:7

Table A.12. Isolated-pentagon fullerene isomers of C_{84}

Isomer	Ring spiral	Point group	NMR pattern	Vibrations	Hexagon indices	Band gap	Transformations
84:1	1 7 9 11 13 18 24 35 38 40 42 44	D_2	21×4	183, 246, 183	8 16 4 4	0.6143	2(2)
84:2	1 7 9 11 13 24 28 30 36 40 42 44	C_2	42×2	246, 246, 246	6 18 6 2	0.3523	1, 5
84:3	1 7 9 11 14 22 27 29 31 35 41 43	C_s	4×1, 40×2	246, 246, 246	3 22 7 0	0.0191	4, 7, 8(2)
84:4	1 7 9 11 14 22 27 30 35 39 41 43	D_{2d}	3×4, 9×8	93, 156, 93	4 20 8 0	0.3519	3(4)
84:5	1 7 9 11 14 23 28 30 36 40 42 44	D_2	21×4	183, 246, 183	4 20 8 0	0.2403	2(2)
84:6	1 7 9 11 22 24 26 28 30 32 36 44	C_{2v}	4×2, 19×4	186, 246, 186	4 20 8 0	0.1892	7, 14(2)
84:7	1 7 9 11 22 24 26 28 30 32 42 44	C_{2v}	4×2, 19×4	186, 246, 186	2 24 6 0	0.1892	3(2), 6, 15(2)
84:8	1 7 9 12 14 20 26 29 33 37 40 42	C_2	42×2	246, 246, 246	2 24 6 0	0.1776	3(2), 9, 15(2)
84:9	1 7 9 12 14 20 27 29 32 35 41 43	C_2	42×2	246, 246, 246	2 24 6 0	0.0556	8, 10(2), 12(2)
84:10	1 7 9 12 14 20 27 29 33 35 40 43	C_s	8×1, 38×2	246, 246, 246	1 26 5 0	0.0916	9(2), 13(2), 21(2)
84:11	1 7 9 12 20 24 26 28 30 33 36 44	C_2	42×2	246, 246, 246	2 24 6 0	0.2540	12(2), 14(2), 16(2)
84:12	1 7 9 12 20 24 26 28 30 33 42 44	C_1	84×1	246, 246, 246	1 26 5 0	0.2164	9, 11, 13, 15, 17, 21, 22
84:13	1 7 9 12 20 24 26 28 30 34 41 44	C_2	42×2	246, 246, 246	2 24 6 0	0.0988	10(2), 12(2)
84:14	1 7 9 12 20 24 26 28 33 36 39 41	C_s	2×1, 41×2	246, 246, 246	3 23 5 1	0.4054	6, 11(2), 15, 18
84:15	1 7 9 12 21 24 26 28 30 32 42 44	C_s	4×1, 40×2	246, 246, 246	1 26 5 0	0.2191	7, 8(2), 12(2), 14, 23
84:16	1 7 9 13 20 22 25 28 30 34 37 44	C_s	2×1, 41×2	246, 246, 246	1 27 3 1	0.3369	11(2), 17, 18(2), 19, 22(2), 24
84:17	1 7 9 13 20 22 25 28 30 37 41 43	C_{2v}	6×2, 18×4	187, 246, 187	2 25 4 1	0.1745	12(4), 16(2)
84:18	1 7 9 13 20 22 25 28 34 37 39 41	C_{2v}	2×2, 20×4	185, 246, 185	2 26 2 2	0.3285	14(2), 16(4), 20, 23
84:19	1 7 9 13 20 22 26 28 30 34 36 44	D_{3d}	2×6, 6×12	62, 63, 0	2 24 6 0	0.1861	16(6)
84:20	1 7 9 13 20 23 25 28 33 37 39 41	T_d	1×12, 3×24	31, 63, 31	4 24 0 4	0.6962	18(6)
84:21	1 7 10 12 14 18 26 31 33 37 39 42	D_2	21×4	183, 246, 183	0 28 4 0	0.1381	10(4), 12(4)
84:22	1 7 10 13 18 22 25 27 31 34 38 44	D_2	21×4	183, 246, 183	0 28 4 0	0.3449	12(4), 16(4), 23(2)
84:23	1 7 10 13 18 22 25 27 31 38 41 43	D_{2d}	1×4, 10×8	92, 155, 92	0 28 4 0	0.3449	15(4), 18(2), 22(4)
84:24	1 7 10 13 19 22 25 28 30 34 37 44	D_{6h}	3×12, 2×24	31, 53, 0	0 30 0 2	0.5293	16(12)

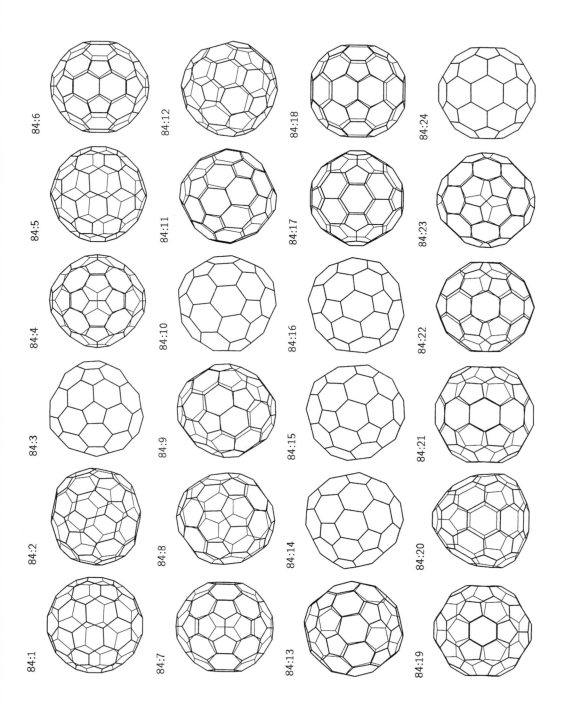

84:6

84:12

84:18

84:24

84:5

84:11

84:17

84:23

84:4

84:10

84:16

84:22

84:3

84:9

84:15

84:21

84:2

84:8

84:14

84:20

84:1

84:7

84:13

84:19

Table A.13. Isolated-pentagon fullerene isomers of C_{86}

Isomer	Ring spiral	Point group	NMR pattern	Vibrations	Hexagon indices	Band gap	Transformations
86:1	1 7 9 11 13 24 28 31 33 37 39 43	C_1	86×1	252, 252, 252	5 18 9 1	0.0289	2, 6
86:2	1 7 9 11 13 24 28 32 37 39 41 43	C_2	43×2	252, 252, 252	6 17 8 2	0.3763	1(2)
86:3	1 7 9 11 13 25 29 31 34 38 43 45	C_2	43×2	252, 252, 252	6 15 12 0	0.0706	
86:4	1 7 9 11 14 22 28 31 34 37 39 43	C_2	43×2	252, 252, 252	4 19 10 0	0.0716	5(2), 6
86:5	1 7 9 11 14 22 28 31 35 37 39 42	C_1	86×1	252, 252, 252	3 21 9 0	0.1619	4, 7, 14
86:6	1 7 9 11 14 23 28 31 33 37 39 43	C_2	43×2	252, 252, 252	4 19 10 0	0.1668	1(2), 4
86:7	1 7 9 11 22 24 26 29 32 38 40 43	C_1	86×1	252, 252, 252	3 21 9 0	0.0357	5, 8, 11
86:8	1 7 9 11 22 24 27 29 32 37 40 43	C_s	8×1, 39×2	252, 252, 252	2 23 8 0	0.0499	7(2), 15
86:9	1 7 9 11 23 25 27 29 33 41 43 45	C_{2v}	7×2, 18×4	192, 252, 192	2 23 8 0	0.0050	
86:10	1 7 9 12 14 20 27 33 35 38 40 45	C_{2v}	5×2, 19×4	191, 252, 191	2 24 6 1	0.0765	13(4), 16(2)
86:11	1 7 9 12 14 20 28 32 35 37 39 42	C_1	86×1	252, 252, 252	2 23 8 0	0.1704	7, 12, 14, 15, 17
86:12	1 7 9 12 14 20 28 33 35 37 39 41	C_1	86×1	252, 252, 252	2 23 8 0	0.1278	11, 13(2), 16, 18
86:13	1 7 9 12 14 20 28 33 35 37 40 45	C_1	86×1	252, 252, 252	1 25 7 0	0.1282	10, 12(2), 15, 17, 19
86:14	1 7 9 12 14 21 28 31 35 37 39 42	C_2	43×2	252, 252, 252	2 23 8 0	0.0403	5(2), 11(2)
86:15	1 7 9 12 14 28 30 32 34 36 38 45	C_s	6×1, 40×2	252, 252, 252	1 25 7 0	0.1514	8, 11(2), 13(2)
86:16	1 7 9 12 20 24 26 28 34 39 41 44	C_s	6×1, 40×2	252, 252, 252	3 22 7 1	0.4087	10, 12(2), 17(2)
86:17	1 7 9 12 20 24 26 29 33 38 40 43	C_2	43×2	252, 252, 252	2 23 8 0	0.2338	11(2), 13(2), 16(2)
86:18	1 7 9 12 20 24 27 30 34 37 39 42	C_3	2×1, 28×3	164, 164, 164	3 21 9 0	0.2481	12(3)
86:19	1 7 10 12 14 19 28 33 35 37 40 45	D_3	1×2, 14×6	125, 127, 84	0 27 6 0	0.1967	13(6)

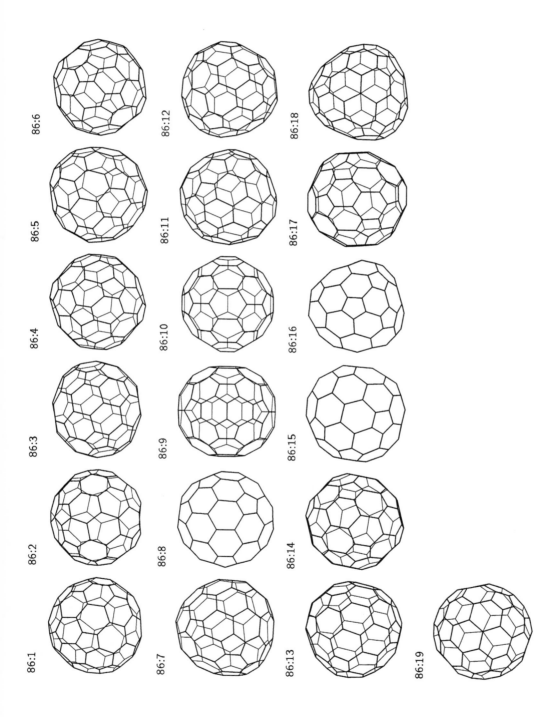

86:6

86:12

86:18

86:5

86:11

86:17

86:4

86:10

86:16

86:3

86:9

86:15

86:2

86:8

86:14

86:1

86:7

86:13

86:19

Table A.14. Isolated-pentagon fullerene isomers of C_{88}

Table A.14. Isolated-pentagon fullerene isomers of C_{88}

Isomer	Ring spiral	Point group	NMR pattern	Vibrations	Hexagon indices	Band gap	Transformations
88:1	1 7 9 11 13 18 33 35 38 40 42 44	D_2	22×4	192, 258, 192	8 14 8 4	0.3119	
88:2	1 7 9 11 13 24 29 31 36 38 43 46	C_1	88×1	258, 258, 258	6 16 10 2	0.3465	7
88:3	1 7 9 11 13 25 29 32 38 42 44 46	C_2	44×2	258, 258, 258	6 14 14 0	0.1586	
88:4	1 7 9 11 13 28 31 34 36 38 40 43	C_s	8×1, 40×2	258, 258, 258	5 17 11 1	0.1903	
88:5	1 7 9 11 14 22 27 35 38 40 42 46	C_{2v}	6×2, 19×4	196, 258, 196	4 20 8 2	0.2516	6(4), 10
88:6	1 7 9 11 14 22 28 35 37 40 42 46	C_1	88×1	258, 258, 258	3 21 9 1	0.0439	5, 9, 11, 28
88:7	1 7 9 11 14 23 29 31 36 38 43 46	C_2	44×2	258, 258, 258	4 18 12 0	0.2456	2(2)
88:8	1 7 9 11 14 28 32 34 36 38 40 42	C_s	4×1, 42×2	258, 258, 258	4 19 10 1	0.0407	9, 23
88:9	1 7 9 11 14 28 32 34 36 38 41 46	C_s	6×1, 41×2	258, 258, 258	2 22 10 0	0.0346	6(2), 8, 17
88:10	1 7 9 11 22 24 26 28 39 41 43 45	C_{2v}	6×2, 19×4	196, 258, 196	4 20 8 2	0.2381	5, 11(4)
88:11	1 7 9 11 22 24 26 29 38 41 43 45	C_1	88×1	258, 258, 258	3 21 9 1	0.3240	6, 10, 13, 17, 33
88:12	1 7 9 11 22 24 27 29 36 41 43 46	C_1	88×1	258, 258, 258	2 23 8 1	0.1021	13, 14, 18, 19, 30
88:13	1 7 9 11 22 24 27 29 37 41 43 45	C_1	88×1	258, 258, 258	3 21 9 1	0.1011	11, 12, 15, 29
88:14	1 7 9 11 22 25 27 29 35 41 43 46	C_s	6×1, 41×2	258, 258, 258	3 22 7 2	0.0310	12(2), 32
88:15	1 7 9 11 23 25 27 29 32 34 40 44	C_1	88×1	258, 258, 258	2 22 10 0	0.1382	13, 16, 17, 18, 21
88:16	1 7 9 11 23 25 27 29 33 40 42 44	C_s	4×1, 42×2	258, 258, 258	3 21 9 1	0.0288	15(2), 22
88:17	1 7 9 11 23 25 27 30 32 34 39 44	C_s	4×1, 42×2	258, 258, 258	2 22 10 0	0.2443	9, 11(2), 15(2), 23
88:18	1 7 9 11 23 25 27 30 33 39 43 46	C_1	88×1	258, 258, 258	3 20 11 0	0.0404	12, 15
88:19	1 7 9 11 23 26 28 30 33 37 43 46	C_s	6×1, 41×2	258, 258, 258	3 20 11 0	0.0118	12(2), 24
88:20	1 7 9 11 23 26 29 32 34 36 39 41	C_2	44×2	258, 258, 258	4 18 12 0	0.2720	21(2)
88:21	1 7 9 11 23 26 29 33 36 39 41 43	C_1	88×1	258, 258, 258	3 21 9 1	0.1141	15, 20, 22, 23
88:22	1 7 9 11 23 27 29 33 36 38 41 43	C_{2v}	6×2, 19×4	196, 258, 196	2 24 6 2	0.0918	16(2), 21(4)
88:23	1 7 9 11 23 27 30 32 34 36 38 40	C_s	2×1, 43×2	258, 258, 258	4 20 8 2	0.3763	8, 17, 21(2)
88:24	1 7 9 12 14 20 27 34 38 40 42 44	C_s	4×1, 42×2	258, 258, 258	2 23 8 1	0.1600	19, 27(2), 30(2)

262

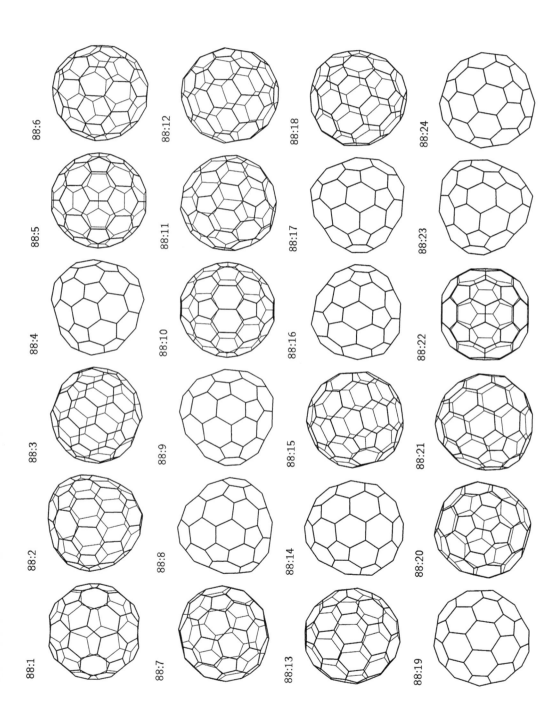

88:6

88:12

88:18

88:24

88:5

88:11

88:17

88:23

88:4

88:10

88:16

88:22

88:3

88:9

88:15

88:21

88:2

88:8

88:14

88:20

88:1

88:7

88:13

88:19

Table A.14. (*Continued*)

Isomer	Ring spiral	Point group	NMR pattern	Vibrations	Hexagon indices	Band gap	Transformations
88:25	1 7 9 12 14 20 28 33 36 40 43 45	C_2	44×2	258, 258, 258	2 22 10 0	0.0631	26(2), 29(2)
88:26	1 7 9 12 14 20 28 34 36 40 42 45	C_1	88×1	258, 258, 258	1 24 9 0	0.0824	25, 27, 30, 35
88:27	1 7 9 12 14 20 28 34 37 40 42 44	C_2	44×2	258, 258, 258	2 22 10 0	0.0466	24(2), 26(2)
88:28	1 7 9 12 14 21 28 35 37 40 42 46	C_2	44×2	258, 258, 258	2 22 10 0	0.0622	6(2), 33
88:29	1 7 9 12 14 28 30 32 34 37 43 45	C_1	88×1	258, 258, 258	2 22 10 0	0.1753	13, 25, 30, 33
88:30	1 7 9 12 14 28 30 32 35 37 42 45	C_1	88×1	258, 258, 258	1 24 9 0	0.1047	12, 24, 26, 29, 32
88:31	1 7 9 12 14 28 30 33 35 37 40 46	C_s	6×1, 41×2	258, 258, 258	2 22 10 0	0.0583	32
88:32	1 7 9 12 14 28 30 33 35 37 41 45	C_s	4×1, 42×2	258, 258, 258	1 24 9 0	0.0160	14, 30(2), 31
88:33	1 7 9 12 21 24 26 29 38 41 43 45	C_2	44×2	258, 258, 258	2 22 10 0	0.2729	11(2), 28, 29(2)
88:34	1 7 9 12 24 27 31 33 35 38 40 42	T	1×4, 7×12	64, 108, 64	4 18 12 0	0.0000	
88:35	1 7 10 12 14 19 28 34 36 40 42 45	D_2	22×4	192, 258, 192	0 26 8 0	0.0553	26(4)

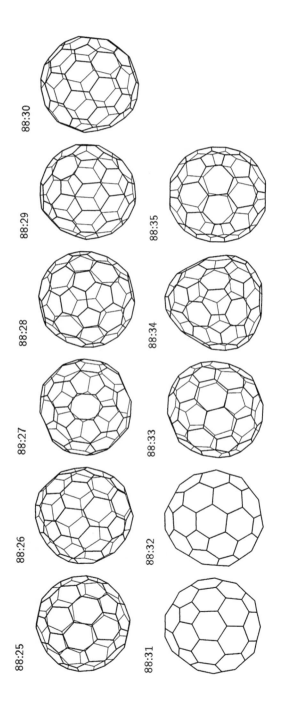

88:30

88:29

88:28

88:27

88:26

88:25

88:35

88:34

88:33

88:32

88:31

Table A.15. Isolated-pentagon fullerene isomers of C_{90}

Isomer	Ring spiral	Point group	NMR pattern	Vibrations	Hexagon indices	Band gap	Transformations
90:1	1 7 9 11 13 15 37 39 41 43 45 47	D_{5h}	3×10, 3×20	40, 68, 0	10 10 10 5	0.4989	
90:2	1 7 9 11 13 24 27 38 40 42 44 47	C_{2v}	7×2, 19×4	201, 264, 201	6 18 6 5	0.5988	3(4)
90:3	1 7 9 11 13 24 28 37 40 42 44 47	C_1	90×1	264, 264, 264	5 18 9 3	0.3085	2, 5, 8
90:4	1 7 9 11 13 25 30 38 41 43 45 47	C_2	45×2	264, 264, 264	6 14 14 1	0.1946	
90:5	1 7 9 11 13 28 34 36 38 41 43 47	C_s	6×1, 42×2	264, 264, 264	4 19 10 2	0.1340	3(2)
90:6	1 7 9 11 14 22 28 31 36 43 45 47	C_2	45×2	264, 264, 264	4 17 14 0	0.1036	7(2)
90:7	1 7 9 11 14 22 28 36 40 42 44 46	C_1	90×1	264, 264, 264	3 20 11 1	0.0733	6, 9, 23
90:8	1 7 9 11 14 23 28 37 40 42 44 47	C_2	45×2	264, 264, 264	4 18 12 1	0.1356	3(2)
90:9	1 7 9 11 14 28 32 34 37 41 44 46	C_1	90×1	264, 264, 264	3 19 13 0	0.0882	7, 10
90:10	1 7 9 11 14 28 32 35 37 41 43 46	C_s	8×1, 41×2	264, 264, 264	2 21 12 0	0.0747	9(2)
90:11	1 7 9 11 22 24 27 30 36 39 43 47	C_1	90×1	264, 264, 264	3 20 11 1	0.1080	12, 15, 18, 20
90:12	1 7 9 11 22 24 28 30 36 38 43 47	C_2	45×2	264, 264, 264	2 23 8 2	0.1568	11(2), 14(2), 26(2)
90:13	1 7 9 11 22 24 28 32 38 40 44 46	C_{2v}	5×2, 20×4	200, 264, 200	4 19 10 2	0.1696	17(2), 36
90:14	1 7 9 11 22 24 28 36 38 40 42 44	C_1	90×1	264, 264, 264	3 22 7 3	0.2981	12, 15, 16, 17, 27, 32
90:15	1 7 9 11 22 25 27 30 35 39 43 47	C_1	90×1	264, 264, 264	4 19 10 2	0.3974	11, 14, 30
90:16	1 7 9 11 22 25 28 35 38 40 42 44	C_{2v}	5×2, 20×4	200, 264, 200	4 22 4 5	0.6499	14(4), 34(2)
90:17	1 7 9 11 22 26 28 30 35 37 43 47	C_s	4×1, 43×2	264, 264, 264	3 21 9 2	0.3389	13, 14(2), 35, 39
90:18	1 7 9 11 24 27 30 33 35 37 40 47	C_2	45×2	264, 264, 264	4 17 14 0	0.2990	11(2)
90:19	1 7 9 12 14 20 28 34 36 39 41 47	C_2	45×2	264, 264, 264	2 21 12 0	0.1173	21(2)
90:20	1 7 9 12 14 20 29 33 36 38 43 46	C_1	90×1	264, 264, 264	2 21 12 0	0.2044	11, 22, 26, 30
90:21	1 7 9 12 14 20 29 34 36 38 41 47	C_1	90×1	264, 264, 264	1 23 11 0	0.1167	19, 22, 44
90:22	1 7 9 12 14 20 29 34 36 38 42 46	C_1	90×1	264, 264, 264	2 21 12 0	0.0617	20, 21, 38
90:23	1 7 9 12 14 21 28 36 40 42 44 46	C_2	45×2	264, 264, 264	2 21 12 0	0.0637	7(2)
90:24	1 7 9 12 14 28 30 33 36 40 43 45	C_1	90×1	264, 264, 264	3 20 11 1	0.1997	25, 28

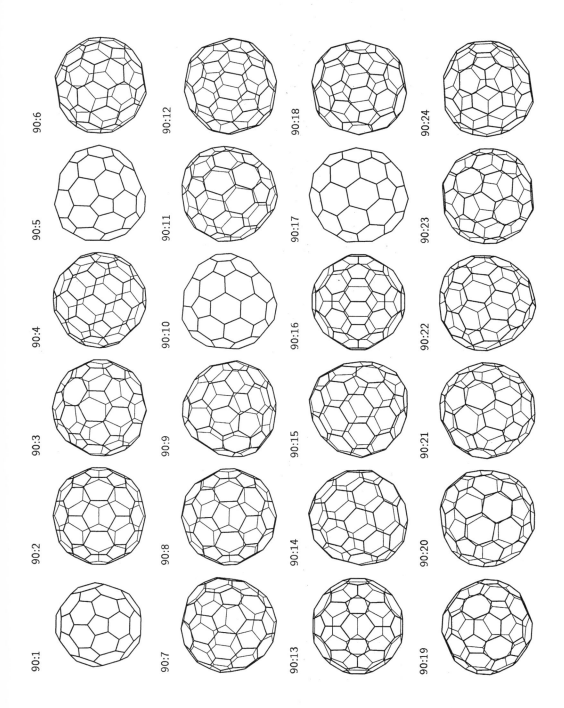

90:6

90:12

90:18

90:24

90:5

90:11

90:17

90:23

90:4

90:10

90:16

90:22

90:3

90:9

90:15

90:21

90:2

90:8

90:14

90:20

90:1

90:7

90:13

90:19

Table A.15. *(Continued)*

Isomer	Ring spiral	Point group	NMR pattern	Vibrations	Hexagon indices	Band gap	Transformations
90:25	1 7 9 12 14 28 30 34 36 40 42 45	C_{2v}	9×2, 18×4	202, 264, 202	2 23 8 2	0.1095	24(4)
90:26	1 7 9 12 21 24 28 30 36 38 43 47	C_1	90×1	264, 264, 264	1 24 9 1	0.2223	12, 20, 27, 32, 38, 40
90:27	1 7 9 12 21 24 28 36 38 40 42 44	C_1	90×1	264, 264, 264	2 23 8 2	0.3535	14, 26, 34, 37, 39, 45
90:28	1 7 9 12 21 25 27 30 34 39 42 45	C_2	45×2	264, 264, 264	2 21 12 0	0.2811	24(2), 29(2)
90:29	1 7 9 12 21 25 27 30 34 39 44 47	C_1	90×1	264, 264, 264	3 20 11 1	0.3541	28, 30, 31
90:30	1 7 9 12 21 25 27 30 35 39 43 47	C_1	90×1	264, 264, 264	2 21 12 0	0.2793	15, 20, 29, 32
90:31	1 7 9 12 21 25 28 30 34 38 44 47	C_2	45×2	264, 264, 264	2 23 8 2	0.3742	29(2), 32(2), 33(2)
90:32	1 7 9 12 21 25 28 30 35 38 43 47	C_1	90×1	264, 264, 264	1 24 9 1	0.3008	14, 26, 30, 31, 34, 35, 45
90:33	1 7 9 12 21 25 28 34 38 40 42 45	C_s	4×1, 43×2	264, 264, 264	3 21 9 2	0.4160	31(2), 34, 35
90:34	1 7 9 12 21 25 28 35 38 40 42 44	C_s	2×1, 44×2	264, 264, 264	2 24 6 3	0.3402	16, 27(2), 32(2), 33, 46
90:35	1 7 9 12 21 26 28 30 35 37 43 47	C_s	2×1, 44×2	264, 264, 264	1 23 11 0	0.3134	17, 32(2), 33, 36, 46
90:36	1 7 9 12 21 26 28 30 35 43 45 47	C_{2v}	5×2, 20×4	200, 264, 200	2 21 12 0	0.2455	13, 35(2)
90:37	1 7 9 13 20 23 25 29 37 41 43 46	C_2	45×2	264, 264, 264	2 23 8 2	0.1959	27(2), 38(2), 40
90:38	1 7 9 13 20 23 26 29 36 41 43 46	C_1	90×1	264, 264, 264	1 24 9 1	0.1345	22, 26, 37, 42, 44
90:39	1 7 9 13 23 25 29 31 33 39 43 45	C_{2v}	3×2, 21×4	199, 264, 199	2 24 6 3	0.4170	17(2), 27(4), 46
90:40	1 7 10 12 14 18 30 33 36 38 43 46	C_2	45×2	264, 264, 264	0 25 10 0	0.1922	26(2), 37, 42(2), 45(2)
90:41	1 7 10 12 14 18 30 34 36 38 41 47	C_2	45×2	264, 264, 264	0 25 10 0	0.0814	42(2), 43(2), 44
90:42	1 7 10 12 14 18 30 34 36 38 42 46	C_2	45×2	264, 264, 264	0 25 10 0	0.1098	38(2), 40(2), 41(2)
90:43	1 7 10 12 14 18 31 34 36 38 40 47	C_2	45×2	264, 264, 264	0 25 10 0	0.0071	41(2), 43(2)
90:44	1 7 10 12 14 19 29 34 36 38 41 47	C_2	45×2	264, 264, 264	0 25 10 0	0.1817	21(2), 38(2), 41
90:45	1 7 10 13 19 25 27 33 35 39 42 47	C_2	45×2	264, 264, 264	0 25 10 0	0.2518	27(2), 32(2), 40(2), 46(2)
90:46	1 7 10 13 19 25 28 33 35 38 42 47	C_{2v}	3×2, 21×4	199, 264, 199	0 26 8 1	0.2368	34(2), 35(2), 39, 45(4)

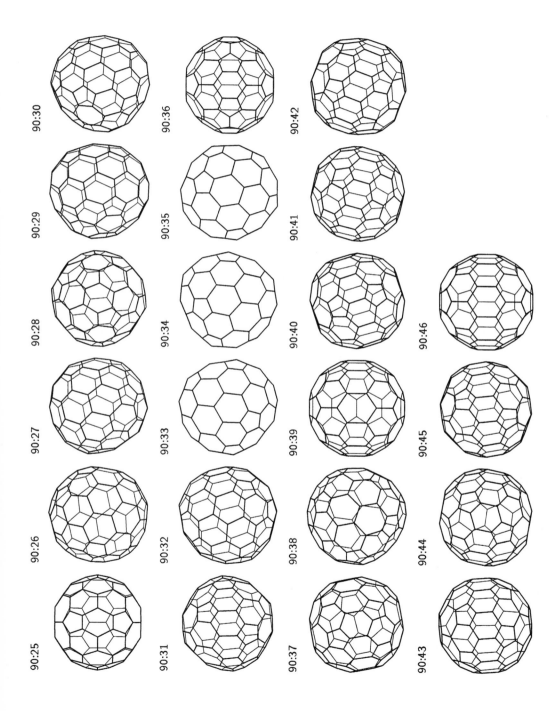

90:30

90:29

90:28

90:27

90:26

90:25

90:36

90:35

90:34

90:33

90:32

90:31

90:42

90:41

90:40

90:39

90:38

90:37

90:46

90:45

90:44

90:43

Table A.16. Isolated-pentagon fullerene isomers of C_{92}

Isomer	Ring spiral	Point group	NMR pattern	Vibrations	Hexagon indices	Band gap	Transformations
92:1	1 7 9 11 13 18 32 34 37 44 46 48	D_2	23×4	201, 270, 201	8 12 12 4	0.1616	
92:2	1 7 9 11 13 24 29 31 37 42 44 46	C_1	92×1	270, 270, 270	5 16 13 2	0.0315	3, 9
92:3	1 7 9 11 13 24 30 37 40 42 44 46	C_2	46×2	270, 270, 270	6 16 10 4	0.3466	2(2)
92:4	1 7 9 11 13 25 28 40 42 44 46 48	C_2	46×2	270, 270, 270	6 14 14 2	0.2861	
92:5	1 7 9 11 13 28 31 34 37 44 46 48	C_s	6×1, 43×2	270, 270, 270	5 15 15 1	0.2293	6
92:6	1 7 9 11 13 28 34 37 41 43 45 47	C_s	4×1, 44×2	270, 270, 270	4 18 12 2	0.0240	5
92:7	1 7 9 11 14 22 29 35 38 42 45 47	C_2	46×2	270, 270, 270	4 18 12 2	0.2543	8(2), 13(2)
92:8	1 7 9 11 14 22 29 36 38 42 44 47	C_1	92×1	270, 270, 270	3 19 13 1	0.1217	7, 12, 34
92:9	1 7 9 11 14 23 29 31 37 42 44 46	C_2	46×2	270, 270, 270	4 16 16 0	0.2059	2(2)
92:10	1 7 9 11 14 28 31 33 35 37 40 48	C_1	92×1	270, 270, 270	2 20 14 0	0.1041	11, 14, 15
92:11	1 7 9 11 14 28 31 33 35 37 46 48	C_1	92×1	270, 270, 270	3 18 15 0	0.0883	10, 12
92:12	1 7 9 11 14 28 31 33 35 38 45 48	C_1	92×1	270, 270, 270	3 18 15 0	0.1599	8, 11, 13
92:13	1 7 9 11 14 28 31 33 36 38 44 48	C_1	92×1	270, 270, 270	4 17 14 1	0.0750	7, 12, 26
92:14	1 7 9 11 14 28 32 35 37 40 42 48	C_s	6×1, 43×2	270, 270, 270	3 19 13 1	0.2047	10(2)
92:15	1 7 9 11 14 29 31 33 35 37 39 48	C_s	6×1, 43×2	270, 270, 270	2 20 14 0	0.0506	10(2), 16
92:16	1 7 9 11 14 30 33 35 37 39 41 48	C_s	8×1, 42×2	270, 270, 270	2 20 14 0	0.1538	15
92:17	1 7 9 11 22 24 27 29 40 42 44 46	C_2	46×2	270, 270, 270	4 18 12 2	0.1075	18(2)
92:18	1 7 9 11 22 24 27 30 39 42 44 46	C_1	92×1	270, 270, 270	3 19 13 1	0.1293	17, 20, 46
92:19	1 7 9 11 22 24 28 30 37 42 44 47	C_2	46×2	270, 270, 270	2 22 10 2	0.0796	20(2), 47(2)
92:20	1 7 9 11 22 24 28 30 37 43 45 47	C_1	92×1	270, 270, 270	3 20 11 2	0.0743	18, 19, 30, 39
92:21	1 7 9 11 22 26 29 35 39 41 44 46	C_s	2×1, 45×2	270, 270, 270	5 17 11 3	0.4162	22, 45
92:22	1 7 9 11 22 27 29 35 38 41 44 46	C_{2v}	4×2, 21×4	204, 270, 204	4 20 8 4	0.0605	21(2), 35
92:23	1 7 9 11 23 25 27 29 41 43 45 48	C_2	46×2	270, 270, 270	4 17 14 1	0.1282	
92:24	1 7 9 11 23 25 27 30 32 44 46 48	C_s	6×1, 43×2	270, 270, 270	3 18 15 0	0.0490	

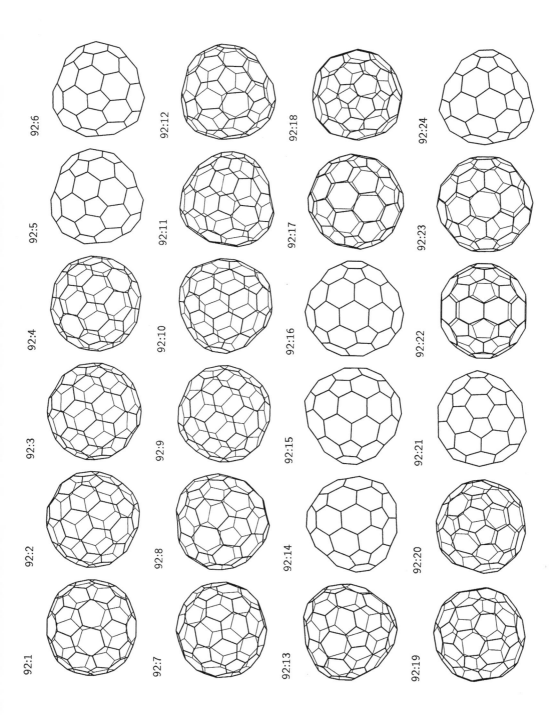

92:6

92:12

92:18

92:24

92:5

92:11

92:17

92:23

92:4

92:10

92:16

92:22

92:3

92:9

92:15

92:21

92:2

92:8

92:14

92:20

92:1

92:7

92:13

92:19

Table A.16. *(Continued)*

Isomer	Ring spiral	Point group	NMR pattern	Vibrations	Hexagon indices	Band gap	Transformations
92:25	1 7 9 11 23 25 29 33 39 41 44 46	C_2	46×2	270, 270, 270	4 18 12 2	0.1553	
92:26	1 7 9 11 23 26 30 32 34 37 39 47	C_2	46×2	270, 270, 270	4 16 16 0	0.3421	13(2), 27
92:27	1 7 9 11 23 26 30 33 36 40 46 48	C_2	46×2	270, 270, 270	4 16 16 0	0.0748	26
92:28	1 7 9 11 24 26 30 32 34 37 39 48	D_3	1×2, 15×6	134, 136, 90	6 12 18 0	0.4595	
92:29	1 7 9 12 14 18 30 34 38 42 44 47	D_{2h}	5×4, 9×8	103, 135, 0	4 18 12 2	0.1344	
92:30	1 7 9 12 14 20 28 35 39 41 43 45	C_1	92×1	270, 270, 270	2 21 12 1	0.1805	20, 33, 38, 47
92:31	1 7 9 12 14 20 29 34 37 41 44 46	C_2	46×2	270, 270, 270	2 20 14 0	0.0827	32(2)
92:32	1 7 9 12 14 20 29 35 37 41 43 46	C_1	92×1	270, 270, 270	1 22 13 0	0.0340	31, 33, 42, 65
92:33	1 7 9 12 14 20 29 35 38 41 43 45	C_1	92×1	270, 270, 270	2 21 12 1	0.0970	30, 32, 43, 56
92:34	1 7 9 12 14 21 29 36 38 42 44 47	C_2	46×2	270, 270, 270	2 20 14 0	0.0682	8(2)
92:35	1 7 9 12 14 28 30 34 36 41 46 48	C_{2v}	4×2, 21×4	204, 270, 204	2 22 10 2	0.1040	22, 45(2)
92:36	1 7 9 12 14 28 31 33 35 38 40 48	C_2	46×2	270, 270, 270	2 20 14 0	0.0745	
92:37	1 7 9 12 20 24 26 29 36 40 42 48	C_1	92×1	270, 270, 270	3 20 11 2	0.4083	38, 40, 48, 49
92:38	1 7 9 12 20 24 26 29 40 42 44 46	C_1	92×1	270, 270, 270	1 22 13 0	0.2983	30, 37, 39, 43, 50, 77
92:39	1 7 9 12 20 24 26 29 41 44 46 48	C_1	92×1	270, 270, 270	2 21 12 1	0.2870	20, 38, 46, 47, 51
92:40	1 7 9 12 20 24 26 30 36 39 42 48	C_1	92×1	270, 270, 270	2 21 12 1	0.1355	37, 41, 43, 52(2)
92:41	1 7 9 12 20 24 26 30 36 39 43 47	C_3	2×1, 30×3	176, 176, 176	3 18 15 0	0.3145	40(3)
92:42	1 7 9 12 20 24 26 30 38 42 44 47	C_1	92×1	270, 270, 270	1 22 13 0	0.1248	32, 43, 44, 53, 59
92:43	1 7 9 12 20 24 26 30 39 42 44 46	C_1	92×1	270, 270, 270	1 22 13 0	0.2413	33, 38, 40, 42, 54, 76
92:44	1 7 9 12 20 24 27 30 37 42 44 47	C_1	92×1	270, 270, 270	1 22 13 0	0.2204	42, 55, 66
92:45	1 7 9 12 20 24 29 34 37 39 41 45	C_s	4×1, 44×2	270, 270, 270	3 19 13 1	0.2246	21, 35
92:46	1 7 9 12 21 24 27 30 39 42 44 46	C_2	46×2	270, 270, 270	2 20 14 0	0.1620	18(2), 39(2)
92:47	1 7 9 12 21 24 28 30 37 42 44 47	C_1	92×1	270, 270, 270	1 23 11 1	0.1383	19, 30, 39, 56, 77
92:48	1 7 9 13 20 23 25 30 37 39 42 48	C_2	46×2	270, 270, 270	2 22 10 2	0.0914	37(2), 52(2), 77

92:30

92:29

92:28

92:27

92:26

92:25

92:36

92:35

92:34

92:33

92:32

92:31

92:42

92:41

92:40

92:39

92:38

92:37

92:48

92:47

92:46

92:45

92:44

92:43

Table A.16. *(Continued)*

Isomer	Ring spiral	Point group	NMR pattern	Vibrations	Hexagon indices	Band gap	Transformations
92:49	1 7 9 13 20 23 26 29 36 40 42 48	C_2	46×2	270, 270, 270	2 22 10 2	0.3092	37(2), 50(2), 52(2)
92:50	1 7 9 13 20 23 26 29 40 42 44 46	C_1	92×1	270, 270, 270	1 23 11 1	0.3610	38, 49, 51, 54, 57, 76, 84
92:51	1 7 9 13 20 23 26 29 41 44 46 48	C_2	46×2	270, 270, 270	2 22 10 2	0.2887	39(2), 50(2), 56(2)
92:52	1 7 9 13 20 23 26 30 36 39 42 48	C_1	92×1	270, 270, 270	1 23 11 1	0.0622	40(2), 48, 49, 54, 76, 78
92:53	1 7 9 13 20 23 26 30 38 42 44 47	C_1	92×1	270, 270, 270	1 23 11 1	0.1673	42, 54, 55, 58, 60, 63
92:54	1 7 9 13 20 23 26 30 39 42 44 46	C_1	92×1	270, 270, 270	1 23 11 1	0.1744	43, 50, 52, 53, 58, 70, 79
92:55	1 7 9 13 20 23 27 30 37 42 44 47	C_1	92×1	270, 270, 270	1 23 11 1	0.2019	44, 53, 57, 60, 66
92:56	1 7 9 13 20 25 27 30 34 42 45 47	C_1	92×1	270, 270, 270	1 23 11 1	0.1611	33, 47, 51, 57, 65, 76
92:57	1 7 9 13 20 26 29 34 36 41 43 46	C_1	92×1	270, 270, 270	1 23 11 1	0.2381	50, 55, 56, 58, 59, 70
92:58	1 7 9 13 20 26 30 34 36 39 43 47	C_1	92×1	270, 270, 270	1 23 11 1	0.1672	53, 54, 57, 58, 63, 74
92:59	1 7 10 12 14 18 30 35 37 41 43 46	C_1	92×1	270, 270, 270	0 24 12 0	0.0176	42, 57, 63, 65, 66, 69, 76
92:60	1 7 10 12 14 18 31 34 37 39 44 47	C_1	92×1	270, 270, 270	0 24 12 0	0.0964	53, 55, 62, 63, 66, 67, 70
92:61	1 7 10 12 14 18 31 35 37 39 42 48	C_2	46×2	270, 270, 270	0 24 12 0	0.1324	62(2), 64(2), 68(2), 71
92:62	1 7 10 12 14 18 31 35 37 39 43 47	C_1	92×1	270, 270, 270	0 24 12 0	0.0016	60, 61, 63, 67, 68, 72, 73, 80
92:63	1 7 10 12 14 18 31 35 37 40 43 46	C_1	92×1	270, 270, 270	0 24 12 0	0.1118	53, 58, 59, 60, 62, 73, 74, 79
92:64	1 7 10 12 14 18 32 35 37 39 41 48	C_2	46×2	270, 270, 270	0 24 12 0	0.0394	61(2), 67(2), 85
92:65	1 7 10 12 14 19 29 35 37 41 43 46	C_2	46×2	270, 270, 270	0 24 12 0	0.0363	32(2), 56(2), 59(2)
92:66	1 7 10 12 14 28 30 32 35 37 46 48	C_1	92×1	270, 270, 270	0 24 12 0	0.1680	44, 55, 59, 60, 67
92:67	1 7 10 12 14 28 30 32 35 38 45 48	C_1	92×1	270, 270, 270	0 24 12 0	0.0868	60, 62, 64, 66, 68, 69
92:68	1 7 10 12 14 28 30 32 36 38 43 47	C_1	92×1	270, 270, 270	0 24 12 0	0.0790	61, 62, 67, 68, 72, 73, 75
92:69	1 7 10 12 18 24 26 31 38 42 44 47	C_2	46×2	270, 270, 270	0 24 12 0	0.0544	59(2), 67(2), 70(2), 73(2)
92:70	1 7 10 12 18 24 26 31 39 42 44 46	C_1	92×1	270, 270, 270	0 24 12 0	0.1580	54, 57, 60, 69, 74, 76, 80, 84
92:71	1 7 10 12 18 24 26 32 38 40 43 45	D_3	1×2, 15×6	134, 136, 90	0 24 12 0	0.2586	61(3), 72(6)
92:72	1 7 10 12 18 24 26 32 38 40 44 48	C_1	92×1	270, 270, 270	0 24 12 0	0.0973	62, 68, 71, 73, 75(2), 81, 82, 83

92:54

92:60

92:66

92:72

92:53

92:59

92:65

92:71

92:52

92:58

92:64

92:70

92:51

92:57

92:63

92:69

92:50

92:56

92:62

92:68

92:49

92:55

92:61

92:67

Table A.16. (*Continued*)

Isomer	Ring spiral	Point group	NMR pattern	Vibrations	Hexagon indices	Band gap	Transformations
92:73	1 7 10 12 18 24 26 32 38 41 44 47	C_1	92×1	270, 270, 270	0 24 12 0	0.1367	62, 63, 68, 69, 72, 73, 74, 80, 82
92:74	1 7 10 12 18 24 26 32 39 41 44 46	C_2	46×2	270, 270, 270	0 24 12 0	0.0335	58(2), 63(2), 70(2), 73(2)
92:75	1 7 10 12 18 24 27 32 37 40 44 48	C_2	46×2	270, 270, 270	0 24 12 0	0.0764	68(2), 72(4), 81, 86
92:76	1 7 10 12 18 25 27 32 34 39 44 47	C_1	92×1	270, 270, 270	0 24 12 0	0.0890	43, 50, 52, 56, 59, 70, 77, 79
92:77	1 7 10 12 18 25 27 32 34 39 45 48	C_2	46×2	270, 270, 270	0 24 12 0	0.1012	38(2), 47(2), 48, 76(2)
92:78	1 7 10 13 18 22 27 34 36 39 42 47	D_3	1×2, 15×6	134, 136, 90	0 24 12 0	0.0623	52(6), 79(3)
92:79	1 7 10 13 18 22 27 34 36 40 42 46	C_2	46×2	270, 270, 270	0 24 12 0	0.0143	54(2), 63(2), 76(2), 78, 80(2)
92:80	1 7 10 13 18 23 26 31 39 42 44 46	C_2	46×2	270, 270, 270	0 24 12 0	0.0189	62(2), 70(2), 73(2), 79(2), 83
92:81	1 7 10 13 18 23 26 32 38 40 44 48	D_2	23×4	201, 270, 201	0 24 12 0	0.2085	72(4), 75(2), 81(2), 82(2)
92:82	1 7 10 13 18 23 26 32 38 41 44 47	D_2	23×4	201, 270, 201	0 24 12 0	0.2632	72(4), 73(4), 81(2)
92:83	1 7 10 13 18 27 32 34 36 38 42 47	D_3	1×2, 15×6	134, 136, 90	0 24 12 0	0.0754	72(6), 80(3)
92:84	1 7 10 13 19 27 30 33 36 38 44 47	D_2	23×4	201, 270, 201	0 24 12 0	0.3145	50(4), 70(4)
92:85	1 7 12 14 19 21 29 34 36 38 42 47	D_3	1×2, 15×6	134, 136, 90	0 24 12 0	0.0394	64(3)
92:86	1 7 12 14 19 22 29 32 37 39 44 48	T	2×4, 7×12	67, 113, 67	0 24 12 0	0.2498	75(6)

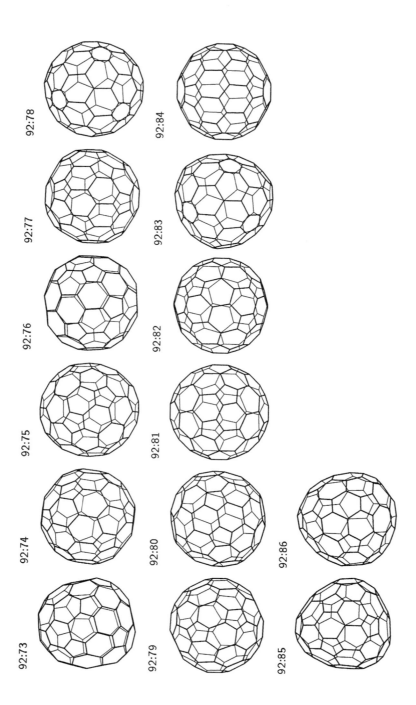

92:78

92:84

92:77

92:83

92:76

92:82

92:75

92:81

92:74

92:80

92:86

92:73

92:79

92:85

Table A.17. Isolated-pentagon fullerene isomers of C_{94}

Isomer	Ring spiral	Point group	NMR pattern	Vibrations	Hexagon indices	Band gap	Transformations
94:1	1 7 9 11 13 24 29 33 38 45 47 49	C_2	47×2	276, 276, 276	6 15 12 4	0.1412	2(2)
94:2	1 7 9 11 13 24 29 38 42 44 46 48	C_1	94×1	276, 276, 276	5 17 11 4	0.3336	1, 6, 9
94:3	1 7 9 11 13 25 37 39 41 43 45 49	C_2	47×2	276, 276, 276	6 14 14 3	0.2354	3
94:4	1 7 9 11 13 28 32 37 43 45 47 49	C_2	47×2	276, 276, 276	6 14 14 3	0.2955	
94:5	1 7 9 11 13 28 33 35 37 41 47 49	C_1	94×1	276, 276, 276	4 17 14 2	0.2685	6
94:6	1 7 9 11 13 28 33 35 38 41 46 49	C_1	94×1	276, 276, 276	5 16 13 3	0.2152	2, 5
94:7	1 7 9 11 14 22 28 36 39 41 43 49	C_2	47×2	276, 276, 276	4 17 14 2	0.2618	8(2)
94:8	1 7 9 11 14 22 29 36 38 41 43 49	C_1	94×1	276, 276, 276	3 18 15 1	0.0792	7, 51
94:9	1 7 9 11 14 23 29 38 42 44 46 48	C_2	47×2	276, 276, 276	4 17 14 2	0.2940	2(2)
94:10	1 7 9 11 14 28 31 33 36 40 44 46	C_1	94×1	276, 276, 276	3 19 13 2	0.0907	11, 13, 17, 34
94:11	1 7 9 11 14 28 31 34 36 40 43 46	C_1	94×1	276, 276, 276	2 20 14 1	0.0306	10, 12, 16, 36
94:12	1 7 9 11 14 28 31 34 37 40 43 45	C_1	94×1	276, 276, 276	3 17 17 0	0.0402	11, 15
94:13	1 7 9 11 14 28 32 36 40 42 44 46	C_s	4×1, 45×2	276, 276, 276	3 20 11 3	0.1334	10(2), 14, 39
94:14	1 7 9 11 14 28 32 36 41 44 46 48	C_s	6×1, 44×2	276, 276, 276	4 17 14 2	0.0162	13, 31
94:15	1 7 9 11 14 29 31 33 35 38 45 47	C_1	94×1	276, 276, 276	3 17 17 0	0.2423	12, 16, 21
94:16	1 7 9 11 14 29 31 33 36 38 44 47	C_1	94×1	276, 276, 276	2 20 14 1	0.1967	11, 15, 17, 18, 22, 43
94:17	1 7 9 11 14 29 31 33 36 39 44 46	C_s	4×1, 45×2	276, 276, 276	3 20 11 3	0.3474	10(2), 16(2), 23, 44
94:18	1 7 9 11 14 29 31 34 36 38 43 47	C_s	4×1, 45×2	276, 276, 276	2 19 16 0	0.1790	16(2), 23, 44, 45
94:19	1 7 9 11 14 30 33 35 37 40 46 48	C_1	94×1	276, 276, 276	3 18 15 1	0.1113	20, 24
94:20	1 7 9 11 14 30 33 35 38 40 45 48	C_2	47×2	276, 276, 276	2 19 16 0	0.0122	19(2), 21(2)
94:21	1 7 9 11 14 30 33 35 38 41 45 47	C_1	94×1	276, 276, 276	3 18 15 1	0.0056	15, 20, 22
94:22	1 7 9 11 14 30 33 36 38 41 44 47	C_2	47×2	276, 276, 276	2 21 12 2	0.0932	16(2), 21(2), 23(2)
94:23	1 7 9 11 14 30 33 36 39 41 44 46	C_s	6×1, 44×2	276, 276, 276	3 19 13 2	0.0238	17, 18, 22(2)
94:24	1 7 9 11 14 30 35 37 40 42 44 47	C_{2v}	9×2, 19×4	211, 276, 211	2 21 12 2	0.0361	19(4)

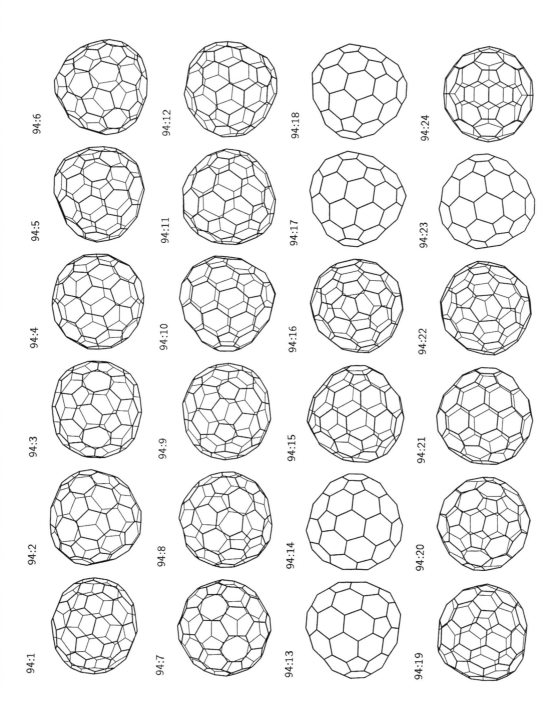

94:6

94:5

94:4

94:3

94:2

94:1

94:12

94:11

94:10

94:9

94:8

94:7

94:18

94:17

94:16

94:15

94:14

94:13

94:24

94:23

94:22

94:21

94:20

94:19

Table A.17. (*Continued*)

Isomer	Ring spiral	Point group	NMR pattern	Vibrations	Hexagon indices	Band gap	Transformations
94:25	1 7 9 11 22 24 28 37 40 42 45 49	C_1	94×1	276, 276, 276	3 20 11 3	0.0150	29, 48, 60
94:26	1 7 9 11 22 24 29 36 39 41 44 47	C_2	47×2	276, 276, 276	4 19 10 4	0.2525	28(2), 30(2)
94:27	1 7 9 11 22 24 29 37 39 41 43 45	C_1	94×1	276, 276, 276	3 20 11 3	0.1541	28, 29, 88
94:28	1 7 9 11 22 24 29 37 39 41 44 46	C_1	94×1	276, 276, 276	3 20 11 3	0.2755	26, 27, 41, 89
94:29	1 7 9 11 22 24 29 37 39 42 45 49	C_1	94×1	276, 276, 276	2 21 12 2	0.1002	25, 27, 90
94:30	1 7 9 11 22 25 29 35 39 41 44 47	C_1	94×1	276, 276, 276	5 17 11 4	0.3902	26, 41
94:31	1 7 9 11 23 25 27 30 32 43 45 47	C_s	4×1, 45×2	276, 276, 276	3 17 17 0	0.2648	14, 39
94:32	1 7 9 11 23 25 30 33 38 40 45 48	C_1	94×1	276, 276, 276	4 16 16 1	0.1811	
94:33	1 7 9 11 23 26 29 32 36 41 44 47	C_2	47×2	276, 276, 276	2 19 16 0	0.1673	36(2)
94:34	1 7 9 11 23 26 30 32 36 39 44 48	C_1	94×1	276, 276, 276	2 19 16 0	0.2599	10, 35, 36, 39, 44
94:35	1 7 9 11 23 26 30 32 36 39 46 49	C_1	94×1	276, 276, 276	3 18 15 1	0.2042	34, 37, 40, 42
94:36	1 7 9 11 23 26 30 32 36 40 44 47	C_1	94×1	276, 276, 276	2 19 16 0	0.1471	11, 33, 34, 43
94:37	1 7 9 11 23 26 30 32 37 39 45 49	C_1	94×1	276, 276, 276	4 15 18 0	0.3308	35
94:38	1 7 9 11 23 27 29 34 36 41 43 47	C_1	94×1	276, 276, 276	4 18 12 3	0.1929	41
94:39	1 7 9 11 23 27 30 32 36 38 44 48	C_s	2×1, 46×2	276, 276, 276	2 20 14 1	0.3002	13, 31, 34(2), 40
94:40	1 7 9 11 23 27 30 32 36 38 46 49	C_s	4×1, 45×2	276, 276, 276	4 18 12 3	0.3786	35(2), 39
94:41	1 7 9 11 23 27 30 34 36 40 43 47	C_1	94×1	276, 276, 276	3 19 13 2	0.2370	28, 30, 38
94:42	1 7 9 11 24 26 30 34 37 40 49	C_s	4×1, 45×2	276, 276, 276	3 17 17 0	0.3940	35(2), 44, 45
94:43	1 7 9 11 24 26 30 32 34 39 45 48	C_2	47×2	276, 276, 276	2 19 16 0	0.3667	16(2), 36(2), 44(2)
94:44	1 7 9 11 24 26 30 32 34 40 45 47	C_s	2×1, 46×2	276, 276, 276	2 20 14 1	0.2478	17, 18, 34(2), 42, 43(2)
94:45	1 7 9 11 24 26 31 34 37 40 42 49	C_s	6×1, 44×2	276, 276, 276	3 17 17 0	0.0789	18, 42
94:46	1 7 9 12 14 18 34 39 41 43 45 47	C_s	6×1, 44×2	276, 276, 276	4 19 10 4	0.1789	52
94:47	1 7 9 12 14 20 29 35 37 42 47 49	C_2	47×2	276, 276, 276	2 19 16 0	0.1145	49(2), 61
94:48	1 7 9 12 14 20 34 37 39 41 45 48	C_1	94×1	276, 276, 276	2 21 12 2	0.1418	25, 50, 59, 90

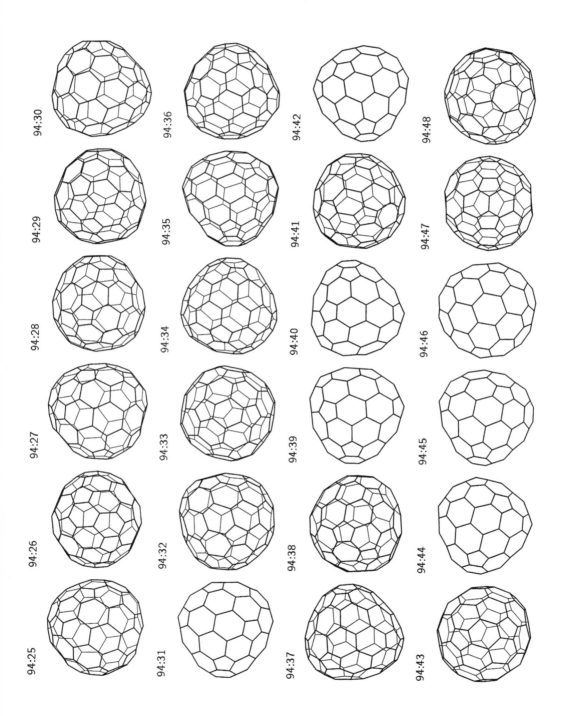

94:30

94:29

94:28

94:27

94:26

94:25

94:36

94:35

94:34

94:33

94:32

94:31

94:42

94:41

94:40

94:39

94:38

94:37

94:48

94:47

94:46

94:45

94:44

94:43

Table A.17. (*Continued*)

Isomer	Ring spiral	Point group	NMR pattern	Vibrations	Hexagon indices	Band gap	Transformations
94:49	1 7 9 12 14 20 35 37 39 41 43 49	C_1	94×1	276, 276, 276	1 22 13 1	0.0861	47, 50, 57, 62, 121
94:50	1 7 9 12 14 20 35 37 39 41 44 48	C_1	94×1	276, 276, 276	2 21 12 2	0.1135	48, 49, 56, 63, 76
94:51	1 7 9 12 14 21 29 36 38 41 43 49	C_2	47×2	276, 276, 276	2 19 16 0	0.1795	8(2)
94:52	1 7 9 12 14 28 30 35 41 43 45 47	C_{2v}	7×2, 20×4	210, 276, 210	2 21 12 2	0.2343	46(2)
94:53	1 7 9 12 14 28 31 34 38 40 43 45	C_1	94×1	276, 276, 276	3 18 15 1	0.1636	55, 94
94:54	1 7 9 12 14 28 32 34 37 41 47 49	C_1	94×1	276, 276, 276	1 21 15 0	0.1029	57, 95
94:55	1 7 9 12 14 28 32 34 38 40 42 45	C_1	94×1	276, 276, 276	2 20 14 1	0.1742	53, 56, 58, 93
94:56	1 7 9 12 14 28 32 34 38 41 45 48	C_1	94×1	276, 276, 276	1 21 15 0	0.0853	50, 55, 57, 59, 92
94:57	1 7 9 12 14 28 32 34 38 41 46 49	C_1	94×1	276, 276, 276	1 21 15 0	0.1470	49, 54, 56, 91
94:58	1 7 9 12 14 28 32 35 38 40 42 44	C_1	94×1	276, 276, 276	3 19 13 2	0.1888	55, 59
94:59	1 7 9 12 14 28 32 35 38 41 44 48	C_1	94×1	276, 276, 276	1 21 15 0	0.1594	48, 56, 58, 60
94:60	1 7 9 12 14 28 33 35 38 41 43 48	C_1	94×1	276, 276, 276	2 20 14 1	0.1010	25, 59
94:61	1 7 9 12 20 24 26 30 38 43 45 48	C_2	47×2	276, 276, 276	2 19 16 0	0.2937	47, 62(2), 66(2)
94:62	1 7 9 12 20 24 26 38 40 42 44 46	C_1	94×1	276, 276, 276	1 22 13 1	0.1275	49, 61, 63, 69, 91, 113, 126
94:63	1 7 9 12 20 24 26 38 40 42 45 49	C_1	94×1	276, 276, 276	2 21 12 2	0.3243	50, 62, 70, 77, 84, 92
94:64	1 7 9 12 20 24 27 30 36 42 46 49	C_1	94×1	276, 276, 276	2 20 14 1	0.0676	64, 65, 74, 87
94:65	1 7 9 12 20 24 27 30 36 43 46 48	C_1	94×1	276, 276, 276	1 21 15 0	0.1821	64, 66, 68, 75, 87, 127
94:66	1 7 9 12 20 24 27 30 37 43 45 48	C_1	94×1	276, 276, 276	2 20 14 1	0.0871	61, 65, 69, 86, 113
94:67	1 7 9 12 20 24 27 36 39 42 44 48	C_1	94×1	276, 276, 276	1 22 13 1	0.1672	68, 71(2), 81, 96, 115
94:68	1 7 9 12 20 24 27 36 40 42 44 47	C_1	94×1	276, 276, 276	1 22 13 1	0.1367	65, 67, 69, 82, 85, 97, 128
94:69	1 7 9 12 20 24 27 37 40 42 44 46	C_1	94×1	276, 276, 276	1 22 13 1	0.0384	62, 66, 68, 70, 83, 98, 129
94:70	1 7 9 12 20 24 27 37 40 42 45 49	C_1	94×1	276, 276, 276	2 21 12 2	0.0253	63, 69, 84, 99, 104
94:71	1 7 9 12 20 24 28 36 38 42 44 48	C_1	94×1	276, 276, 276	1 23 11 2	0.1410	67(2), 85, 114, 116, 119
94:72	1 7 9 12 20 25 27 30 34 44 46 48	C_1	94×1	276, 276, 276	1 21 15 0	0.1939	75, 80, 112, 116

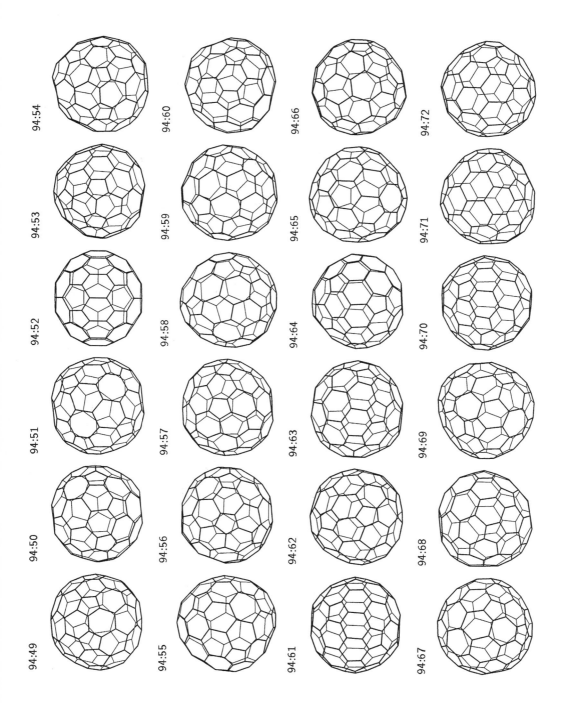

94:54

94:60

94:66

94:72

94:53

94:59

94:65

94:71

94:52

94:58

94:64

94:70

94:51

94:57

94:63

94:69

94:50

94:56

94:62

94:68

94:49

94:55

94:61

94:67

Table A.17. (Continued)

Isomer	Ring spiral	Point group	NMR pattern	Vibrations	Hexagon indices	Band gap	Transformations
94:73	1 7 9 12 20 25 27 30 35 42 45 47	C_2	47×2	276, 276, 276	2 19 16 0	0.1097	74(2)
94:74	1 7 9 12 20 25 27 30 35 42 46 49	C_1	94×1	276, 276, 276	1 21 15 0	0.0277	64, 73, 75, 130
94:75	1 7 9 12 20 25 27 30 35 43 46 48	C_1	94×1	276, 276, 276	1 21 15 0	0.0677	65, 72, 74, 82, 123
94:76	1 7 9 12 20 25 27 33 39 41 44 46	C_1	94×1	276, 276, 276	1 22 13 1	0.0891	50, 77, 90, 108, 121
94:77	1 7 9 12 20 25 27 34 39 41 43 46	C_1	94×1	276, 276, 276	1 22 13 1	0.1680	63, 76, 79, 104, 109, 126
94:78	1 7 9 12 20 25 27 34 39 42 45 48	C_1	94×1	276, 276, 276	1 22 13 1	0.2539	79, 80, 81, 83, 98, 103, 128
94:79	1 7 9 12 20 25 27 34 39 42 46 49	C_1	94×1	276, 276, 276	1 22 13 1	0.2159	77, 78, 84, 99, 105, 106, 129
94:80	1 7 9 12 20 25 27 34 40 42 45 47	C_1	94×1	276, 276, 276	1 22 13 1	0.2463	72, 78, 82, 102, 111, 115
94:81	1 7 9 12 20 25 27 35 39 42 44 48	C_1	94×1	276, 276, 276	1 22 13 1	0.1582	67, 78, 82, 85, 97, 125
94:82	1 7 9 12 20 25 27 35 40 42 44 47	C_1	94×1	276, 276, 276	1 22 13 1	0.0918	68, 75, 80, 81, 101, 125
94:83	1 7 9 12 20 25 28 34 38 42 45 48	C_1	94×1	276, 276, 276	1 23 11 2	0.2977	69, 78, 84, 85, 86, 113, 127, 133
94:84	1 7 9 12 20 25 28 34 38 42 46 49	C_1	94×1	276, 276, 276	2 22 10 3	0.2373	63, 70, 79, 83, 84, 105, 113
94:85	1 7 9 12 20 25 28 35 38 42 44 48	C_1	94×1	276, 276, 276	1 23 11 2	0.1107	68, 71, 81, 83, 87, 122, 123
94:86	1 7 9 12 20 26 28 34 37 42 45 48	C_2	47×2	276, 276, 276	2 21 12 2	0.2165	66(2), 83(2), 87(2)
94:87	1 7 9 12 20 26 28 35 37 42 44 48	C_1	94×1	276, 276, 276	1 22 13 1	0.0669	64, 65, 85, 86, 127, 130
94:88	1 7 9 12 21 24 29 37 39 41 44 46	C_1	94×1	276, 276, 276	2 21 12 2	0.1234	27, 89, 90, 108
94:89	1 7 9 12 21 24 29 37 39 41 43 45	C_2	47×2	276, 276, 276	1 22 13 1	0.2041	28(2), 88(2)
94:90	1 7 9 12 21 24 29 37 39 42 45 49	C_1	94×1	276, 276, 276	1 22 13 1	0.0697	29, 48, 76, 88
94:91	1 7 9 12 24 26 31 33 39 41 44 46	C_1	94×1	276, 276, 276	1 21 15 0	0.2781	57, 62, 92, 95, 98, 103
94:92	1 7 9 12 24 26 31 33 39 41 45 49	C_1	94×1	276, 276, 276	1 21 15 0	0.1952	56, 63, 91, 93, 99, 105
94:93	1 7 9 12 24 26 31 33 39 42 45 48	C_1	94×1	276, 276, 276	2 20 14 1	0.2340	55, 92, 94, 100, 107
94:94	1 7 9 12 24 26 31 33 40 42 45 47	C_2	47×2	276, 276, 276	2 19 16 0	0.1164	53(2), 93(2)
94:95	1 7 9 12 24 26 31 34 39 41 43 46	C_1	94×1	276, 276, 276	1 21 15 0	0.1271	54, 91, 102, 111
94:96	1 7 9 12 24 27 31 33 37 40 44 48	C_3	1×1, 31×3	182, 182, 182	1 21 15 0	0.0057	67(3), 97(3)

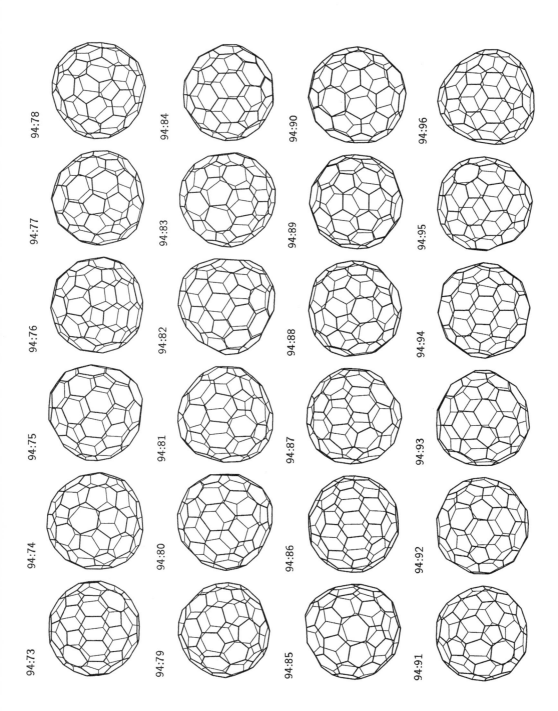

94:78

94:84

94:90

94:96

94:77

94:83

94:89

94:95

94:76

94:82

94:88

94:94

94:75

94:81

94:87

94:93

94:74

94:80

94:86

94:92

94:73

94:79

94:85

94:91

Table A.17. (*Continued*)

Isomer	Ring spiral	Point group	NMR pattern	Vibrations	Hexagon indices	Band gap	Transformations
94:97	1 7 9 12 24 27 31 33 37 41 44 47	C_1	94×1	276, 276, 276	1 21 15 0	0.1434	68, 81, 96, 98, 101(2)
94:98	1 7 9 12 24 27 31 33 38 41 44 46	C_1	94×1	276, 276, 276	1 21 15 0	0.0700	69, 78, 91, 97, 99, 102
94:99	1 7 9 12 24 27 31 33 38 41 45 49	C_1	94×1	276, 276, 276	1 21 15 0	0.0269	70, 79, 92, 98, 100
94:100	1 7 9 12 24 27 31 33 38 42 45 48	C_1	94×1	276, 276, 276	2 20 14 1	0.0797	93, 99, 106
94:101	1 7 9 12 24 27 31 34 37 41 43 47	C_1	94×1	276, 276, 276	1 21 15 0	0.0096	82, 97(2), 102, 110
94:102	1 7 9 12 24 27 31 34 38 41 43 46	C_1	94×1	276, 276, 276	1 21 15 0	0.0757	80, 95, 98, 101
94:103	1 7 9 12 24 28 30 32 34 37 39 49	C_1	94×1	276, 276, 276	1 22 13 1	0.1175	78, 91, 105, 111, 113, 122
94:104	1 7 9 12 24 28 30 32 34 38 45 48	C_1	94×1	276, 276, 276	1 22 13 1	0.0369	70, 77, 105, 106, 129
94:105	1 7 9 12 24 28 30 32 34 39 45 47	C_1	94×1	276, 276, 276	1 22 13 1	0.2152	79, 84, 92, 103, 104, 107, 133
94:106	1 7 9 12 24 28 30 32 35 38 44 48	C_1	94×1	276, 276, 276	2 21 12 2	0.0790	79, 100, 104, 107, 109
94:107	1 7 9 12 24 28 30 32 35 39 44 47	C_2	47×2	276, 276, 276	2 21 12 2	0.3076	93(2), 105(2), 106(2)
94:108	1 7 9 12 24 28 31 35 38 41 44 49	C_2	47×2	276, 276, 276	2 21 12 2	0.0364	76(2), 88(2), 109
94:109	1 7 9 12 24 28 31 35 38 42 44 48	C_2	47×2	276, 276, 276	2 21 12 2	0.1610	77(2), 106(2), 108
94:110	1 7 9 12 25 27 31 34 36 41 43 47	C_3	1×1, 31×3	182, 182, 182	1 21 15 0	0.0000	101(3)
94:111	1 7 9 12 25 28 30 32 34 36 39 49	C_1	94×1	276, 276, 276	1 22 13 1	0.0805	80, 95, 103, 112, 119
94:112	1 7 9 12 26 28 30 32 34 36 38 49	C_1	94×1	276, 276, 276	1 21 15 0	0.0287	72, 111, 131
94:113	1 7 9 13 20 23 26 38 40 42 44 46	C_1	94×1	276, 276, 276	1 23 11 2	0.0542	62, 66, 83, 84, 103, 127, 129
94:114	1 7 9 13 20 23 28 36 38 42 44 48	C_3	1×1, 31×3	182, 182, 182	1 24 9 3	0.0000	71(3), 131(3)
94:115	1 7 10 12 14 18 31 36 38 42 44 47	C_1	94×1	276, 276, 276	0 23 14 0	0.1054	67, 80, 116, 119, 125, 128
94:116	1 7 10 12 14 18 32 35 38 40 45 48	C_1	94×1	276, 276, 276	0 24 12 1	0.1434	71, 72, 115, 118, 123, 131
94:117	1 7 10 12 14 18 32 36 38 40 43 49	C_2	47×2	276, 276, 276	0 23 14 0	0.0300	118(2), 120(2)
94:118	1 7 10 12 14 18 32 36 38 40 44 48	C_1	94×1	276, 276, 276	0 23 14 0	0.0558	116, 117, 119, 124, 132
94:119	1 7 10 12 14 18 32 36 38 41 44 47	C_1	94×1	276, 276, 276	0 23 14 0	0.0755	71, 111, 115, 118, 122, 131
94:120	1 7 10 12 14 18 33 36 38 40 42 49	C_s	8×1, 43×2	276, 276, 276	0 23 14 0	0.0737	117(2), 132(2)

94:102

94:108

94:114

94:120

94:101

94:107

94:113

94:119

94:100

94:106

94:112

94:118

94:99

94:105

94:111

94:117

94:98

94:104

94:110

94:116

94:97

94:103

94:109

94:115

Table A.17. (*Continued*)

Isomer	Ring spiral	Point group	NMR pattern	Vibrations	Hexagon indices	Band gap	Transformations
94:121	1 7 10 12 12 14 19 35 37 39 41 43 49	C_2	47×2	276, 276, 276	0 23 14 0	0.0883	49(2), 76(2), 126
94:122	1 7 10 12 18 24 27 31 36 43 46 48	C_1	94×1	276, 276, 276	0 23 14 0	0.1296	85, 103, 119, 124, 127, 128, 133
94:123	1 7 10 12 18 25 27 31 35 42 46 49	C_1	94×1	276, 276, 276	0 24 12 1	0.0449	75, 85, 116, 124, 125, 127, 130
94:124	1 7 10 12 18 25 27 31 35 43 46 48	C_2	47×2	276, 276, 276	0 23 14 0	0.0186	118(2), 122(2), 123(2)
94:125	1 7 10 12 18 25 27 32 35 41 46 49	C_1	94×1	276, 276, 276	0 23 14 0	0.0135	81, 82, 115, 123, 125, 128
94:126	1 7 10 12 18 26 32 34 36 39 41 46	C_2	47×2	276, 276, 276	0 23 14 0	0.1245	62(2), 77(2), 121, 129(2)
94:127	1 7 10 12 19 24 27 30 36 43 46 48	C_1	94×1	276, 276, 276	0 24 12 1	0.1148	65, 83, 87, 113, 122, 123, 127, 128
94:128	1 7 10 12 19 24 27 36 40 42 44 47	C_1	94×1	276, 276, 276	0 23 14 0	0.1116	68, 78, 115, 122, 125, 127, 129
94:129	1 7 10 12 19 24 27 37 40 42 44 46	C_1	94×1	276, 276, 276	0 23 14 0	0.0535	69, 79, 104, 113, 126, 128, 133
94:130	1 7 10 12 19 25 27 30 35 42 46 49	C_2	47×2	276, 276, 276	0 23 14 0	0.0869	74(2), 87(2), 123(2)
94:131	1 7 10 12 25 28 30 32 34 36 39 48	C_1	94×1	276, 276, 276	0 24 12 1	0.0543	112, 114, 116, 119, 132(2)
94:132	1 7 10 12 26 28 30 32 34 36 38 48	C_1	94×1	276, 276, 276	0 23 14 0	0.0657	118, 120, 131(2), 134
94:133	1 7 10 13 18 22 26 38 41 43 45 48	C_2	47×2	276, 276, 276	0 23 14 0	0.2529	83(2), 105(2), 122(2), 129(2)
94:134	1 7 10 13 25 28 30 32 34 36 39 49	C_{3v}	1×1, 7×3, 12×6	142, 142, 142	0 24 12 1	0.0000	132(6)

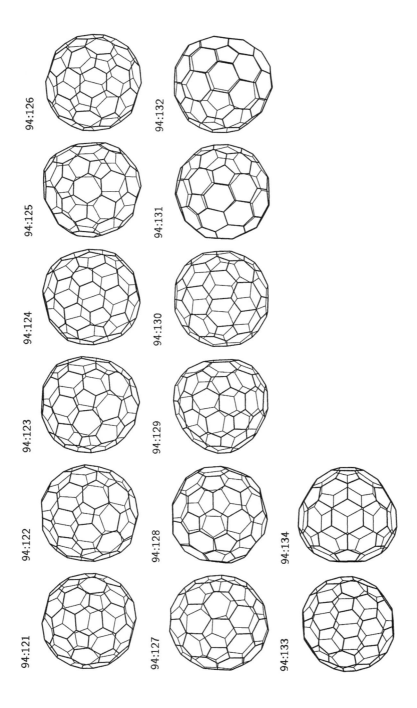

94:126

94:132

94:125

94:131

94:124

94:130

94:123

94:129

94:122

94:128

94:134

94:121

94:127

94:133

Table A.18. Isolated-pentagon fullerene isomers of C_{96}

Isomer	Ring spiral	Point group	NMR pattern	Vibrations	Hexagon indices	Band gap	Transformations
96:1	1 7 9 11 13 18 36 41 43 45 47 49	D_2	24×4	210, 282, 210	8 12 12 6	0.5374	
96:2	1 7 9 11 13 20 34 37 40 45 47 49	D_3	16×6	140, 142, 94	8 12 12 6	0.2105	
96:3	1 7 9 11 13 20 36 39 43 45 47 49	D_{3d}	4×6, 6×12	72, 73, 0	8 12 12 6	0.5173	
96:4	1 7 9 11 13 24 36 39 41 43 47 49	C_2	48×2	282, 282, 282	6 16 10 6	0.5738	5(2), 6
96:5	1 7 9 11 13 24 37 39 41 43 46 49	C_1	96×1	282, 282, 282	5 16 13 4	0.3053	4, 14
96:6	1 7 9 11 13 25 35 39 41 43 47 49	C_2	48×2	282, 282, 282	6 14 14 4	0.2629	4
96:7	1 7 9 11 13 28 33 35 37 40 42 50	C_1	96×1	282, 282, 282	4 17 14 3	0.0388	8, 10
96:8	1 7 9 11 13 28 34 37 40 42 44 50	C_s	6×1, 45×2	282, 282, 282	5 16 13 4	0.3451	7(2)
96:9	1 7 9 11 13 29 32 35 38 45 47 49	C_2	48×2	282, 282, 282	6 12 18 2	0.0342	
96:10	1 7 9 11 13 29 33 35 37 39 42 50	C_s	6×1, 45×2	282, 282, 282	3 18 15 2	0.1221	7(2), 11
96:11	1 7 9 11 13 29 33 35 37 39 48 50	C_s	4×1, 46×2	282, 282, 282	5 15 15 3	0.1425	10
96:12	1 7 9 11 14 22 29 37 41 43 45 47	C_1	96×1	282, 282, 282	3 17 17 1	0.1468	13, 55
96:13	1 7 9 11 14 22 29 37 42 45 47 49	C_2	48×2	282, 282, 282	4 16 16 2	0.1994	12(2), 15
96:14	1 7 9 11 14 23 37 39 41 43 46 49	C_2	48×2	282, 282, 282	4 16 16 2	0.0541	5(2)
96:15	1 7 9 11 14 28 31 34 38 43 45 48	C_2	48×2	282, 282, 282	4 14 20 0	0.0331	13
96:16	1 7 9 11 14 28 34 36 41 43 47 49	C_1	96×1	282, 282, 282	3 18 15 2	0.0204	17, 18, 43
96:17	1 7 9 11 14 28 34 36 41 43 48 50	C_1	96×1	282, 282, 282	2 20 14 2	0.0633	16, 20, 44
96:18	1 7 9 11 14 28 34 37 41 43 46 49	C_1	96×1	282, 282, 282	3 17 17 1	0.0429	16, 19
96:19	1 7 9 11 14 28 35 37 41 43 45 49	C_1	96×1	282, 282, 282	4 16 16 2	0.1402	18
96:20	1 7 9 11 14 29 32 36 38 43 48 50	C_1	96×1	282, 282, 282	3 17 17 1	0.0058	17, 47
96:21	1 7 9 11 22 24 28 30 36 46 48 50	D_2	24×4	210, 282, 210	4 18 12 4	0.1873	22(4)
96:22	1 7 9 11 22 24 28 36 40 42 47 50	C_1	96×1	282, 282, 282	3 20 11 4	0.3766	21, 24, 31, 99, 105
96:23	1 7 9 11 22 24 29 36 38 42 48 50	C_1	96×1	282, 282, 282	2 20 14 2	0.1868	24, 25, 29, 100
96:24	1 7 9 11 22 24 29 36 39 42 47 50	C_1	96×1	282, 282, 282	3 19 13 3	0.2073	22, 23, 27, 87

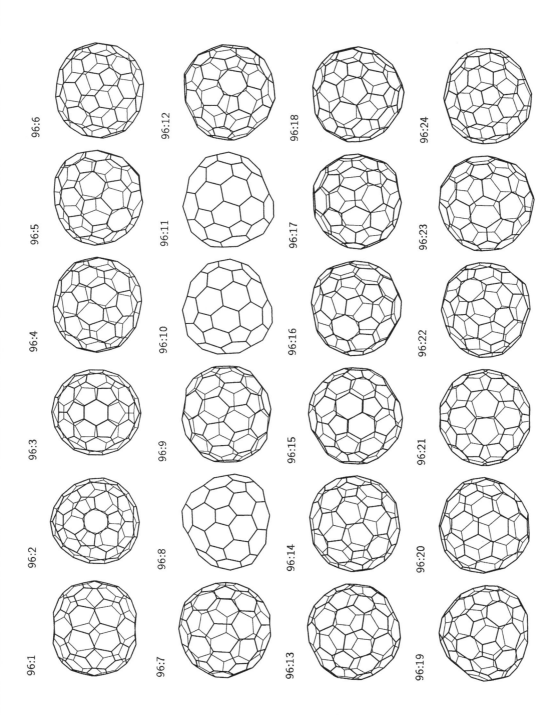

96:6

96:5

96:4

96:3

96:2

96:1

96:12

96:11

96:10

96:9

96:8

96:7

96:18

96:17

96:16

96:15

96:14

96:13

96:24

96:23

96:22

96:21

96:20

96:19

Table A.18. (*Continued*)

Isomer	Ring spiral	Point group	NMR pattern	Vibrations	Hexagon indices	Band gap	Transformations
96:25	1 7 9 11 22 24 29 38 42 45 47 49	C_1	96×1	282, 282, 282	3 18 15 2	0.0713	23, 51
96:26	1 7 9 11 22 25 27 30 43 45 47 49	C_1	96×1	282, 282, 282	2 20 14 2	0.1338	29, 37, 101
96:27	1 7 9 11 22 25 27 39 42 44 46 49	C_1	96×1	282, 282, 282	2 21 12 3	0.3465	24, 28, 29, 31, 41, 102
96:28	1 7 9 11 22 25 27 39 42 44 47 50	C_1	96×1	282, 282, 282	3 20 11 4	0.2934	27, 30, 32, 40, 103
96:29	1 7 9 11 22 25 27 40 42 44 46 48	C_1	96×1	282, 282, 282	2 21 12 3	0.2090	23, 26, 27, 38, 104
96:30	1 7 9 11 22 25 28 35 40 42 46 49	C_2	48×2	282, 282, 282	4 20 8 6	0.6223	28(2), 31, 106(2)
96:31	1 7 9 11 22 25 28 35 40 42 47 50	C_2	48×2	282, 282, 282	2 22 10 4	0.3392	22(2), 27(2), 30, 107(2)
96:32	1 7 9 11 22 25 29 35 39 41 43 46	C_1	96×1	282, 282, 282	4 18 12 4	0.3925	28, 39, 108
96:33	1 7 9 11 22 28 35 39 41 45 47 49	D_{3h}	2×6, 7×12	71, 119, 48	6 18 6 8	0.6319	110(3)
96:34	1 7 9 11 23 25 29 32 38 41 44 46	C_1	96×1	282, 282, 282	3 17 17 1	0.0283	35, 43
96:35	1 7 9 11 23 25 29 32 38 41 45 50	C_1	96×1	282, 282, 282	2 19 16 1	0.0716	34, 36, 44, 45
96:36	1 7 9 11 23 25 29 32 38 45 48 50	C_1	96×1	282, 282, 282	2 19 16 1	0.1218	35, 37, 42, 46
96:37	1 7 9 11 23 25 29 32 39 45 47 50	C_1	96×1	282, 282, 282	2 19 16 1	0.1868	26, 36, 38
96:38	1 7 9 11 23 25 29 34 38 42 48 50	C_1	96×1	282, 282, 282	2 20 14 2	0.2684	29, 37, 41, 42
96:39	1 7 9 11 23 25 29 34 39 41 43 46	C_1	96×1	282, 282, 282	3 19 13 3	0.1527	32, 40, 48
96:40	1 7 9 11 23 25 29 34 39 42 46 49	C_2	48×2	282, 282, 282	2 20 14 2	0.2341	28(2), 39(2), 41
96:41	1 7 9 11 23 25 29 34 39 42 47 50	C_2	48×2	282, 282, 282	2 20 14 2	0.2961	27(2), 38(2), 40
96:42	1 7 9 11 23 25 29 38 42 44 46 49	C_2	48×2	282, 282, 282	2 20 14 2	0.1240	36(2), 38(2)
96:43	1 7 9 11 23 25 30 32 38 40 44 46	C_1	96×1	282, 282, 282	3 17 17 1	0.2595	16, 34, 44
96:44	1 7 9 11 23 25 30 32 38 40 45 50	C_1	96×1	282, 282, 282	2 19 16 1	0.1000	17, 35, 43, 47
96:45	1 7 9 11 23 26 29 32 36 46 48 50	C_1	96×1	282, 282, 282	3 17 17 1	0.1344	35, 46, 47
96:46	1 7 9 11 23 26 29 32 37 45 48 50	C_2	48×2	282, 282, 282	2 18 18 0	0.1894	36(2), 45(2)
96:47	1 7 9 11 23 26 30 32 37 40 45 50	C_1	96×1	282, 282, 282	3 16 19 0	0.3119	20, 44, 45
96:48	1 7 9 11 23 29 34 36 38 40 42 44	C_2	48×2	282, 282, 282	4 18 12 4	0.2151	39(2)

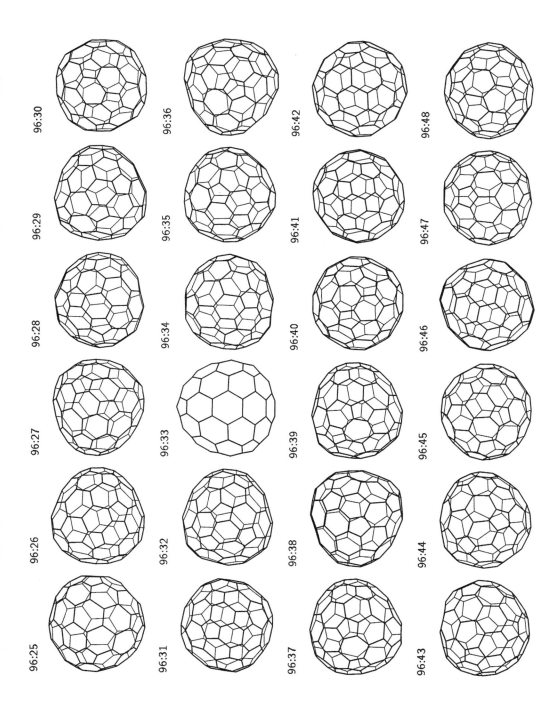

96:30

96:36

96:42

96:48

96:29

96:35

96:41

96:47

96:28

96:34

96:40

96:46

96:27

96:33

96:39

96:45

96:26

96:32

96:38

96:44

96:25

96:31

96:37

96:43

Table A.18. (*Continued*)

Isomer	Ring spiral	Point group	NMR pattern	Vibrations	Hexagon indices	Band gap	Transformations
96:49	1 7 9 11 24 26 31 34 37 41 47 49	C_1	96×1	282, 282, 282	4 15 18 1	0.2586	50
96:50	1 7 9 11 24 26 31 37 41 43 45 48	C_s	8×1, 44×2	282, 282, 282	3 18 15 2	0.0948	49(2)
96:51	1 7 9 12 14 20 29 36 42 44 46 48	C_1	96×1	282, 282, 282	2 19 16 1	0.1769	25, 54, 100
96:52	1 7 9 12 14 20 35 38 41 43 46 48	C_2	48×2	282, 282, 282	2 20 14 2	0.1268	53(2), 63(2)
96:53	1 7 9 12 14 20 36 38 41 43 45 48	C_1	96×1	282, 282, 282	1 21 15 1	0.1001	52, 54, 60, 158
96:54	1 7 9 12 14 20 36 39 41 43 45 47	C_1	96×1	282, 282, 282	2 20 14 2	0.1333	51, 53, 66, 84
96:55	1 7 9 12 14 21 29 37 41 43 45 47	C_2	48×2	282, 282, 282	2 18 18 0	0.0615	12(2)
96:56	1 7 9 12 14 28 31 33 36 40 46 49	C_1	96×1	282, 282, 282	1 21 15 1	0.1007	58, 69, 112
96:57	1 7 9 12 14 28 31 33 37 40 44 50	C_1	96×1	282, 282, 282	1 20 17 0	0.0851	58, 59, 61, 70
96:58	1 7 9 12 14 28 31 33 37 40 45 49	C_1	96×1	282, 282, 282	1 20 17 0	0.0823	56, 57, 70
96:59	1 7 9 12 14 28 31 33 37 44 47 49	C_1	96×1	282, 282, 282	1 20 17 0	0.0847	57, 60, 62, 65
96:60	1 7 9 12 14 28 31 33 38 44 46 49	C_1	96×1	282, 282, 282	1 20 17 0	0.0828	53, 59, 63, 66
96:61	1 7 9 12 14 28 31 34 37 40 43 50	C_1	96×1	282, 282, 282	1 21 15 1	0.1624	57, 62, 64, 69, 116
96:62	1 7 9 12 14 28 31 34 37 43 47 49	C_1	96×1	282, 282, 282	2 20 14 2	0.1150	59, 61, 63, 117
96:63	1 7 9 12 14 28 31 34 38 43 46 49	C_1	96×1	282, 282, 282	2 19 16 1	0.1532	52, 60, 62, 115
96:64	1 7 9 12 14 28 32 34 37 40 42 50	C_1	96×1	282, 282, 282	2 19 16 1	0.1551	61, 67, 88
96:65	1 7 9 12 14 28 33 37 41 43 45 48	C_1	96×1	282, 282, 282	2 19 16 1	0.0614	59, 66
96:66	1 7 9 12 14 28 33 38 41 43 45 47	C_1	96×1	282, 282, 282	2 19 16 1	0.1935	54, 60, 65
96:67	1 7 9 12 14 29 31 33 36 38 47 49	C_1	96×1	282, 282, 282	1 20 17 0	0.1172	64, 68, 69, 113
96:68	1 7 9 12 14 29 31 33 36 38 48 50	C_1	96×1	282, 282, 282	2 18 18 0	0.1212	67, 94
96:69	1 7 9 12 14 29 31 33 36 39 46 49	C_1	96×1	282, 282, 282	1 21 15 1	0.2238	56, 61, 67, 70, 114
96:70	1 7 9 12 14 29 31 33 37 39 44 50	C_1	96×1	282, 282, 282	1 20 17 0	0.2309	57, 58, 69, 70
96:71	1 7 9 12 20 24 27 36 39 41 43 50	C_1	96×1	282, 282, 282	2 20 14 2	0.0318	72, 74, 77, 85, 95
96:72	1 7 9 12 20 24 27 39 41 43 45 47	C_1	96×1	282, 282, 282	1 21 15 1	0.0670	71, 73, 78, 79, 167

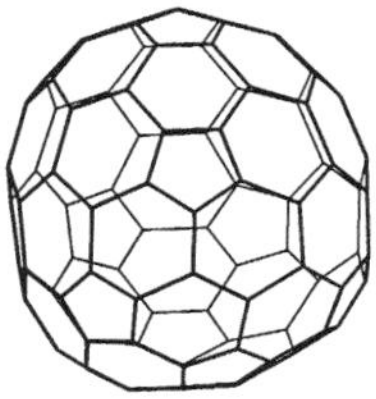
96:49

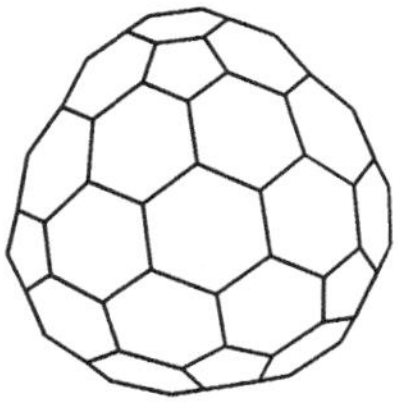
96:50

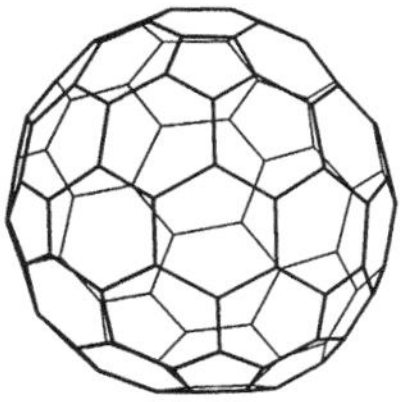
96:51

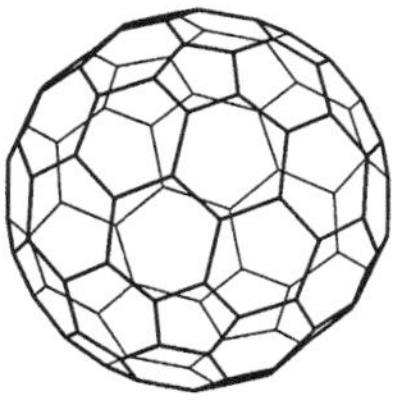
96:52

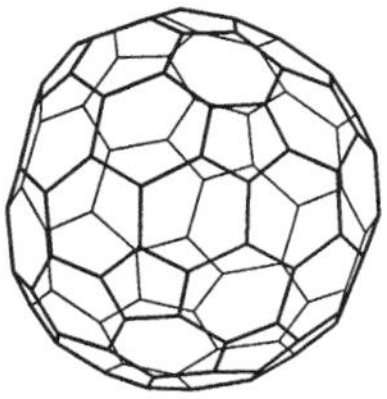
96:53

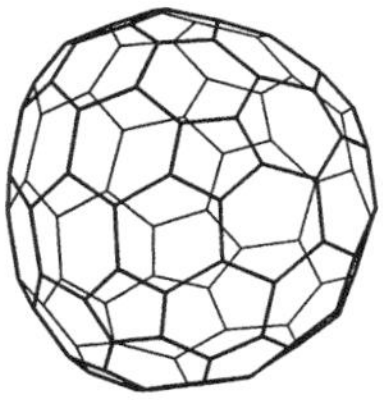
96:54

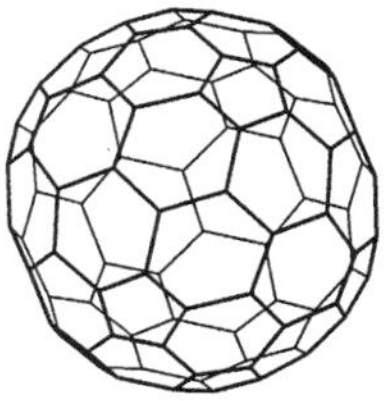
96:55

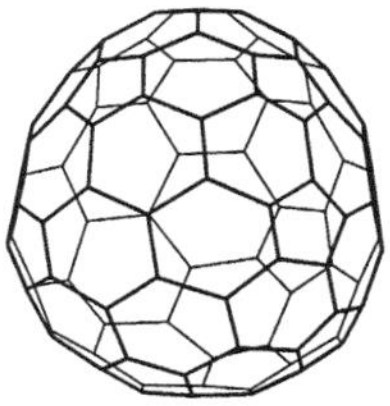
96:56

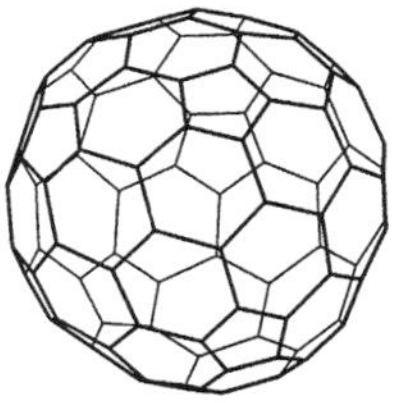
96:57

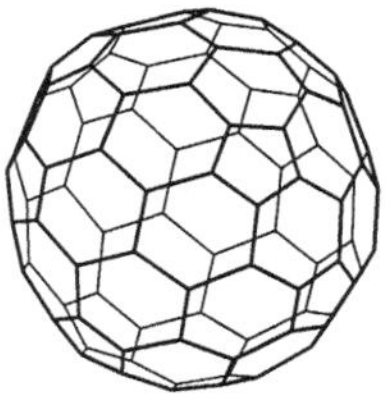
96:58

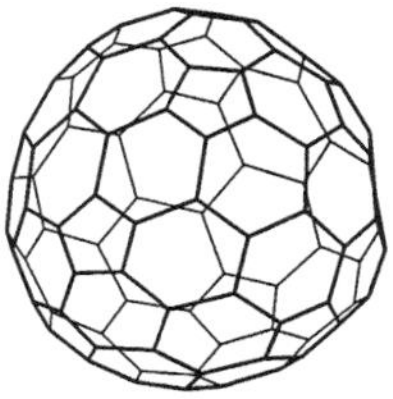
96:59

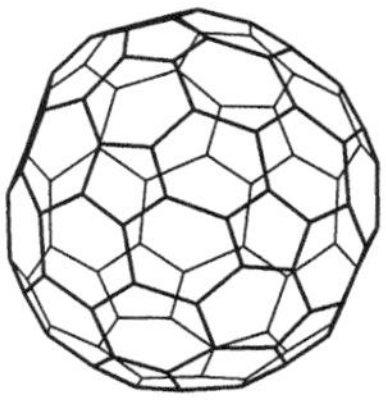
96:60

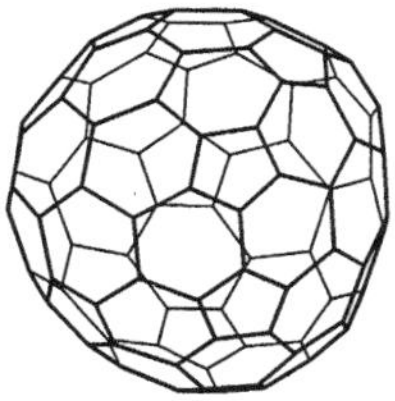
96:61

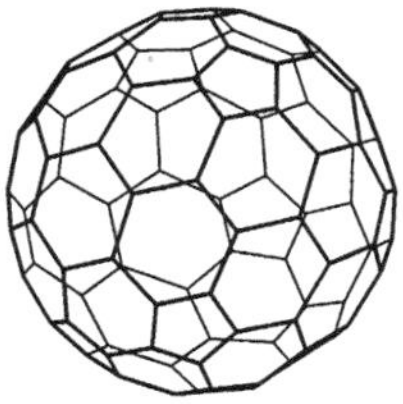
96:62

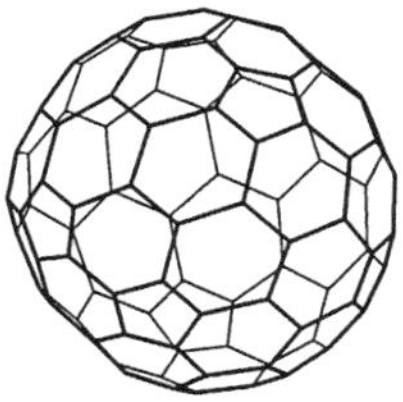
96:63

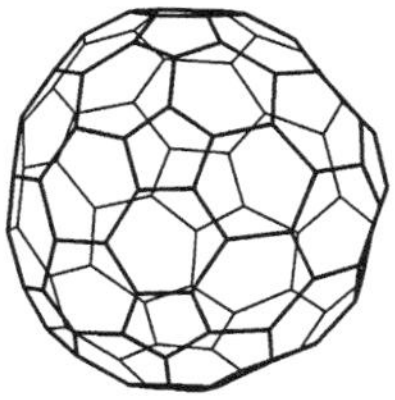
96:64

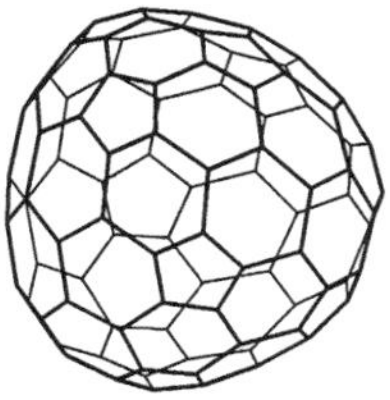
96:65

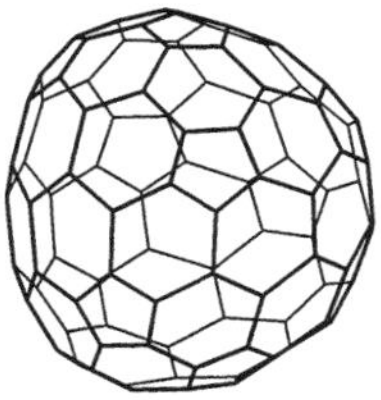
96:66

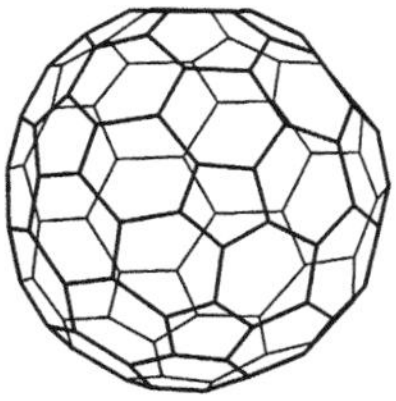
96:67

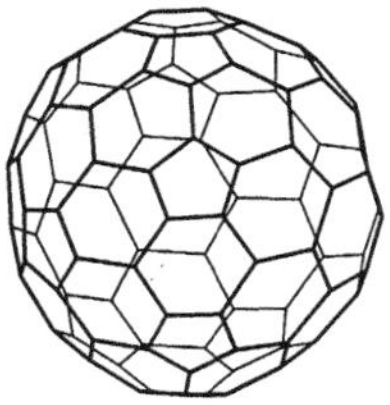
96:68

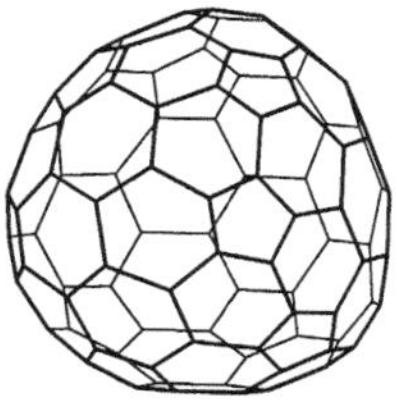
96:69

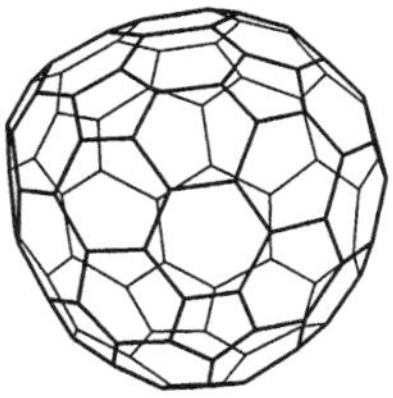
96:70

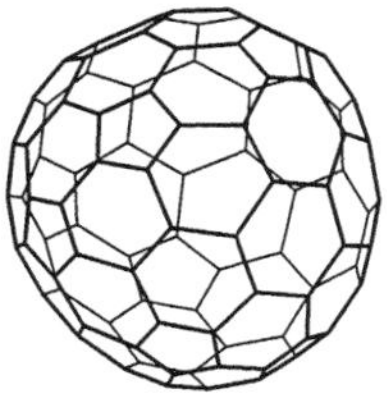
96:71

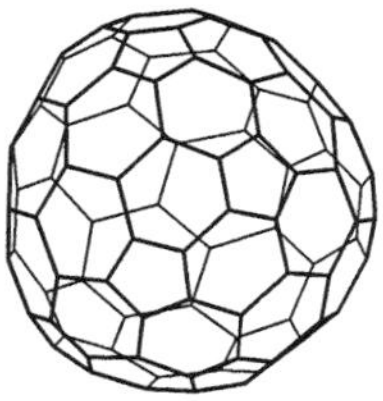
96:72

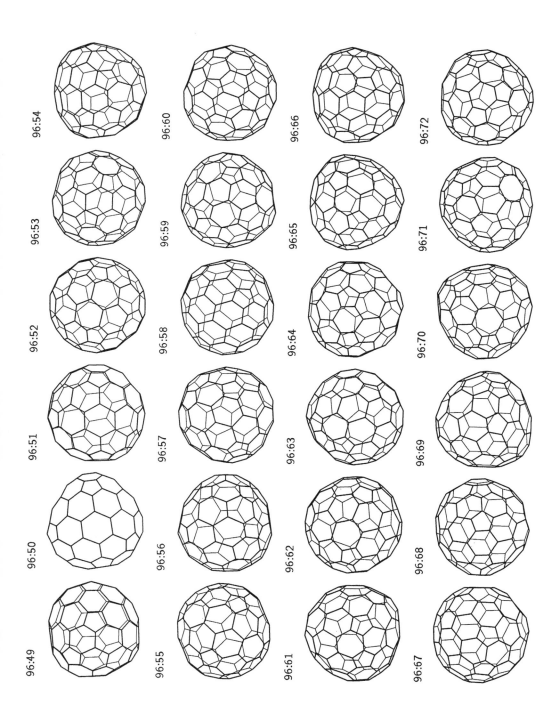

Table A.18. (*Continued*)

Isomer	Ring spiral	Point group	NMR pattern	Vibrations	Hexagon indices	Band gap	Transformations
96:73	1 7 9 12 20 24 27 39 42 45 47 50	C_2	48×2	282, 282, 282	2 20 14 2	0.1172	72(2), 80(2), 121
96:74	1 7 9 12 20 24 28 36 38 41 43 50	C_1	96×1	282, 282, 282	2 21 12 3	0.3214	71, 79, 88, 125, 129, 139
96:75	1 7 9 12 20 24 28 37 40 43 45 49	C_1	96×1	282, 282, 282	1 22 13 2	0.1521	76, 81, 133, 152, 159
96:76	1 7 9 12 20 24 28 37 41 43 45 48	C_1	96×1	282, 282, 282	1 22 13 2	0.1932	75, 78, 79, 134, 164, 166
96:77	1 7 9 12 20 24 28 37 42 44 46 48	C_1	96×1	282, 282, 282	2 20 14 2	0.0811	71, 78, 90, 98, 129
96:78	1 7 9 12 20 24 28 37 42 45 48 50	C_1	96×1	282, 282, 282	1 23 11 3	0.1177	72, 76, 77, 80, 135, 168, 169
96:79	1 7 9 12 20 24 28 38 41 43 45 47	C_1	96×1	282, 282, 282	1 22 13 2	0.2835	72, 74, 76, 80, 135, 169, 176
96:80	1 7 9 12 20 24 28 38 42 45 47 50	C_2	48×2	282, 282, 282	2 22 10 4	0.2720	73(2), 78(2), 79(2), 136
96:81	1 7 9 12 20 24 29 37 39 43 45 49	C_s	6×1, 45×2	282, 282, 282	1 21 15 1	0.0200	75(2), 170
96:82	1 7 9 12 20 25 27 30 34 45 47 49	C_2	48×2	282, 282, 282	2 18 18 0	0.1266	98(2), 127
96:83	1 7 9 12 20 25 27 33 39 45 48 50	C_2	48×2	282, 282, 282	2 20 14 2	0.2450	84(2), 87(2)
96:84	1 7 9 12 20 25 27 33 40 45 47 50	C_1	96×1	282, 282, 282	1 21 15 1	0.2108	54, 83, 100, 158
96:85	1 7 9 12 20 25 27 35 39 41 43 50	C_1	96×1	282, 282, 282	2 19 16 1	0.0953	71, 88, 119
96:86	1 7 9 12 20 25 27 35 40 42 46 49	C_1	96×1	282, 282, 282	3 18 15 2	0.1191	122, 123
96:87	1 7 9 12 20 25 28 33 38 45 48 50	C_1	96×1	282, 282, 282	2 21 12 3	0.3253	24, 83, 99, 100, 102
96:88	1 7 9 12 20 25 28 35 38 41 43 50	C_1	96×1	282, 282, 282	2 20 14 2	0.3333	64, 74, 85, 113, 116
96:89	1 7 9 12 20 25 28 36 41 47 50	C_s	6×1, 45×2	282, 282, 282	1 20 17 0	0.0240	92(2), 170
96:90	1 7 9 12 20 26 30 34 36 39 46 49	C_1	96×1	282, 282, 282	1 20 17 0	0.1937	77, 91, 93, 95, 143, 169
96:91	1 7 9 12 20 26 30 34 36 39 48 50	C_1	96×1	282, 282, 282	1 20 17 0	0.0930	90, 92, 98, 145, 166
96:92	1 7 9 12 20 26 30 34 36 40 47 50	C_1	96×1	282, 282, 282	1 20 17 0	0.2080	89, 91, 147, 152
96:93	1 7 9 12 20 26 30 34 37 39 45 49	C_1	96×1	282, 282, 282	2 19 16 1	0.1286	90, 94, 96, 139, 142
96:94	1 7 9 12 20 26 30 35 37 39 44 49	C_1	96×1	282, 282, 282	2 18 18 0	0.2705	68, 93, 97, 113
96:95	1 7 9 12 20 26 34 36 39 41 43 47	C_1	96×1	282, 282, 282	1 21 15 1	0.0396	71, 90, 96, 119, 139, 167
96:96	1 7 9 12 20 26 34 37 39 41 43 46	C_2	48×2	282, 282, 282	2 20 14 2	0.1578	93(2), 95(2), 97(2)

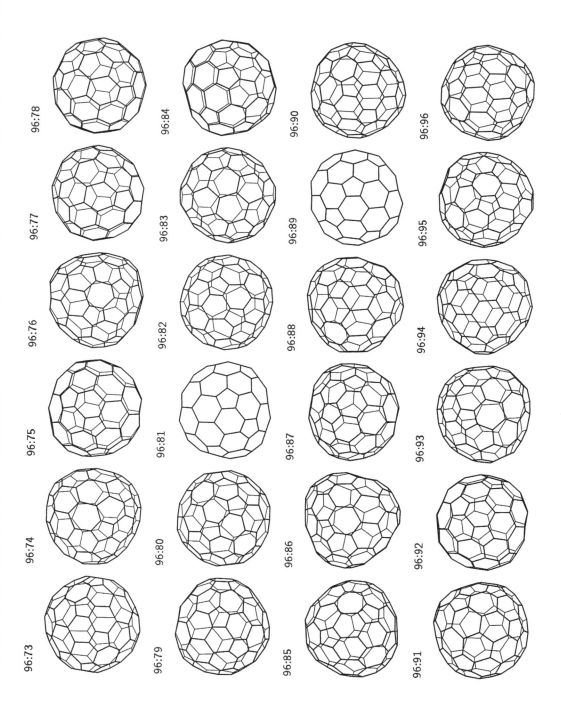

96:78

96:77

96:76

96:75

96:74

96:73

96:84

96:83

96:82

96:81

96:80

96:79

96:90

96:89

96:88

96:87

96:86

96:85

96:96

96:95

96:94

96:93

96:92

96:91

297

Table A.18. (*Continued*)

Isomer	Ring spiral	Point group	NMR pattern	Vibrations	Hexagon indices	Band gap	Transformations
96:97	1 7 9 12 20 26 35 37 39 41 43 45	C_1	96×1	282, 282, 282	2 19 16 1	0.0905	94, 96, 118, 119
96:98	1 7 9 12 20 27 30 34 36 38 48 50	C_1	96×1	282, 282, 282	1 21 15 1	0.0479	77, 82, 91, 128, 168
96:99	1 7 9 12 21 24 28 36 40 42 47 50	C_2	48×2	282, 282, 282	2 22 10 4	0.3500	22(2), 87(2), 107(2)
96:100	1 7 9 12 21 24 29 36 38 42 48 50	C_1	96×1	282, 282, 282	1 22 13 2	0.2181	23, 51, 84, 87, 104
96:101	1 7 9 12 21 25 27 30 43 45 47 49	C_1	96×1	282, 282, 282	1 20 17 0	0.1417	26, 104
96:102	1 7 9 12 21 25 27 39 42 44 46 49	C_1	96×1	282, 282, 282	1 21 15 1	0.3097	27, 87, 103, 104, 107
96:103	1 7 9 12 21 25 27 39 42 44 47 50	C_1	96×1	282, 282, 282	2 20 14 2	0.2701	28, 102, 106, 108
96:104	1 7 9 12 21 25 27 40 42 44 46 48	C_1	96×1	282, 282, 282	1 21 15 1	0.1972	29, 100, 101, 102
96:105	1 7 9 12 21 25 28 30 35 46 48 50	C_2	48×2	282, 282, 282	2 20 14 2	0.2931	22(2), 107(2)
96:106	1 7 9 12 21 25 28 35 40 42 46 49	C_1	96×1	282, 282, 282	3 20 11 4	0.2938	30, 103, 107, 151
96:107	1 7 9 12 21 25 28 35 40 42 47 50	C_1	96×1	282, 282, 282	1 22 13 2	0.2691	31, 99, 102, 105, 106, 183
96:108	1 7 9 12 21 25 29 35 39 41 43 46	C_1	96×1	282, 282, 282	3 18 15 2	0.1734	32, 103
96:109	1 7 9 12 21 25 30 34 38 40 46 49	D_2	24×4	210, 282, 210	4 16 16 2	0.2529	
96:110	1 7 9 12 21 28 35 39 41 45 47 49	C_{2v}	4×2, 22×4	213, 282, 213	4 20 8 6	0.3019	33, 149(2)
96:111	1 7 9 12 21 29 34 36 38 40 42 47	D_2	24×4	210, 282, 210	4 18 12 4	0.1086	
96:112	1 7 9 12 24 26 29 32 34 41 47 50	C_1	96×1	282, 282, 282	1 20 17 0	0.1853	56, 114, 140
96:113	1 7 9 12 24 26 30 32 34 39 48 50	C_1	96×1	282, 282, 282	1 21 15 1	0.1640	67, 88, 94, 114, 119, 139
96:114	1 7 9 12 24 26 30 32 34 40 47 50	C_1	96×1	282, 282, 282	1 20 17 0	0.3024	69, 112, 113, 116, 138
96:115	1 7 9 12 24 26 31 33 40 44 47 49	C_2	48×2	282, 282, 282	2 18 18 0	0.0441	63(2), 117(2)
96:116	1 7 9 12 24 26 31 34 40 42 47 50	C_1	96×1	282, 282, 282	1 20 17 0	0.1859	61, 88, 114, 117, 125
96:117	1 7 9 12 24 26 31 34 40 43 47 49	C_1	96×1	282, 282, 282	2 19 16 1	0.2266	62, 115, 116, 126
96:118	1 7 9 12 24 27 30 32 34 37 47 49	C_2	48×2	282, 282, 282	2 18 18 0	0.2293	97(2), 124
96:119	1 7 9 12 24 27 30 32 34 38 48 50	C_1	96×1	282, 282, 282	1 20 17 0	0.1459	85, 95, 97, 113
96:120	1 7 9 12 24 27 31 33 37 42 47 50	C_1	96×1	282, 282, 282	3 18 15 2	0.1595	120, 123

96:102

96:108

96:114

96:120

96:101

96:107

96:113

96:119

96:100

96:106

96:112

96:118

96:99

96:105

96:111

96:117

96:98

96:104

96:110

96:116

96:97

96:103

96:109

96:115

Table A.18. (*Continued*)

Isomer	Ring spiral	Point group	NMR pattern	Vibrations	Hexagon indices	Band gap	Transformations
96:121	1 7 9 12 24 27 31 33 40 45 47 50	D_3	16×6	140, 142, 94	2 18 18 0	0.0304	73(3)
96:122	1 7 9 12 24 27 31 34 41 45 48 50	C_2	48×2	282, 282, 282	2 18 18 0	0.0951	86(2)
96:123	1 7 9 12 24 27 31 35 41 43 46 48	C_1	96×1	282, 282, 282	2 19 16 1	0.0693	86, 120
96:124	1 7 9 12 25 27 30 32 34 36 47 49	C_2	48×2	282, 282, 282	2 18 18 0	0.0634	118
96:125	1 7 9 12 25 28 30 33 35 39 43 46	C_1	96×1	282, 282, 282	1 21 15 1	0.2311	74, 116, 126, 131, 138, 176
96:126	1 7 9 12 25 28 30 33 36 39 43 45	C_2	48×2	282, 282, 282	2 20 14 2	0.2798	117(2), 125(2), 132
96:127	1 7 9 12 26 28 30 32 34 37 45 47	C_2	48×2	282, 282, 282	2 18 18 0	0.2379	82, 128(2)
96:128	1 7 9 12 26 28 30 32 35 37 44 47	C_1	96×1	282, 282, 282	1 21 15 1	0.3055	98, 127, 129, 130, 145, 173
96:129	1 7 9 12 26 28 30 32 35 38 44 46	C_1	96×1	282, 282, 282	2 21 12 3	0.3941	74, 77, 128, 131, 135, 143
96:130	1 7 9 12 26 28 30 33 35 37 43 47	C_1	96×1	282, 282, 282	1 20 17 0	0.2578	128, 131, 141, 146, 177
96:131	1 7 9 12 26 28 30 33 35 38 43 46	C_1	96×1	282, 282, 282	1 21 15 1	0.2436	125, 129, 130, 132, 144, 174
96:132	1 7 9 12 26 28 30 33 36 38 43 45	C_2	48×2	282, 282, 282	2 18 18 0	0.3009	126, 131(2)
96:133	1 7 9 13 20 23 28 37 40 43 45 49	C_s	6×1, 45×2	282, 282, 282	1 23 11 3	0.1923	75(2), 134(2), 153(2), 156
96:134	1 7 9 13 20 23 28 37 41 43 45 48	C_1	96×1	282, 282, 282	1 23 11 3	0.3569	76, 133, 135(2), 137, 172, 175, 177
96:135	1 7 9 13 20 23 28 37 42 45 48 50	C_1	96×1	282, 282, 282	1 24 9 4	0.3090	78, 79, 129, 134(2), 136, 171, 173, 174
96:136	1 7 9 13 20 23 28 38 42 45 47 50	D_3	16×6	140, 142, 94	2 24 6 6	0.6418	80(3), 135(6)
96:137	1 7 9 13 20 28 34 36 38 42 44 48	C_{3v}	6×3, 13×6	144, 144, 144	1 24 9 4	0.4201	134(6), 178(3)
96:138	1 7 9 13 22 25 30 32 38 40 44 50	C_1	96×1	282, 282, 282	1 21 15 1	0.2141	114, 125, 139, 140, 144, 164
96:139	1 7 9 13 22 25 30 33 38 40 43 50	C_1	96×1	282, 282, 282	1 22 13 2	0.0669	74, 93, 95, 113, 138, 143, 169
96:140	1 7 9 13 22 25 31 38 40 42 44 50	C_1	96×1	282, 282, 282	1 21 15 1	0.1390	112, 138, 141, 159
96:141	1 7 9 13 22 26 29 32 35 41 47 50	C_1	96×1	282, 282, 282	1 20 17 0	0.1627	130, 140, 144, 153
96:142	1 7 9 13 22 26 30 32 35 39 46 49	C_2	48×2	282, 282, 282	2 18 18 0	0.3012	93(2), 143(2)
96:143	1 7 9 13 22 26 30 32 35 39 48 50	C_1	96×1	282, 282, 282	1 21 15 1	0.2064	90, 129, 139, 142, 144, 145, 171
96:144	1 7 9 13 22 26 30 32 35 40 47 50	C_1	96×1	282, 282, 282	1 20 17 0	0.3341	131, 138, 141, 143, 146, 175

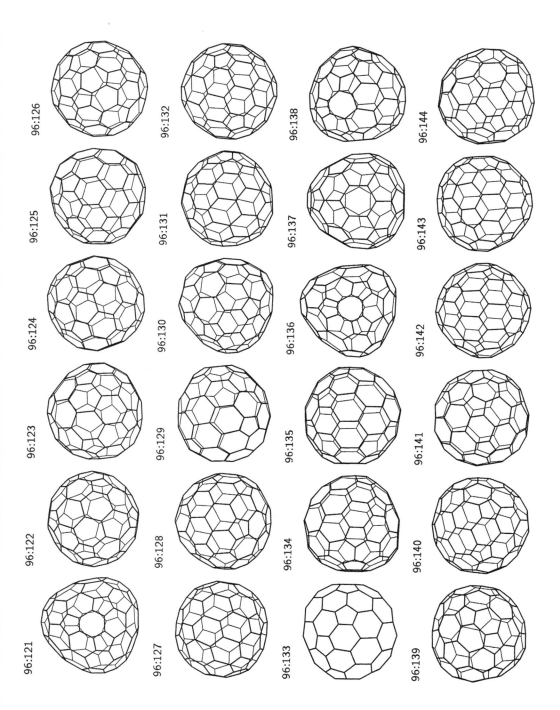

96:126

96:132

96:138

96:144

96:125

96:131

96:137

96:143

96:124

96:130

96:136

96:142

96:123

96:129

96:135

96:141

96:122

96:128

96:134

96:140

96:121

96:127

96:133

96:139

Table A.18. (*Continued*)

Isomer	Ring spiral	Point group	NMR pattern	Vibrations	Hexagon indices	Band gap	Transformations
96:145	1 7 9 13 22 26 30 32 39 44 46 49	C_1	96×1	282, 282, 282	1 20 17 0	0.2713	91, 128, 143, 146, 147, 172
96:146	1 7 9 13 22 26 30 32 40 44 46 48	C_s	6×1, 45×2	282, 282, 282	1 21 15 1	0.2897	130(2), 144(2), 145(2), 178
96:147	1 7 9 13 22 26 30 33 39 43 46 49	C_s	6×1, 45×2	282, 282, 282	1 20 17 0	0.1695	92(2), 145(2), 156
96:148	1 7 9 13 23 25 29 31 41 44 46 49	D_{2h}	4×4, 10×8	107, 141, 0	4 20 8 6	0.3466	149(2)
96:149	1 7 9 13 23 25 29 31 42 44 46 48	C_{2v}	2×2, 23×4	212, 282, 212	2 22 10 4	0.2006	110(2), 148, 184
96:150	1 7 9 13 23 25 29 33 39 41 43 45	D_2	24×4	210, 282, 210	4 18 12 4	0.3419	151(2)
96:151	1 7 9 13 23 25 29 33 39 42 45 49	C_2	48×2	282, 282, 282	2 20 14 2	0.1950	106(2), 150, 183
96:152	1 7 10 12 14 18 32 37 39 43 45 48	C_1	96×1	282, 282, 282	0 23 14 1	0.1149	75, 92, 156, 160, 166, 170
96:153	1 7 10 12 14 18 33 36 39 41 46 49	C_1	96×1	282, 282, 282	0 23 14 1	0.1065	133, 141, 155, 159, 161, 175, 177
96:154	1 7 10 12 14 18 33 37 39 41 44 50	D_2	24×4	210, 282, 210	0 22 16 0	0.0061	155(4), 157(2)
96:155	1 7 10 12 14 18 33 37 39 41 45 49	C_1	96×1	282, 282, 282	0 22 16 0	0.1160	153, 154, 156, 160, 162, 179, 181
96:156	1 7 10 12 14 18 33 37 39 42 45 48	C_s	4×1, 46×2	282, 282, 282	0 23 14 1	0.0400	133, 147, 152(2), 155(2), 172(2)
96:157	1 7 10 12 14 18 34 37 39 41 43 50	C_2	48×2	282, 282, 282	0 22 16 0	0.0276	154, 160(2), 186
96:158	1 7 10 12 14 19 36 38 41 43 45 48	C_2	48×2	282, 282, 282	0 23 14 1	0.1472	53(2), 84(2)
96:159	1 7 10 12 14 28 30 33 37 44 47 49	C_1	96×1	282, 282, 282	0 22 16 0	0.1166	75, 140, 153, 160, 164
96:160	1 7 10 12 14 28 30 33 38 44 46 49	C_1	96×1	282, 282, 282	0 22 16 0	0.1280	152, 155, 157, 159, 165
96:161	1 7 10 12 14 28 30 34 37 42 47 50	C_s	6×1, 45×2	282, 282, 282	0 22 16 0	0.1446	153(2), 162, 163, 182(2)
96:162	1 7 10 12 14 28 30 34 38 42 46 50	C_{2v}	6×2, 21×4	214, 282, 214	0 23 14 1	0.0484	155(4), 161(2), 180(2)
96:163	1 7 10 12 14 28 31 34 37 41 47 50	D_{2d}	4×4, 10×8	107, 179, 107	0 22 16 0	0.1834	161(4)
96:164	1 7 10 12 18 26 31 34 36 39 46 49	C_1	96×1	282, 282, 282	0 22 16 0	0.1986	76, 138, 159, 165, 169, 175, 176
96:165	1 7 10 12 18 26 31 34 36 39 48 50	C_2	48×2	282, 282, 282	0 22 16 0	0.1405	160(2), 164(2), 166(2), 181
96:166	1 7 10 12 18 27 31 34 36 38 48 50	C_1	96×1	282, 282, 282	0 23 14 1	0.0396	76, 91, 152, 165, 168, 169, 172
96:167	1 7 10 12 19 24 27 39 41 43 45 47	C_2	48×2	282, 282, 282	0 22 16 0	0.0189	72(2), 95(2), 169(2)
96:168	1 7 10 12 19 24 28 37 42 45 48 50	C_2	48×2	282, 282, 282	0 24 12 2	0.0628	78(2), 98(2), 166(2), 173

96:150

96:149

96:148

96:147

96:146

96:145

96:156

96:155

96:154

96:153

96:152

96:151

96:162

96:161

96:160

96:159

96:158

96:157

96:168

96:167

96:166

96:165

96:164

96:163

303

Table A.18. (*Continued*)

Isomer	Ring spiral	Point group	NMR pattern	Vibrations	Hexagon indices	Band gap	Transformations
96:169	1 7 10 12 19 24 28 38 41 43 45 47	C_1	96×1	282, 282, 282	0 23 14 1	0.0582	78, 79, 90, 139, 164, 166, 167, 171
96:170	1 7 10 12 19 24 29 37 39 43 45 49	C_s	4×1, 46×2	282, 282, 282	0 23 14 1	0.0407	81, 89, 152(2)
96:171	1 7 10 13 18 22 25 37 40 42 44 47	C_2	48×2	282, 282, 282	0 23 14 1	0.0833	135(2), 143(2), 169(2), 172(2), 175(2)
96:172	1 7 10 13 18 22 25 38 40 42 44 46	C_1	96×1	282, 282, 282	0 23 14 1	0.1555	134, 145, 156, 166, 171, 173, 178, 179, 181
96:173	1 7 10 13 18 22 25 38 40 43 46 49	C_2	48×2	282, 282, 282	0 24 12 2	0.3003	128(2), 135(2), 168, 172(2), 177(2)
96:174	1 7 10 13 19 25 28 33 41 44 46 48	C_2	48×2	282, 282, 282	0 24 12 2	0.1843	131(2), 135(2), 175(2), 176, 177(2)
96:175	1 7 10 13 19 25 28 34 41 43 46 48	C_1	96×1	282, 282, 282	0 23 14 1	0.1836	134, 144, 153, 164, 171, 174, 178, 181, 182
96:176	1 7 10 13 19 27 34 36 40 43 47 50	C_2	48×2	282, 282, 282	0 22 16 0	0.2001	79(2), 125(2), 164(2), 174
96:177	1 7 10 13 19 28 33 36 38 42 45 48	C_1	96×1	282, 282, 282	0 23 14 1	0.2162	130, 134, 153, 173, 174, 178, 179, 182
96:178	1 7 10 13 19 28 34 36 38 42 44 48	C_s	4×1, 46×2	282, 282, 282	0 24 12 2	0.1232	137, 146, 172(2), 175(2), 177(2), 180(2)
96:179	1 7 10 13 25 27 31 33 35 40 42 46	C_2	48×2	282, 282, 282	0 22 16 0	0.1749	155(2), 172(2), 177(2), 180(2)
96:180	1 7 10 13 25 28 31 33 35 39 42 46	C_s	6×1, 45×2	282, 282, 282	0 23 14 1	0.1309	162, 178(2), 179(2), 181(2), 182(2), 185
96:181	1 7 10 14 18 22 24 38 40 42 44 46	C_2	48×2	282, 282, 282	0 22 16 0	0.2911	155(2), 165, 172(2), 175(2), 180(2)
96:182	1 7 10 14 18 22 31 35 37 39 44 48	C_2	48×2	282, 282, 282	0 22 16 0	0.2461	161(2), 175(2), 177(2), 180(2)
96:183	1 7 10 14 19 23 27 33 40 45 47 50	D_2	24×4	210, 282, 210	0 22 16 0	0.1932	107(4), 151(2)
96:184	1 7 10 14 19 27 33 35 37 43 48 50	D_{6h}	4×12, 2×24	36, 61, 0	0 24 12 2	0.1293	149(6)
96:185	1 7 11 18 21 25 27 32 36 40 45 49	D_{6d}	2×12, 3×24	35, 60, 0	0 24 12 2	0.0728	180(12)
96:186	1 7 12 14 19 21 29 36 38 42 44 50	D_2	24×4	210, 282, 210	0 22 16 0	0.0276	157(2)
96:187	1 8 10 12 16 29 33 35 37 39 48 50	D_{6d}	4×12, 2×24	36, 61, 0	0 24 12 2	0.0000	

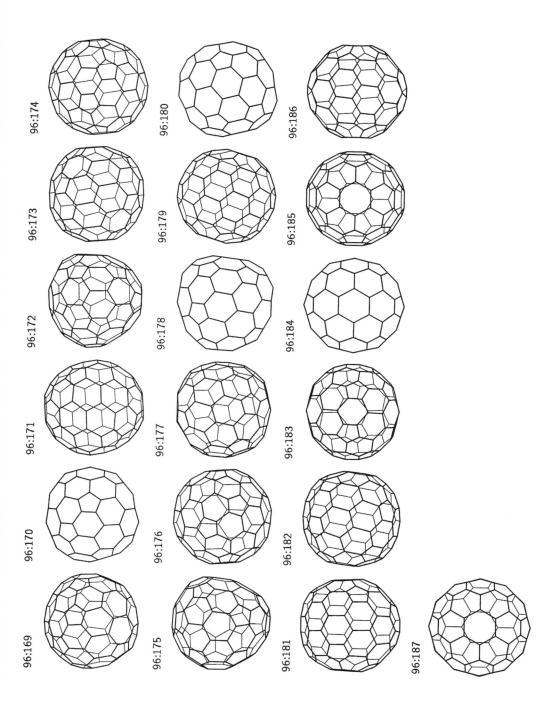

96:174

96:180

96:186

96:173

96:179

96:185

96:172

96:178

96:184

96:171

96:177

96:183

96:170

96:176

96:182

96:169

96:175

96:181

96:187

Table A.19. Isolated-pentagon fullerene isomers of C_{98}

Isomer	Ring spiral	Point group	NMR pattern	Vibrations	Hexagon indices	Band gap	Transformations
98:1	1 7 9 11 13 24 28 39 41 43 45 51	C_2	49×2	288, 288, 288	6 16 10 7	0.3304	2(2)
98:2	1 7 9 11 13 24 29 38 41 43 45 51	C_1	98×1	288, 288, 288	5 16 13 5	0.0288	1, 11
98:3	1 7 9 11 13 25 33 39 41 45 47 49	C_2	49×2	288, 288, 288	6 12 18 3	0.2320	
98:4	1 7 9 11 13 28 33 36 40 42 45 47	C_1	98×1	288, 288, 288	4 17 14 4	0.1412	5, 6
98:5	1 7 9 11 13 28 33 36 41 45 47 49	C_1	98×1	288, 288, 288	5 14 17 3	0.0733	4
98:6	1 7 9 11 13 29 33 35 38 42 46 48	C_1	98×1	288, 288, 288	4 16 16 3	0.0607	4, 7
98:7	1 7 9 11 13 29 33 36 38 42 45 48	C_s	8×1, 45×2	288, 288, 288	3 17 17 2	0.0261	6(2)
98:8	1 7 9 11 13 30 35 38 42 46 48 50	C_2	49×2	288, 288, 288	6 13 16 4	0.2143	
98:9	1 7 9 11 14 22 36 39 41 44 47 49	C_2	49×2	288, 288, 288	4 16 16 3	0.3039	10(2)
98:10	1 7 9 11 14 22 37 39 41 44 46 49	C_1	98×1	288, 288, 288	3 17 17 2	0.0429	9, 72
98:11	1 7 9 11 14 23 29 38 41 43 45 51	C_2	49×2	288, 288, 288	4 16 16 3	0.1600	2(2)
98:12	1 7 9 11 14 28 31 35 40 44 46 50	C_1	98×1	288, 288, 288	4 16 16 3	0.0787	17, 19
98:13	1 7 9 11 14 28 32 36 43 47 49 51	C_{2v}	5×2, 22×4	218, 288, 218	4 16 16 3	0.1123	62
98:14	1 7 9 11 14 28 33 35 41 46 48 50	C_1	98×1	288, 288, 288	2 19 16 2	0.0628	15, 16
98:15	1 7 9 11 14 28 33 36 41 45 48 50	C_1	98×1	288, 288, 288	3 18 15 3	0.1523	14, 49
98:16	1 7 9 11 14 28 35 41 43 45 47 49	C_1	98×1	288, 288, 288	3 18 15 3	0.0919	14, 19
98:17	1 7 9 11 14 29 31 33 37 45 47 50	C_1	98×1	288, 288, 288	4 15 18 2	0.2941	12, 22
98:18	1 7 9 11 14 29 31 35 37 43 48 51	C_s	6×1, 46×2	288, 288, 288	3 17 17 2	0.0755	23
98:19	1 7 9 11 14 29 32 37 43 45 47 49	C_1	98×1	288, 288, 288	3 17 17 2	0.1074	12, 16
98:20	1 7 9 11 14 29 34 36 39 43 49 51	C_1	98×1	288, 288, 288	3 17 17 2	0.0454	67
98:21	1 7 9 11 14 30 33 35 39 46 48 50	C_1	98×1	288, 288, 288	4 14 20 1	0.0730	
98:22	1 7 9 11 14 30 33 37 41 45 47 50	C_1	98×1	288, 288, 288	4 16 16 3	0.0680	17
98:23	1 7 9 11 14 30 35 37 41 43 48 51	C_s	8×1, 45×2	288, 288, 288	4 17 14 4	0.2071	18
98:24	1 7 9 11 14 34 36 38 40 42 44 50	C_s	6×1, 46×2	288, 288, 288	4 15 18 2	0.0100	65

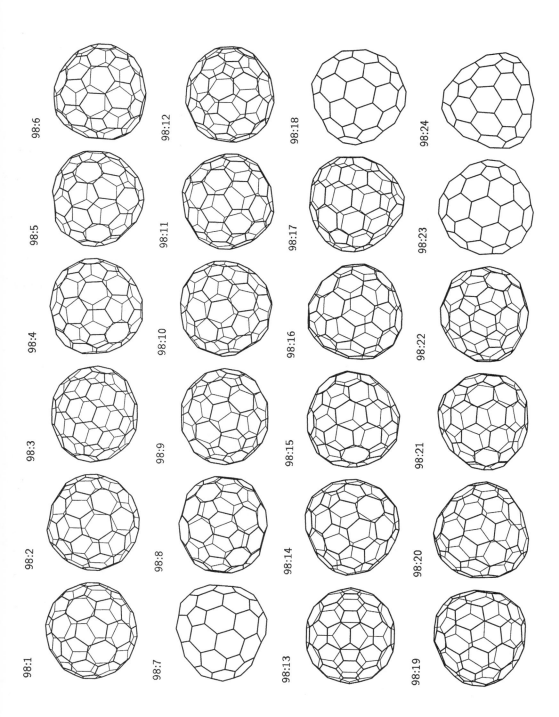

98:6

98:12

98:18

98:24

98:5

98:11

98:17

98:23

98:4

98:10

98:16

98:22

98:3

98:9

98:15

98:21

98:2

98:8

98:14

98:20

98:1

98:7

98:13

98:19

Table A.19. (*Continued*)

Isomer	Ring spiral	Point group	NMR pattern	Vibrations	Hexagon indices	Band gap	Transformations
98:25	1 7 9 11 22 24 29 36 38 41 43 51	C_1	98×1	288, 288, 288	3 18 15 3	0.0995	26, 33, 68
98:26	1 7 9 11 22 24 30 36 38 40 43 51	C_1	98×1	288, 288, 288	2 19 16 2	0.1539	25, 27, 34, 129
98:27	1 7 9 11 22 24 30 36 38 40 49 51	C_1	98×1	288, 288, 288	3 18 15 3	0.2037	26, 29, 35, 126
98:28	1 7 9 11 22 24 30 37 40 45 48 50	C_2	49×2	288, 288, 288	4 17 14 4	0.0178	29(2)
98:29	1 7 9 11 22 24 30 38 40 45 47 50	C_1	98×1	288, 288, 288	3 18 15 3	0.1718	27, 28, 130
98:30	1 7 9 11 22 25 27 30 42 44 46 48	C_1	98×1	288, 288, 288	2 19 16 2	0.1307	31, 32, 38, 131
98:31	1 7 9 11 22 25 27 30 42 44 47 51	C_1	98×1	288, 288, 288	3 17 17 2	0.0894	30, 124
98:32	1 7 9 11 22 25 27 30 43 46 48 50	C_1	98×1	288, 288, 288	3 18 15 3	0.2511	30, 37, 45, 115
98:33	1 7 9 11 22 25 29 35 38 41 43 51	C_1	98×1	288, 288, 288	3 19 13 4	0.3809	25, 34, 36, 46, 90
98:34	1 7 9 11 22 25 30 35 38 40 43 51	C_1	98×1	288, 288, 288	2 20 14 3	0.1133	26, 33, 35, 40, 50, 132
98:35	1 7 9 11 22 25 30 35 38 40 49 51	C_1	98×1	288, 288, 288	3 19 13 4	0.3669	27, 34, 41, 51, 128
98:36	1 7 9 11 22 26 29 35 37 41 43 51	C_1	98×1	288, 288, 288	3 19 13 4	0.2663	33, 37, 40, 53, 133
98:37	1 7 9 11 22 26 29 35 41 43 46 48	C_1	98×1	288, 288, 288	2 20 14 3	0.3134	32, 36, 38, 43, 54, 134
98:38	1 7 9 11 22 26 29 36 41 43 45 48	C_1	98×1	288, 288, 288	2 20 14 3	0.1977	30, 37, 39, 44, 135
98:39	1 7 9 11 22 26 29 36 41 44 48 51	C_1	98×1	288, 288, 288	2 20 14 3	0.1832	38, 42, 63, 136
98:40	1 7 9 11 22 26 30 35 37 40 43 51	C_1	98×1	288, 288, 288	2 19 16 2	0.0597	34, 36, 41, 43, 55, 137
98:41	1 7 9 11 22 26 30 35 37 40 49 51	C_1	98×1	288, 288, 288	3 18 15 3	0.1082	35, 40, 56, 138
98:42	1 7 9 11 22 26 30 35 39 43 46 49	C_1	98×1	288, 288, 288	2 19 16 2	0.1613	39, 43, 44, 57, 139
98:43	1 7 9 11 22 26 30 35 40 43 46 48	C_1	98×1	288, 288, 288	2 19 16 2	0.1762	37, 40, 42, 44, 58, 140
98:44	1 7 9 11 22 26 30 36 39 43 45 49	C_1	98×1	288, 288, 288	2 19 16 2	0.1447	38, 42, 43, 44, 141
98:45	1 7 9 11 23 25 29 32 39 44 46 49	C_1	98×1	288, 288, 288	3 16 19 1	0.0850	32, 54
98:46	1 7 9 11 23 25 29 34 38 41 43 51	C_1	98×1	288, 288, 288	3 18 15 3	0.0717	33, 50, 53
98:47	1 7 9 11 23 25 29 38 42 44 47 50	C_2	49×2	288, 288, 288	4 15 18 2	0.1157	
98:48	1 7 9 11 23 25 29 39 41 44 46 51	C_1	98×1	288, 288, 288	4 16 16 3	0.2159	

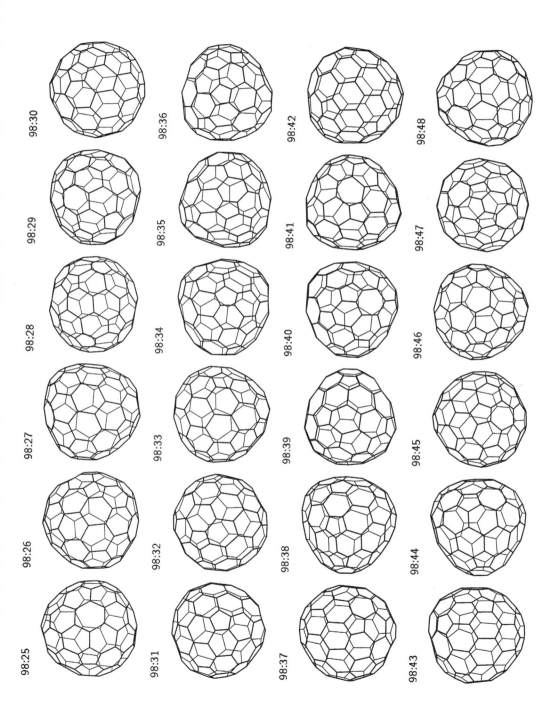

98:30

98:36

98:42

98:48

98:29

98:35

98:41

98:47

98:28

98:34

98:40

98:46

98:27

98:33

98:39

98:45

98:26

98:32

98:38

98:44

98:25

98:31

98:37

98:43

Table A.19. (*Continued*)

Isomer	Ring spiral	Point group	NMR pattern	Vibrations	Hexagon indices	Band gap	Transformations
98:49	1 7 9 11 23 25 30 32 39 45 47 49	C_1	98×1	288, 288, 288	3 16 19 1	0.1523	15
98:50	1 7 9 11 23 25 30 34 38 40 43 51	C_1	98×1	288, 288, 288	2 19 16 2	0.0788	34, 46, 51, 55
98:51	1 7 9 11 23 25 30 34 38 40 49 51	C_1	98×1	288, 288, 288	3 18 15 3	0.1692	35, 50, 52, 56
98:52	1 7 9 11 23 25 30 38 40 44 46 50	C_1	98×1	288, 288, 288	3 17 17 2	0.1230	51, 60
98:53	1 7 9 11 23 26 29 34 37 41 43 51	C_1	98×1	288, 288, 288	3 18 15 3	0.0736	36, 46, 54, 55
98:54	1 7 9 11 23 26 29 34 41 43 46 48	C_1	98×1	288, 288, 288	2 19 16 2	0.0620	37, 45, 53, 58
98:55	1 7 9 11 23 26 30 34 37 40 43 51	C_1	98×1	288, 288, 288	2 18 18 1	0.1925	40, 50, 53, 56, 58
98:56	1 7 9 11 23 26 30 34 37 40 49 51	C_1	98×1	288, 288, 288	3 17 17 2	0.1647	41, 51, 55, 60
98:57	1 7 9 11 23 26 30 34 39 43 46 49	C_1	98×1	288, 288, 288	2 18 18 1	0.1630	42, 58, 63
98:58	1 7 9 11 23 26 30 34 40 43 46 48	C_1	98×1	288, 288, 288	2 18 18 1	0.0566	43, 54, 55, 57
98:59	1 7 9 11 23 26 30 36 40 44 47 50	C_1	98×1	288, 288, 288	3 16 19 1	0.1566	60, 67
98:60	1 7 9 11 23 26 30 37 40 44 46 50	C_1	98×1	288, 288, 288	3 16 19 1	0.2419	52, 56, 59
98:61	1 7 9 11 23 26 36 39 41 43 45 51	C_2	49×2	288, 288, 288	4 15 18 2	0.0418	65(2)
98:62	1 7 9 11 23 27 30 32 37 45 48 51	C_{2v}	5×2, 22×4	218, 288, 218	4 14 20 1	0.3262	13
98:63	1 7 9 11 23 27 30 34 38 43 46 49	C_1	98×1	288, 288, 288	2 19 16 2	0.0469	39, 57
98:64	1 7 9 11 23 27 34 36 41 43 49 51	C_1	98×1	288, 288, 288	5 14 17 3	0.3186	
98:65	1 7 9 11 23 27 36 38 41 43 45 51	C_s	4×1, 47×2	288, 288, 288	4 16 16 3	0.2953	24, 61(2)
98:66	1 7 9 11 23 29 34 36 40 42 47 49	D_3	1×2, 16×6	143, 145, 96	6 15 12 6	0.2512	
98:67	1 7 9 11 24 26 30 35 40 44 47 50	C_1	98×1	288, 288, 288	3 16 19 1	0.2608	20, 59
98:68	1 7 9 12 14 20 35 38 40 42 46 50	C_1	98×1	288, 288, 288	2 20 14 3	0.1496	25, 70, 90, 129
98:69	1 7 9 12 14 20 36 38 40 42 44 51	C_1	98×1	288, 288, 288	1 20 17 1	0.0175	70, 71, 166
98:70	1 7 9 12 14 20 36 38 40 42 45 50	C_1	98×1	288, 288, 288	2 19 16 2	0.0442	68, 69, 125
98:71	1 7 9 12 14 20 36 38 41 44 49 51	C_2	49×2	288, 288, 288	2 19 16 2	0.1080	69(2), 74
98:72	1 7 9 12 14 21 37 39 41 44 46 49	C_2	49×2	288, 288, 288	2 18 18 1	0.0792	10(2)

98:54

98:60

98:66

98:72

98:53

98:59

98:65

98:71

98:52

98:58

98:64

98:70

98:51

98:57

98:63

98:69

98:50

98:56

98:62

98:68

98:49

98:55

98:61

98:67

Table A.19. *(Continued)*

Isomer	Ring spiral	Point group	NMR pattern	Vibrations	Hexagon indices	Band gap	Transformations
98:73	1 7 9 12 14 28 31 33 36 47 49 51	C_1	98×1	288, 288, 288	2 18 18 1	0.0850	83, 122
98:74	1 7 9 12 14 28 31 33 38 45 48 50	C_2	49×2	288, 288, 288	2 17 20 0	0.0625	71
98:75	1 7 9 12 14 28 32 35 41 46 48 50	C_1	98×1	288, 288, 288	1 20 17 1	0.0481	79
98:76	1 7 9 12 14 28 32 36 40 42 44 47	C_1	98×1	288, 288, 288	2 19 16 2	0.1569	77, 80, 87, 142
98:77	1 7 9 12 14 28 32 36 41 44 47 49	C_1	98×1	288, 288, 288	1 20 17 1	0.0579	76, 78, 81, 82, 144
98:78	1 7 9 12 14 28 32 36 41 44 48 51	C_1	98×1	288, 288, 288	1 20 17 1	0.0661	77, 79, 83, 143
98:79	1 7 9 12 14 28 32 36 41 45 48 50	C_1	98×1	288, 288, 288	1 20 17 1	0.0036	75, 78, 145
98:80	1 7 9 12 14 28 32 37 40 42 44 46	C_1	98×1	288, 288, 288	2 18 18 1	0.1675	76, 81, 86
98:81	1 7 9 12 14 28 32 37 41 44 46 49	C_1	98×1	288, 288, 288	1 19 19 0	0.0857	77, 80, 84
98:82	1 7 9 12 14 28 33 36 41 43 47 49	C_1	98×1	288, 288, 288	2 20 14 3	0.1150	77, 83, 84, 106
98:83	1 7 9 12 14 28 33 36 41 43 48 51	C_1	98×1	288, 288, 288	1 21 15 2	0.0619	73, 78, 82, 146
98:84	1 7 9 12 14 28 33 37 41 43 46 49	C_1	98×1	288, 288, 288	2 18 18 1	0.1033	81, 82
98:85	1 7 9 12 14 29 31 33 37 45 47 49	C_1	98×1	288, 288, 288	3 16 19 1	0.2022	
98:86	1 7 9 12 14 29 31 33 38 44 46 48	C_1	98×1	288, 288, 288	2 17 20 0	0.0430	80, 87
98:87	1 7 9 12 14 29 31 34 38 43 46 48	C_1	98×1	288, 288, 288	2 18 18 1	0.1791	76, 86, 89, 147
98:88	1 7 9 12 14 29 31 35 39 44 46 49	C_{2v}	7×2, 21×4	219, 288, 219	4 15 18 2	0.3066	
98:89	1 7 9 12 14 29 32 34 38 42 46 48	C_2	49×2	288, 288, 288	2 17 20 0	0.1496	87(2)
98:90	1 7 9 12 20 24 26 39 41 43 47 51	C_1	98×1	288, 288, 288	2 19 16 2	0.1649	33, 68, 132, 133
98:91	1 7 9 12 20 24 28 36 40 43 47 51	C_2	49×2	288, 288, 288	2 21 12 4	0.2821	92(2), 97(2), 104(2)
98:92	1 7 9 12 20 24 28 36 40 44 47 50	C_1	98×1	288, 288, 288	1 22 13 3	0.0404	91, 93, 98, 105, 149, 200, 204
98:93	1 7 9 12 20 24 28 37 40 44 46 50	C_1	98×1	288, 288, 288	2 19 16 2	0.0930	92, 99, 112, 117, 121
98:94	1 7 9 12 20 24 29 36 38 41 44 51	C_1	98×1	288, 288, 288	1 20 17 1	0.1916	96, 101, 102, 109, 160
98:95	1 7 9 12 20 24 29 36 38 43 48 51	C_1	98×1	288, 288, 288	1 21 15 2	0.3253	96, 97, 100, 104, 110, 201
98:96	1 7 9 12 20 24 29 36 38 44 48 50	C_1	98×1	288, 288, 288	1 20 17 1	0.2549	94, 95, 98, 105, 111, 202

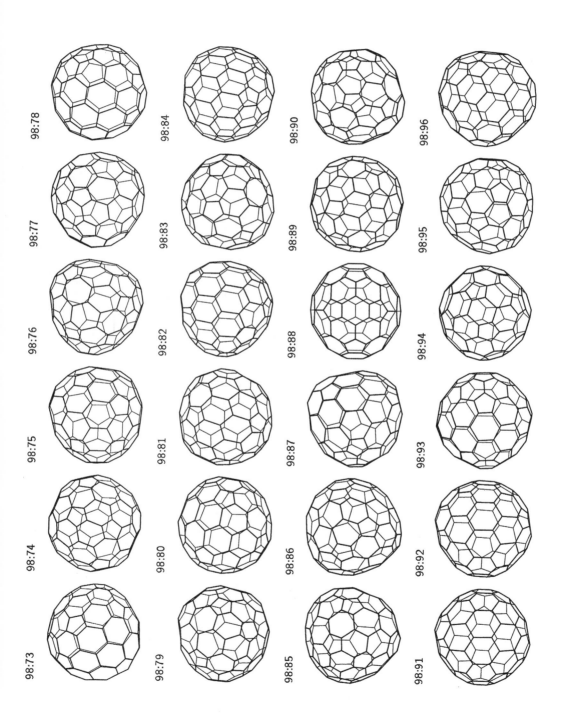

98:78

98:84

98:90

98:96

98:77

98:83

98:89

98:95

98:76

98:82

98:88

98:94

98:75

98:81

98:87

98:93

98:74

98:80

98:86

98:92

98:73

98:79

98:85

98:91

Table A.19. *(Continued)*

Isomer	Ring spiral	Point group	NMR pattern	Vibrations	Hexagon indices	Band gap	Transformations
98:97	1 7 9 12 20 24 29 36 39 43 47 51	C_s	4×1, 47×2	288, 288, 288	2 21 12 4	0.4023	91(2), 95(2), 98, 149
98:98	1 7 9 12 20 24 29 36 39 44 47 50	C_s	2×1, 48×2	288, 288, 288	1 21 15 2	0.1904	92(2), 96(2), 97, 99, 203
98:99	1 7 9 12 20 24 29 37 39 44 46 50	C_s	4×1, 47×2	288, 288, 288	2 18 18 1	0.2567	93(2), 98, 158
98:100	1 7 9 12 20 24 29 38 43 45 47 49	C_1	98×1	288, 288, 288	2 18 18 1	0.2760	95, 103, 157
98:101	1 7 9 12 20 24 30 36 38 40 44 51	C_1	98×1	288, 288, 288	1 20 17 1	0.1914	94, 114, 150, 164
98:102	1 7 9 12 20 25 28 34 40 45 47 50	C_1	98×1	288, 288, 288	1 21 15 2	0.1557	94, 105, 150, 161, 164
98:103	1 7 9 12 20 25 28 35 40 43 46 49	C_1	98×1	288, 288, 288	2 20 14 3	0.2849	100, 104, 106, 152, 155
98:104	1 7 9 12 20 25 28 35 40 43 47 51	C_1	98×1	288, 288, 288	1 21 15 2	0.3295	91, 95, 103, 105, 204, 248
98:105	1 7 9 12 20 25 28 35 40 44 47 50	C_1	98×1	288, 288, 288	1 21 15 2	0.1670	92, 96, 102, 104, 181, 184
98:106	1 7 9 12 20 25 28 35 41 43 46 48	C_1	98×1	288, 288, 288	2 19 16 2	0.2774	82, 103, 144, 146
98:107	1 7 9 12 20 25 29 34 38 41 45 51	C_s	2×1, 48×2	288, 288, 288	1 21 15 2	0.3363	108, 109(2), 112(2), 117, 203
98:108	1 7 9 12 20 25 29 34 38 41 46 50	C_s	4×1, 47×2	288, 288, 288	2 18 18 1	0.2227	107, 113(2), 158
98:109	1 7 9 12 20 25 29 34 38 45 48 50	C_1	98×1	288, 288, 288	1 20 17 1	0.2684	94, 107, 111, 114, 118, 202
98:110	1 7 9 12 20 25 29 35 38 43 48 51	C_s	4×1, 47×2	288, 288, 288	1 20 17 1	0.3207	95(2), 111, 119, 205
98:111	1 7 9 12 20 25 29 35 38 44 48 50	C_s	2×1, 48×2	288, 288, 288	1 21 15 2	0.2929	96(2), 109(2), 110, 120, 206
98:112	1 7 9 12 20 25 30 34 38 40 45 51	C_1	98×1	288, 288, 288	1 21 15 2	0.0430	93, 107, 113, 114, 158, 200
98:113	1 7 9 12 20 25 30 34 38 40 46 50	C_2	49×2	288, 288, 288	2 18 18 1	0.1135	108(2), 112(2)
98:114	1 7 9 12 20 25 30 35 38 40 44 51	C_1	98×1	288, 288, 288	1 20 17 1	0.0900	101, 109, 112, 121, 181
98:115	1 7 9 12 20 26 28 36 41 43 45 47	C_1	98×1	288, 288, 288	2 18 18 1	0.2465	32, 131, 134
98:116	1 7 9 12 20 26 29 34 36 46 48 50	C_1	98×1	288, 288, 288	2 17 20 0	0.2903	118, 123, 157
98:117	1 7 9 12 20 26 29 34 37 41 45 51	C_s	4×1, 47×2	288, 288, 288	2 20 14 3	0.3838	93(2), 107, 118(2), 149
98:118	1 7 9 12 20 26 29 34 37 45 48 50	C_1	98×1	288, 288, 288	1 20 17 1	0.2340	109, 116, 117, 120, 121, 201
98:119	1 7 9 12 20 26 29 35 37 43 48 51	C_s	6×1, 46×2	288, 288, 288	1 19 19 0	0.1871	110, 120, 167
98:120	1 7 9 12 20 26 29 35 37 44 48 50	C_s	4×1, 47×2	288, 288, 288	1 19 19 0	0.2929	111, 118(2), 119, 205

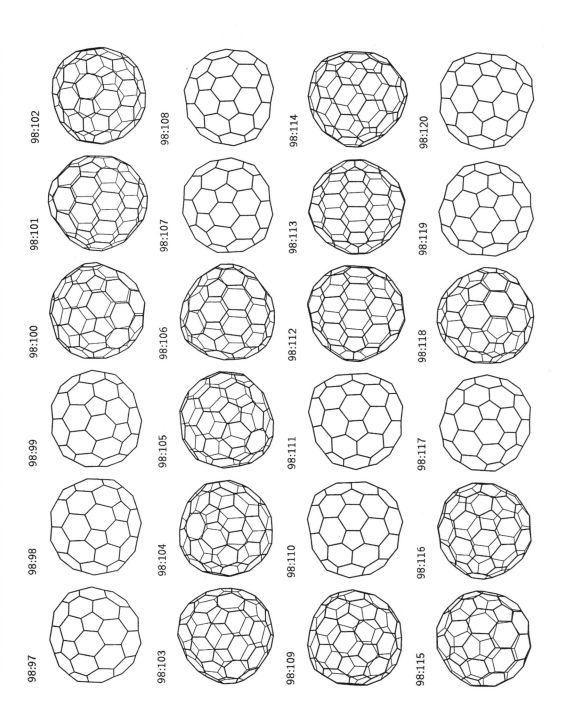

315

98:102

98:108

98:114

98:120

98:101

98:107

98:113

98:119

98:100

98:106

98:112

98:118

98:99

98:105

98:111

98:117

98:98

98:104

98:110

98:116

98:97

98:103

98:109

98:115

Table A.19. (*Continued*)

Isomer	Ring spiral	Point group	NMR pattern	Vibrations	Hexagon indices	Band gap	Transformations
98:121	1 7 9 12 20 26 30 35 37 40 44 51	C_1	98×1	288, 288, 288	1 19 19 0	0.1676	93, 114, 118, 123, 204
98:122	1 7 9 12 20 26 30 35 39 44 46 49	C_1	98×1	288, 288, 288	2 17 20 0	0.2200	73, 123, 146
98:123	1 7 9 12 20 26 30 35 40 44 46 48	C_1	98×1	288, 288, 288	2 18 18 1	0.1403	116, 121, 122, 152
98:124	1 7 9 12 20 26 30 36 39 43 45 47	C_1	98×1	288, 288, 288	2 17 20 0	0.1799	31, 131
98:125	1 7 9 12 20 26 33 36 39 41 47 51	C_1	98×1	288, 288, 288	1 20 17 1	0.1623	70, 127, 129, 166
98:126	1 7 9 12 20 26 33 37 39 41 45 50	C_1	98×1	288, 288, 288	2 20 14 3	0.3071	27, 127, 128, 129, 130
98:127	1 7 9 12 20 26 33 37 39 41 46 51	C_2	49×2	288, 288, 288	2 19 16 2	0.1825	125(2), 126(2)
98:128	1 7 9 12 20 26 34 37 39 41 44 50	C_1	98×1	288, 288, 288	2 19 16 2	0.2364	35, 126, 132, 138
98:129	1 7 9 12 21 24 30 36 38 40 43 51	C_1	98×1	288, 288, 288	1 21 15 2	0.2313	26, 68, 125, 126, 132
98:130	1 7 9 12 21 24 30 38 40 45 47 50	C_2	49×2	288, 288, 288	2 19 16 2	0.2564	29(2), 126(2)
98:131	1 7 9 12 21 25 27 30 42 44 46 48	C_1	98×1	288, 288, 288	1 19 19 0	0.0378	30, 115, 124, 135
98:132	1 7 9 12 21 25 30 35 38 40 43 51	C_1	98×1	288, 288, 288	1 20 17 1	0.0801	34, 90, 128, 129, 137
98:133	1 7 9 12 21 26 29 35 37 41 43 51	C_1	98×1	288, 288, 288	2 19 16 2	0.1714	36, 90, 134, 137
98:134	1 7 9 12 21 26 29 35 41 43 46 48	C_1	98×1	288, 288, 288	1 20 17 1	0.1594	37, 115, 133, 135, 140
98:135	1 7 9 12 21 26 29 36 41 43 45 48	C_1	98×1	288, 288, 288	1 20 17 1	0.0772	38, 131, 134, 136, 141
98:136	1 7 9 12 21 26 29 36 41 44 48 51	C_1	98×1	288, 288, 288	1 20 17 1	0.0564	39, 135, 139
98:137	1 7 9 12 21 26 30 35 37 40 43 51	C_1	98×1	288, 288, 288	1 19 19 0	0.0926	40, 132, 133, 138, 140
98:138	1 7 9 12 21 26 30 35 37 40 49 51	C_1	98×1	288, 288, 288	2 18 18 1	0.1485	41, 128, 137
98:139	1 7 9 12 21 26 30 35 39 43 46 49	C_1	98×1	288, 288, 288	1 19 19 0	0.1192	42, 136, 140, 141
98:140	1 7 9 12 21 26 30 35 40 43 46 48	C_1	98×1	288, 288, 288	1 19 19 0	0.0833	43, 134, 137, 139, 141
98:141	1 7 9 12 21 26 30 36 39 43 45 49	C_1	98×1	288, 288, 288	1 19 19 0	0.0904	44, 135, 139, 140, 141
98:142	1 7 9 12 24 26 29 33 41 43 45 48	C_1	98×1	288, 288, 288	2 18 18 1	0.1790	76, 144, 147, 156
98:143	1 7 9 12 24 26 29 33 42 45 47 49	C_1	98×1	288, 288, 288	1 19 19 0	0.2357	78, 144, 145, 146, 153
98:144	1 7 9 12 24 26 29 33 42 45 48 50	C_1	98×1	288, 288, 288	1 19 19 0	0.0988	77, 106, 142, 143, 155

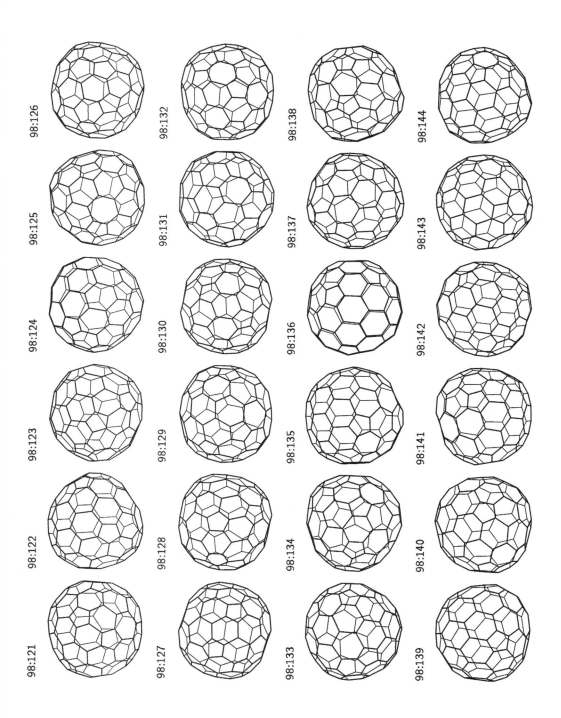

98:126

98:125

98:124

98:123

98:122

98:121

98:132

98:131

98:130

98:129

98:128

98:127

98:138

98:137

98:136

98:135

98:134

98:133

98:144

98:143

98:142

98:141

98:140

98:139

Table A.19. (*Continued*)

Isomer	Ring spiral	Point group	NMR pattern	Vibrations	Hexagon indices	Band gap	Transformations
98:145	1 7 9 12 24 26 29 33 42 46 49 51	C_1	98×1	288, 288, 288	1 19 19 0	0.0793	79, 143, 154
98:146	1 7 9 12 24 26 29 34 42 44 47 49	C_1	98×1	288, 288, 288	1 20 17 1	0.1213	83, 106, 122, 143, 152
98:147	1 7 9 12 24 26 30 33 40 43 45 48	C_2	49×2	288, 288, 288	2 17 20 0	0.0352	87(2), 142(2)
98:148	1 7 9 12 26 30 32 34 36 38 40 50	C_3	2×1, 32×3	188, 188, 188	3 15 21 0	0.2944	
98:149	1 7 9 13 20 23 28 36 40 44 47 50	C_s	2×1, 48×2	288, 288, 288	1 23 11 4	0.0702	92(2), 97, 117, 201(2), 203
98:150	1 7 9 13 20 23 30 36 38 40 44 51	C_1	98×1	288, 288, 288	1 21 15 2	0.1959	101, 102, 232, 247
98:151	1 7 9 13 20 26 30 36 43 45 47 49	C_s	6×1, 46×2	288, 288, 288	1 22 13 3	0.0959	191(2)
98:152	1 7 9 13 22 25 31 38 41 43 45 48	C_1	98×1	288, 288, 288	1 21 15 2	0.0664	103, 123, 146, 153, 157, 204
98:153	1 7 9 13 22 25 31 39 41 43 45 47	C_1	98×1	288, 288, 288	1 20 17 1	0.1472	143, 152, 154, 155, 184
98:154	1 7 9 13 22 25 31 39 41 43 46 51	C_1	98×1	288, 288, 288	1 20 17 1	0.0822	145, 153, 159, 161
98:155	1 7 9 13 22 25 31 39 41 44 47 50	C_1	98×1	288, 288, 288	1 20 17 1	0.2471	103, 144, 153, 156, 248
98:156	1 7 9 13 22 25 31 39 42 44 47 49	C_2	49×2	288, 288, 288	2 19 16 2	0.2832	142(2), 155(2)
98:157	1 7 9 13 22 26 29 35 42 44 47 49	C_1	98×1	288, 288, 288	1 20 17 1	0.0663	100, 116, 152, 201
98:158	1 7 9 13 22 26 33 39 41 43 46 48	C_s	2×1, 48×2	288, 288, 288	1 21 15 2	0.0267	99, 108, 112(2), 203
98:159	1 7 9 13 22 31 36 38 40 42 44 51	C_1	98×1	288, 288, 288	1 20 17 1	0.1493	154, 247
98:160	1 7 10 12 14 18 33 38 40 44 46 49	C_2	49×2	288, 288, 288	0 23 14 2	0.0500	94(2), 164(2), 202(2)
98:161	1 7 10 12 14 18 34 37 40 42 47 50	C_1	98×1	288, 288, 288	0 21 18 0	0.1701	102, 154, 163, 184, 247
98:162	1 7 10 12 14 18 34 38 40 42 45 51	C_2	49×2	288, 288, 288	0 22 16 1	0.0966	163(2), 165(2)
98:163	1 7 10 12 14 18 34 38 40 42 46 50	C_1	98×1	288, 288, 288	0 22 16 1	0.1094	161, 162, 164, 182, 234
98:164	1 7 10 12 14 18 34 38 40 43 46 49	C_1	98×1	288, 288, 288	0 23 14 2	0.1219	101, 102, 160, 163, 181, 232
98:165	1 7 10 12 14 18 35 38 40 42 44 51	C_1	98×1	288, 288, 288	0 22 16 1	0.1411	162, 168, 183, 234
98:166	1 7 10 12 14 19 36 38 40 42 44 51	C_2	49×2	288, 288, 288	0 21 18 0	0.1463	69(2), 125(2)
98:167	1 7 10 12 14 29 31 34 37 39 49 51	C_{2v}	7×2, 21×4	219, 288, 219	0 21 18 0	0.1283	119(2), 205(2)
98:168	1 7 10 12 18 24 26 39 41 43 45 51	C_1	98×1	288, 288, 288	0 21 18 0	0.1081	165, 175, 233, 252

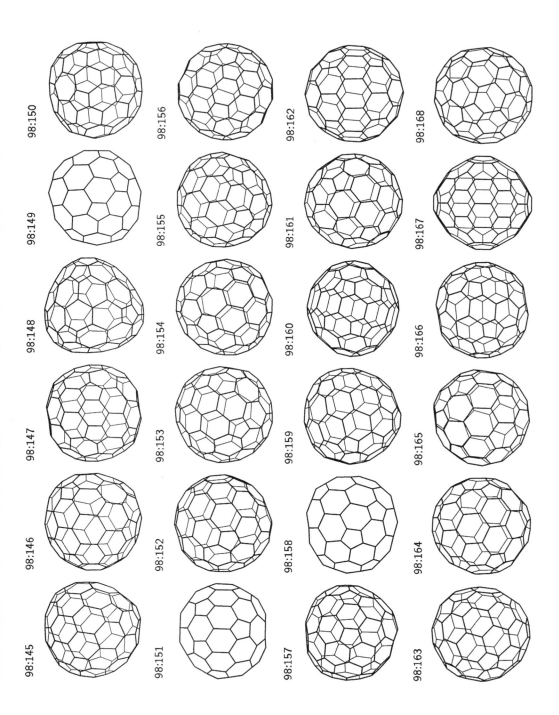

98:150

98:156

98:162

98:168

98:149

98:155

98:161

98:167

98:148

98:154

98:160

98:166

98:147

98:153

98:159

98:165

98:146

98:152

98:158

98:164

98:145

98:151

98:157

98:163

Table A.19. (*Continued*)

Isomer	Ring spiral	Point group	NMR pattern	Vibrations	Hexagon indices	Band gap	Transformations
98:169	1 7 10 12 18 24 27 36 40 42 44 51	C_1	98×1	288, 288, 288	0 21 18 0	0.0572	173, 175, 186, 189, 233
98:170	1 7 10 12 18 24 32 37 39 41 45 51	C_1	98×1	288, 288, 288	0 22 16 1	0.1506	185, 194, 195, 246, 259
98:171	1 7 10 12 18 25 27 33 40 44 46 49	C_1	98×1	288, 288, 288	0 22 16 1	0.0763	172, 178, 195, 196, 197, 221
98:172	1 7 10 12 18 25 27 33 41 44 46 48	C_1	98×1	288, 288, 288	0 21 18 0	0.0722	171, 173, 177, 179, 186, 187
98:173	1 7 10 12 18 25 27 33 41 44 47 51	C_1	98×1	288, 288, 288	0 21 18 0	0.0756	169, 172, 174, 176, 180
98:174	1 7 10 12 18 25 27 33 41 45 47 50	C_2	49×2	288, 288, 288	0 21 18 0	0.1098	173(2), 175(2)
98:175	1 7 10 12 18 25 27 33 42 45 47 49	C_1	98×1	288, 288, 288	0 21 18 0	0.0766	168, 169, 174, 183
98:176	1 7 10 12 18 25 27 34 40 42 45 51	C_2	49×2	288, 288, 288	0 21 18 0	0.1053	173(2), 177(2), 208(2)
98:177	1 7 10 12 18 25 27 34 40 42 46 50	C_1	98×1	288, 288, 288	0 21 18 0	0.0056	172, 176, 178, 197, 207, 242, 244
98:178	1 7 10 12 18 25 27 34 40 43 46 49	C_1	98×1	288, 288, 288	0 22 16 1	0.0441	171, 177, 177, 179, 187, 190, 198, 237, 245
98:179	1 7 10 12 18 25 27 34 41 43 46 48	C_1	98×1	288, 288, 288	0 21 18 0	0.1176	172, 178, 180, 188, 199, 244, 251
98:180	1 7 10 12 18 25 27 34 41 43 47 51	C_1	98×1	288, 288, 288	0 21 18 0	0.0754	173, 179, 189, 208, 225
98:181	1 7 10 12 18 25 31 34 38 40 45 51	C_1	98×1	288, 288, 288	0 23 14 2	0.0350	105, 114, 164, 182, 200, 202, 204
98:182	1 7 10 12 18 25 31 35 38 40 44 51	C_2	49×2	288, 288, 288	0 22 16 1	0.0756	163(2), 181(2), 184(2)
98:183	1 7 10 12 18 25 33 35 38 40 42 51	C_2	49×2	288, 288, 288	0 22 16 1	0.1530	165(2), 175(2)
98:184	1 7 10 12 18 26 31 35 37 40 44 51	C_1	98×1	288, 288, 288	0 21 18 0	0.1883	105, 153, 161, 182, 204, 248
98:185	1 7 10 12 18 26 33 35 39 41 46 50	C_s	8×1, 45×2	288, 288, 288	0 22 16 1	0.0612	170(2), 186(2), 187(2), 193
98:186	1 7 10 12 18 26 33 35 39 42 46 49	C_1	98×1	288, 288, 288	0 21 18 0	0.0307	169, 172, 185, 188, 195, 246
98:187	1 7 10 12 18 26 33 36 39 41 45 50	C_1	98×1	288, 288, 288	0 22 16 1	0.0991	172, 178, 185, 188, 195, 196, 199
98:188	1 7 10 12 18 26 33 36 39 42 45 49	C_2	49×2	288, 288, 288	0 21 18 0	0.2692	179(2), 186(2), 187(2), 189
98:189	1 7 10 12 18 26 33 36 40 42 45 48	C_2	49×2	288, 288, 288	0 21 18 0	0.0889	169(2), 180(2), 188
98:190	1 7 10 12 18 26 34 36 39 41 43 51	C_1	98×1	288, 288, 288	0 21 18 0	0.0815	178, 197, 207, 212, 221, 256
98:191	1 7 10 12 18 27 33 35 37 40 49 51	C_2	49×2	288, 288, 288	0 23 14 2	0.0397	151(2), 192(2)
98:192	1 7 10 12 18 27 33 35 37 41 48 51	C_s	6×1, 46×2	288, 288, 288	0 23 14 2	0.0056	191(2), 194(2), 259

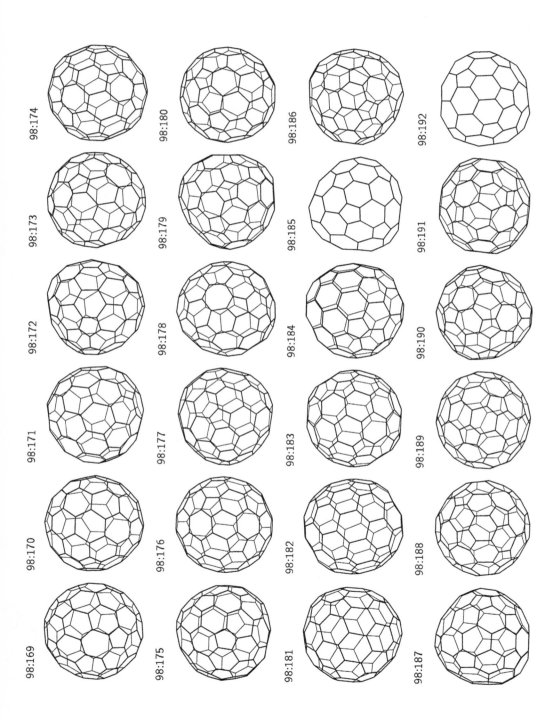

98:174

98:180

98:186

98:192

98:173

98:179

98:185

98:191

98:172

98:178

98:184

98:190

98:171

98:177

98:183

98:189

98:170

98:176

98:182

98:188

98:169

98:175

98:181

98:187

Table A.19. (*Continued*)

Isomer	Ring spiral	Point group	NMR pattern	Vibrations	Hexagon indices	Band gap	Transformations
98:193	1 7 10 12 18 27 33 35 38 41 46 50	C_s	6×1, 46×2	288, 288, 288	0 23 14 2	0.0398	185, 194(2), 195(2), 196(2)
98:194	1 7 10 12 18 27 33 35 38 41 47 51	C_2	49×2	288, 288, 288	0 23 14 2	0.0431	170(2), 192(2), 193(2)
98:195	1 7 10 12 18 27 33 35 38 42 46 49	C_1	98×1	288, 288, 288	0 22 16 1	0.0730	170, 171, 186, 187, 193
98:196	1 7 10 12 18 27 33 36 38 41 45 50	C_2	49×2	288, 288, 288	0 23 14 2	0.0882	171(2), 187(2), 193(2), 198(2)
98:197	1 7 10 12 18 27 34 36 38 41 43 51	C_1	98×1	288, 288, 288	0 22 16 1	0.1069	171, 177, 190, 198, 209, 214, 237
98:198	1 7 10 12 18 27 34 36 38 41 44 50	C_s	6×1, 46×2	288, 288, 288	0 23 14 2	0.2183	178(2), 196(2), 197(2), 199, 243(2)
98:199	1 7 10 12 18 27 34 36 38 42 44 49	C_s	8×1, 45×2	288, 288, 288	0 22 16 1	0.0016	179(2), 187(2), 198, 245(2)
98:200	1 7 10 12 19 24 28 36 40 44 47 50	C_2	49×2	288, 288, 288	0 24 12 3	0.0795	92(2), 112(2), 181(2), 203(2)
98:201	1 7 10 12 19 24 29 36 38 43 48 51	C_1	98×1	288, 288, 288	0 23 14 2	0.0627	95, 118, 149, 157, 202, 204, 205
98:202	1 7 10 12 19 24 29 36 38 44 48 50	C_1	98×1	288, 288, 288	0 23 14 2	0.0743	96, 109, 160, 181, 201, 203, 206
98:203	1 7 10 12 19 24 29 36 39 44 47 50	C_s	49×2	288, 288, 288	0 24 12 3	0.0146	98, 107, 149, 158, 200(2), 202(2)
98:204	1 7 10 12 19 25 28 35 40 43 47 51	C_1	98×1	288, 288, 288	0 22 16 1	0.1186	92, 104, 121, 152, 181, 184, 201
98:205	1 7 10 12 19 25 29 35 38 43 48 51	C_s	2×1, 48×2	288, 288, 288	0 22 16 1	0.0838	110, 120, 167, 201(2), 206
98:206	1 7 10 12 19 25 29 35 38 44 48 50	C_{2v}	5×2, 22×4	218, 288, 218	0 24 12 3	0.0431	111(2), 202(4), 205(2)
98:207	1 7 10 12 24 27 29 32 37 41 44 51	C_1	98×1	288, 288, 288	0 22 16 1	0.0269	177, 190, 208, 209, 210, 224, 235
98:208	1 7 10 12 24 27 29 32 37 41 45 50	C_1	98×1	288, 288, 288	0 21 18 0	0.1068	176, 180, 207, 211, 223, 244
98:209	1 7 10 12 24 27 29 33 37 41 43 51	C_s	6×1, 46×2	288, 288, 288	0 23 14 2	0.1098	197(2), 207(2), 213(2), 218
98:210	1 7 10 12 24 27 30 32 37 40 44 51	C_1	98×1	288, 288, 288	0 22 16 1	0.0795	207, 211, 212, 213, 216, 227, 244, 257
98:211	1 7 10 12 24 27 30 32 37 40 45 50	C_2	49×2	288, 288, 288	0 21 18 0	0.0328	208(2), 210(2), 217(2)
98:212	1 7 10 12 24 27 30 32 40 44 46 48	C_1	98×1	288, 288, 288	0 21 18 0	0.0531	190, 210, 214, 219, 245, 250
98:213	1 7 10 12 24 27 30 33 37 40 43 51	C_1	98×1	288, 288, 288	0 23 14 2	0.0776	209, 210, 214(2), 219, 220, 235, 237, 239
98:214	1 7 10 12 24 27 30 33 40 43 46 48	C_1	98×1	288, 288, 288	0 22 16 1	0.0906	197, 212, 213(2), 215, 236, 242, 243
98:215	1 7 10 12 24 27 31 33 40 42 46 48	D_3	1×2, 16×6	143, 145, 96	0 21 18 0	0.1618	214(6)
98:216	1 7 10 12 24 28 30 32 37 39 44 51	C_1	98×1	288, 288, 288	0 23 14 2	0.0821	210, 217, 219, 220, 229(2), 240, 249

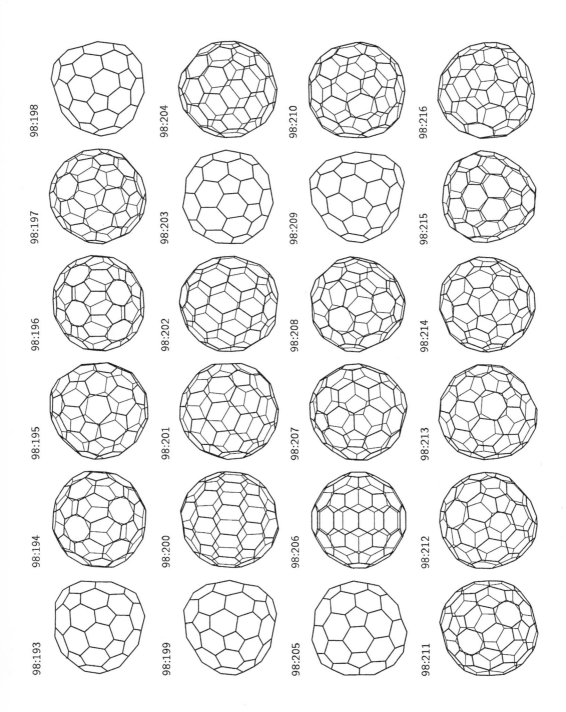

98:198

98:204

98:210

98:216

98:197

98:203

98:209

98:215

98:196

98:202

98:208

98:214

98:195

98:201

98:207

98:213

98:194

98:200

98:206

98:212

98:193

98:199

98:205

98:211

Table A.19. (*Continued*)

Isomer	Ring spiral	Point group	NMR pattern	Vibrations	Hexagon indices	Band gap	Transformations
98:217	1 7 10 12 24 28 30 32 37 39 45 50	C_1	98×1	288, 288, 288	0 22 16 1	0.1542	211, 216, 223, 226, 227, 230
98:218	1 7 10 12 24 28 30 32 38 44 46 49	C_s	8×1, 45×2	288, 288, 288	0 22 16 1	0.0611	209, 219(2), 224(2)
98:219	1 7 10 12 24 28 30 32 39 44 46 48	C_1	98×1	288, 288, 288	0 22 16 1	0.0978	212, 213, 216, 218, 227, 241, 255
98:220	1 7 10 12 24 28 30 33 37 39 43 51	C_3	2×1, 32×3	188, 188, 188	0 24 12 3	0.0625	213(3), 216(3), 238(3)
98:221	1 7 10 12 25 27 29 31 34 45 47 50	C_2	49×2	288, 288, 288	0 21 18 0	0.0884	171(2), 190(2)
98:222	1 7 10 12 25 27 29 32 35 41 46 50	C_2	49×2	288, 288, 288	0 21 18 0	0.1410	223(2), 226(2), 254
98:223	1 7 10 12 25 27 29 32 35 42 46 49	C_1	98×1	288, 288, 288	0 22 16 1	0.0860	208, 217, 222, 224, 225, 240
98:224	1 7 10 12 25 27 29 32 35 42 47 51	C_1	98×1	288, 288, 288	0 21 18 0	0.1054	207, 218, 223, 227, 255
98:225	1 7 10 12 25 27 29 32 36 42 45 49	C_2	49×2	288, 288, 288	0 21 18 0	0.0674	180(2), 223(2), 251
98:226	1 7 10 12 25 27 30 32 35 40 46 50	C_1	98×1	288, 288, 288	0 21 18 0	0.1838	217, 222, 228, 229, 231
98:227	1 7 10 12 25 27 30 32 36 40 44 51	C_1	98×1	288, 288, 288	0 21 18 0	0.1829	210, 217, 219, 224, 229, 249
98:228	1 7 10 12 25 28 30 32 35 39 46 50	C_1	98×1	288, 288, 288	0 22 16 1	0.2305	226, 228(2), 230, 231, 253
98:229	1 7 10 12 25 28 30 32 36 39 44 51	C_1	98×1	288, 288, 288	0 22 16 1	0.1872	216(2), 226, 227, 230, 253, 254
98:230	1 7 10 12 25 28 30 32 36 39 45 50	C_2	49×2	288, 288, 288	0 23 14 2	0.1860	217(2), 228(2), 229(2), 231
98:231	1 7 10 12 26 28 30 32 36 38 45 50	C_2	49×2	288, 288, 288	0 21 18 0	0.2289	226(2), 228(2), 230
98:232	1 7 10 13 18 22 34 36 38 40 42 51	C_2	49×2	288, 288, 288	0 23 14 2	0.1494	150(2), 164(2), 234(2)
98:233	1 7 10 13 18 23 26 39 41 43 45 51	C_2	49×2	288, 288, 288	0 21 18 0	0.0235	168(2), 169(2), 246
98:234	1 7 10 13 18 23 31 36 38 40 44 51	C_1	98×1	288, 288, 288	0 22 16 1	0.1380	163, 165, 232, 247, 252
98:235	1 7 10 13 18 25 32 34 38 40 45 49	C_1	98×1	288, 288, 288	0 22 16 1	0.1068	207, 213, 236, 238, 242, 244, 255, 256, 257
98:236	1 7 10 13 18 25 32 34 38 41 45 48	C_2	49×2	288, 288, 288	0 23 14 2	0.0134	214(2), 235(2), 237(2), 239(2), 250(2)
98:237	1 7 10 13 18 25 32 34 39 41 45 47	C_1	98×1	288, 288, 288	0 22 16 1	0.0740	178, 197, 213, 236, 238, 242, 243, 244, 256
98:238	1 7 10 13 18 25 32 35 38 40 44 49	C_1	98×1	288, 288, 288	0 23 14 2	0.0030	220, 235, 237, 239, 240, 245, 249, 250(2)
98:239	1 7 10 13 18 25 32 35 38 41 44 48	C_s	6×1, 46×2	288, 288, 288	0 24 12 3	0.2762	213(2), 236(2), 238(2), 241, 243(2), 257(2)
98:240	1 7 10 13 18 25 33 35 38 40 43 49	C_1	98×1	288, 288, 288	0 22 16 1	0.1803	216, 223, 238, 241, 244, 251, 254, 255

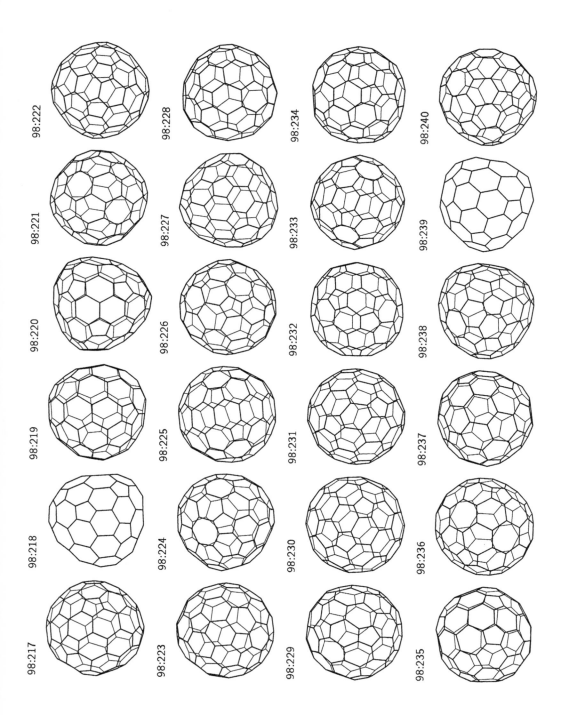

98:222

98:228

98:234

98:240

98:221

98:227

98:233

98:239

98:220

98:226

98:232

98:238

98:219

98:225

98:231

98:237

98:218

98:224

98:230

98:236

98:217

98:223

98:229

98:235

Table A.19. (*Continued*)

Isomer	Ring spiral	Point group	NMR pattern	Vibrations	Hexagon indices	Band gap	Transformations
98:241	1 7 10 13 18 25 33 35 38 41 43 48	C_s	8×1, 45×2	288, 288, 288	0 23 14 2	0.0074	219(2), 239, 240(2), 245(2), 249(2)
98:242	1 7 10 13 18 26 32 34 37 40 45 49	C_2	49×2	288, 288, 288	0 21 18 0	0.1156	177(2), 214(2), 235(2), 237(2)
98:243	1 7 10 13 18 26 32 35 37 41 44 48	C_2	49×2	288, 288, 288	0 23 14 2	0.2206	198(2), 214(2), 237(2), 239(2), 245(2)
98:244	1 7 10 13 18 26 33 35 37 40 43 49	C_1	98×1	288, 288, 288	0 21 18 0	0.1882	177, 179, 208, 210, 235, 237, 240, 245
98:245	1 7 10 13 18 26 33 35 37 41 43 48	C_1	98×1	288, 288, 288	0 22 16 1	0.0561	178, 199, 212, 238, 241, 243, 244, 251
98:246	1 7 10 13 18 27 32 34 38 46 49 51	C_2	49×2	288, 288, 288	0 21 18 0	0.0936	170(2), 186(2), 233
98:247	1 7 10 13 19 26 33 35 39 41 47 51	C_1	98×1	288, 288, 288	0 21 18 0	0.1707	150, 159, 161, 234
98:248	1 7 10 13 19 27 33 36 41 46 48 51	C_2	49×2	288, 288, 288	0 21 18 0	0.3015	104(2), 155(2), 184(2)
98:249	1 7 10 13 25 28 31 33 35 41 46 48	C_1	98×1	288, 288, 288	0 22 16 1	0.0928	216, 227, 238, 241, 254, 255(2), 257
98:250	1 7 10 13 25 29 31 33 35 39 46 49	C_1	98×1	288, 288, 288	0 22 16 1	0.0340	212, 236, 238(2), 255, 256, 257, 258
98:251	1 7 10 13 26 29 31 34 36 38 44 49	C_2	49×2	288, 288, 288	0 21 18 0	0.1924	179(2), 225, 240(2), 245(2)
98:252	1 7 10 13 27 30 32 34 37 39 49 51	C_2	49×2	288, 288, 288	0 21 18 0	0.1245	168(2), 234(2)
98:253	1 7 10 14 18 22 34 36 38 40 43 49	C_3	2×1, 32×3	188, 188, 188	0 21 18 0	0.2911	228(3), 229(3)
98:254	1 7 10 14 18 22 34 36 39 41 43 47	C_2	49×2	288, 288, 288	0 21 18 0	0.2445	222, 229(2), 240(2), 249(2)
98:255	1 7 10 14 18 22 35 37 39 41 43 45	C_1	98×1	288, 288, 288	0 21 18 0	0.2148	219, 224, 235, 240, 249(2), 250
98:256	1 7 10 14 18 24 32 34 39 41 45 47	C_2	49×2	288, 288, 288	0 21 18 0	0.2081	190(2), 235(2), 237(2), 250(2)
98:257	1 7 10 14 18 24 32 35 38 41 44 48	C_2	49×2	288, 288, 288	0 23 14 2	0.1383	210(2), 235(2), 239(2), 249(2), 250(2)
98:258	1 7 10 14 22 29 32 36 38 40 43 46	D_3	1×2, 16×6	143, 145, 96	0 21 18 0	0.0948	250(6)
98:259	1 7 12 14 19 22 29 36 42 44 47 49	C_s	8×1, 45×2	288, 288, 288	0 22 16 1	0.1933	170(2), 192

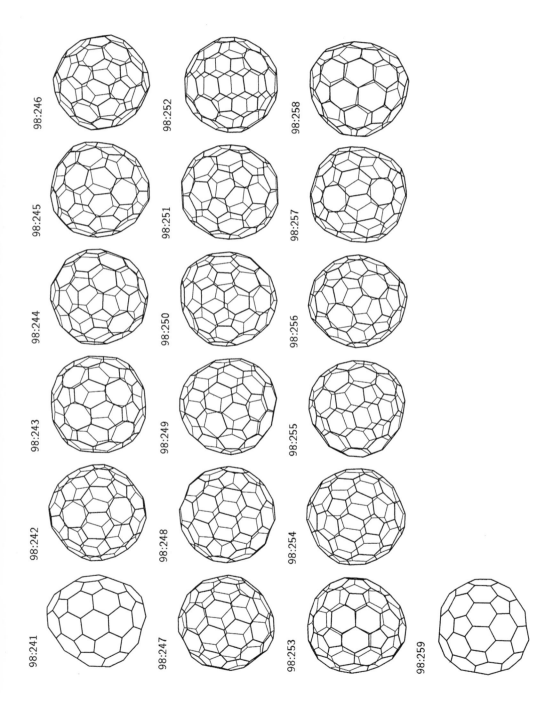

98:246

98:245

98:244

98:243

98:242

98:241

98:252

98:251

98:250

98:249

98:248

98:247

98:258

98:257

98:256

98:255

98:254

98:253

98:259

Table A.20. Isolated-pentagon fullerene isomers of C_{100}

Isomer	Ring spiral	Point group	NMR pattern	Vibrations	Hexagon indices	Band gap	Transformations
100:1	1 7 9 11 13 15 38 40 42 44 46 52	D_{5d}	2×10, 4×20	44, 75, 0	10 10 10 10	0.3501	
100:2	1 7 9 11 13 18 35 40 42 44 46 52	D_2	25×4	219, 294, 219	8 12 12 8	0.2738	
100:3	1 7 9 11 13 24 38 41 43 47 49 51	C_1	100×1	294, 294, 294	5 15 15 5	0.3249	18, 37
100:4	1 7 9 11 13 25 31 39 43 45 47 49	C_2	50×2	294, 294, 294	6 12 18 4	0.0886	
100:5	1 7 9 11 13 28 31 36 43 45 47 49	C_2	50×2	294, 294, 294	6 12 18 4	0.0610	
100:6	1 7 9 11 13 28 36 40 43 45 48 51	C_1	100×1	294, 294, 294	4 17 14 5	0.0490	7, 11
100:7	1 7 9 11 13 28 37 40 43 45 47 51	C_1	100×1	294, 294, 294	5 15 15 5	0.2634	6, 12
100:8	1 7 9 11 13 29 31 37 44 46 48 51	C_2	50×2	294, 294, 294	6 12 18 4	0.1742	13
100:9	1 7 9 11 13 29 36 38 43 45 49 51	C_1	100×1	294, 294, 294	4 15 18 3	0.1134	10, 11
100:10	1 7 9 11 13 29 36 38 43 45 50 52	C_1	100×1	294, 294, 294	3 17 17 3	0.1026	9, 15, 39
100:11	1 7 9 11 13 29 36 39 43 45 48 51	C_1	100×1	294, 294, 294	4 16 16 4	0.2891	6, 9, 12
100:12	1 7 9 11 13 29 37 39 43 45 47 51	C_1	100×1	294, 294, 294	5 14 17 4	0.1961	7, 11
100:13	1 7 9 11 13 30 37 41 44 46 48 51	C_2	50×2	294, 294, 294	6 14 14 6	0.2822	8
100:14	1 7 9 11 13 31 35 37 39 41 44 52	C_s	8×1, 46×2	294, 294, 294	3 17 17 3	0.0810	
100:15	1 7 9 11 13 36 38 40 42 44 46 52	C_s	6×1, 47×2	294, 294, 294	3 18 15 4	0.0352	10(2), 38
100:16	1 7 9 11 14 22 29 37 44 48 50 52	C_2	50×2	294, 294, 294	4 14 20 2	0.0448	17(2)
100:17	1 7 9 11 14 22 37 39 41 45 50 52	C_1	100×1	294, 294, 294	3 16 19 2	0.1530	16, 103
100:18	1 7 9 11 14 23 38 41 43 47 49 51	C_2	50×2	294, 294, 294	4 16 16 4	0.3692	3(2)
100:19	1 7 9 11 14 28 33 35 40 42 46 52	C_1	100×1	294, 294, 294	3 18 15 4	0.0315	20, 22, 27
100:20	1 7 9 11 14 28 33 35 40 42 47 51	C_1	100×1	294, 294, 294	2 19 16 3	0.0839	19, 21, 28
100:21	1 7 9 11 14 28 33 35 41 47 49 51	C_1	100×1	294, 294, 294	3 16 19 2	0.0677	20
100:22	1 7 9 11 14 28 33 36 40 42 45 52	C_1	100×1	294, 294, 294	3 19 13 5	0.1596	19, 24, 30, 91
100:23	1 7 9 11 14 28 33 37 41 45 50 52	C_2	50×2	294, 294, 294	4 14 20 2	0.2326	
100:24	1 7 9 11 14 28 34 36 40 42 44 52	C_1	100×1	294, 294, 294	4 17 14 5	0.0395	22, 32, 47

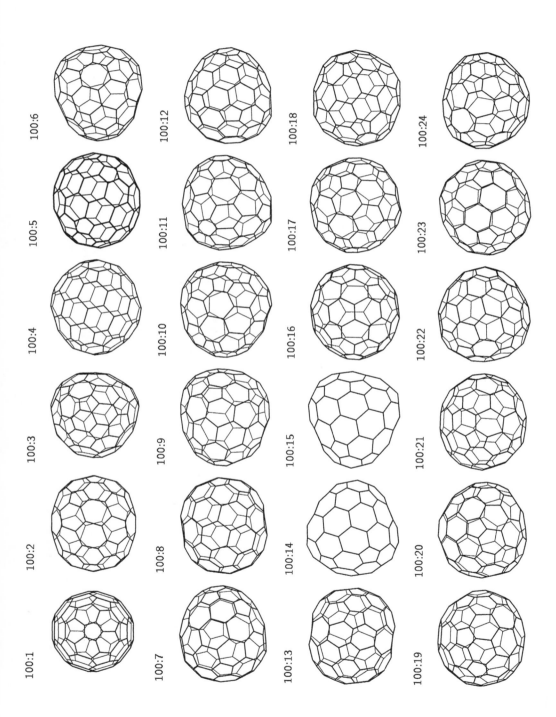

100:6

100:12

100:18

100:24

100:5

100:11

100:17

100:23

100:4

100:10

100:16

100:22

100:3

100:9

100:15

100:21

100:2

100:8

100:14

100:20

100:1

100:7

100:13

100:19

330

Table A.20. (*Continued*)

Isomer	Ring spiral	Point group	NMR pattern	Vibrations	Hexagon indices	Band gap	Transformations
100:25	1 7 9 11 14 29 31 36 43 45 47 49	C_s	4×1, 48×2	294, 294, 294	3 18 15 4	0.2416	36, 64, 94
100:26	1 7 9 11 14 29 33 35 38 42 48 51	C_1	100×1	294, 294, 294	2 17 20 1	0.0381	28
100:27	1 7 9 11 14 29 33 35 39 42 46 52	C_1	100×1	294, 294, 294	2 18 18 2	0.1950	19, 28, 29, 30
100:28	1 7 9 11 14 29 33 35 39 42 47 51	C_1	100×1	294, 294, 294	2 18 18 2	0.0846	20, 26, 27
100:29	1 7 9 11 14 29 33 35 39 46 49 51	C_1	100×1	294, 294, 294	3 17 17 3	0.2022	27, 31, 33
100:30	1 7 9 11 14 29 33 36 39 42 45 52	C_1	100×1	294, 294, 294	2 19 16 3	0.2154	22, 27, 31, 32, 95
100:31	1 7 9 11 14 29 33 36 39 45 49 51	C_1	100×1	294, 294, 294	4 17 14 5	0.3531	29, 30, 96
100:32	1 7 9 11 14 29 34 36 39 42 44 52	C_1	100×1	294, 294, 294	3 17 17 3	0.1969	24, 30, 97
100:33	1 7 9 11 14 29 35 39 43 45 47 50	C_1	100×1	294, 294, 294	4 15 18 3	0.2983	29
100:34	1 7 9 11 14 30 33 38 44 46 48 50	C_2	50×2	294, 294, 294	4 14 20 2	0.1361	
100:35	1 7 9 11 14 30 35 40 44 46 48 50	C_2	50×2	294, 294, 294	4 16 16 4	0.0595	
100:36	1 7 9 11 14 30 36 41 43 45 47 49	C_{2v}	8×2, 21×4	224, 224, 224	4 18 12 6	0.1807	25(2)
100:37	1 7 9 11 15 22 29 36 44 48 50 52	C_2	50×2	294, 294, 294	6 14 14 6	0.1343	3(2)
100:38	1 7 9 11 15 29 34 36 43 48 50 52	C_s	8×1, 46×2	294, 294, 294	3 17 17 3	0.0951	15
100:39	1 7 9 11 18 24 33 36 39 42 47 51	C_s	6×1, 47×2	294, 294, 294	4 15 18 3	0.0134	10(2)
100:40	1 7 9 11 22 24 29 37 41 43 46 48	C_1	100×1	294, 294, 294	3 18 15 4	0.2063	41, 44, 159
100:41	1 7 9 11 22 24 29 37 42 46 48 50	C_2	50×2	294, 294, 294	4 16 16 4	0.1445	40(2)
100:42	1 7 9 11 22 24 30 36 39 43 47 49	C_1	100×1	294, 294, 294	3 17 17 3	0.0406	43, 54, 102
100:43	1 7 9 11 22 24 30 37 39 43 46 49	C_1	100×1	294, 294, 294	2 18 18 2	0.0703	42, 44, 160
100:44	1 7 9 11 22 24 30 37 40 43 46 48	C_1	100×1	294, 294, 294	3 17 17 3	0.0705	40, 43, 154
100:45	1 7 9 11 22 25 27 39 41 43 45 52	C_1	100×1	294, 294, 294	4 16 16 4	0.0941	47, 70
100:46	1 7 9 11 22 25 27 40 42 45 47 50	C_1	100×1	294, 294, 294	3 17 17 3	0.1094	49, 128
100:47	1 7 9 11 22 25 28 35 41 43 48 52	C_1	100×1	294, 294, 294	4 17 14 5	0.3704	24, 45, 91, 97
100:48	1 7 9 11 22 25 28 40 44 46 48 51	C_s	6×1, 47×2	294, 294, 294	2 21 12 5	0.0174	53(2), 161(2)

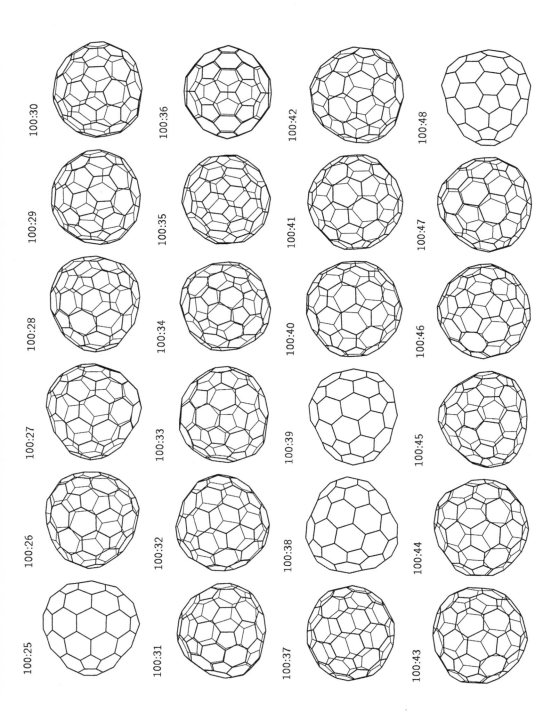

100:30

100:36

100:42

100:48

100:29

100:35

100:41

100:47

100:28

100:34

100:40

100:46

100:27

100:33

100:39

100:45

100:26

100:32

100:38

100:44

100:25

100:31

100:37

100:43

Table A.20. (*Continued*)

Isomer	Ring spiral	Point group	NMR pattern	Vibrations	Hexagon indices	Band gap	Transformations
100:49	1 7 9 11 22 25 29 38 41 43 45 47	C_1	100×1	294, 294, 294	3 18 15 4	0.3348	46, 51, 56, 162
100:50	1 7 9 11 22 25 29 38 41 44 46 52	C_1	100×1	294, 294, 294	2 20 14 4	0.3196	51, 52, 57, 60, 88, 163
100:51	1 7 9 11 22 25 29 38 41 44 47 51	C_1	100×1	294, 294, 294	3 19 13 5	0.2517	49, 50, 58, 87, 164
100:52	1 7 9 11 22 25 29 38 44 46 49 51	C_1	100×1	294, 294, 294	2 20 14 4	0.1410	50, 53, 61, 86, 165
100:53	1 7 9 11 22 25 29 39 44 46 48 51	C_1	100×1	294, 294, 294	2 20 14 4	0.1615	48, 52, 85, 166
100:54	1 7 9 11 22 25 30 35 39 43 47 49	C_1	100×1	294, 294, 294	3 17 17 3	0.1771	42, 65, 116
100:55	1 7 9 11 22 25 30 35 40 46 48 50	C_1	100×1	294, 294, 294	5 16 13 6	0.3755	66, 92
100:56	1 7 9 11 22 25 30 38 40 43 45 47	C_1	100×1	294, 294, 294	3 17 17 3	0.0094	49, 58, 167
100:57	1 7 9 11 22 25 30 38 40 44 46 52	C_1	100×1	294, 294, 294	2 19 16 3	0.0272	50, 58, 62, 76, 168
100:58	1 7 9 11 22 25 30 38 40 44 47 51	C_1	100×1	294, 294, 294	3 18 15 4	0.2054	51, 56, 57, 75, 148
100:59	1 7 9 11 22 26 29 36 44 47 49 51	C_1	100×1	294, 294, 294	3 17 17 3	0.0547	61, 78, 153
100:60	1 7 9 11 22 26 29 37 41 44 46 52	C_1	100×1	294, 294, 294	3 19 13 5	0.2216	50, 61, 62, 80, 146
100:61	1 7 9 11 22 26 29 37 44 46 49 51	C_1	100×1	294, 294, 294	2 20 14 4	0.0605	52, 59, 60, 79, 169
100:62	1 7 9 11 22 26 30 35 39 47 50 52	C_1	100×1	294, 294, 294	3 17 17 3	0.0504	57, 60, 69, 129
100:63	1 7 9 11 22 26 30 36 40 45 47 50	C_1	100×1	294, 294, 294	4 15 18 3	0.1493	170
100:64	1 7 9 11 22 28 35 38 42 45 48 50	C_s	6×1, 47×2	294, 294, 294	5 16 13 6	0.3684	25, 93
100:65	1 7 9 11 23 25 30 34 39 43 47 49	C_1	100×1	294, 294, 294	3 16 19 2	0.1457	54
100:66	1 7 9 11 23 25 30 34 40 46 48 50	C_2	50×2	294, 294, 294	4 16 16 4	0.2033	55(2)
100:67	1 7 9 11 23 26 29 32 44 46 48 50	D_2	25×4	219, 294, 219	4 12 24 0	0.0205	
100:68	1 7 9 11 23 26 29 41 44 46 49 52	C_1	100×1	294, 294, 294	3 16 19 2	0.1007	73
100:69	1 7 9 11 23 26 30 34 39 47 50 52	C_1	100×1	294, 294, 294	3 17 17 3	0.1601	62, 71, 76, 80
100:70	1 7 9 11 23 26 30 36 40 43 45 52	C_1	100×1	294, 294, 294	3 16 19 2	0.1589	45, 97
100:71	1 7 9 11 23 26 30 39 44 46 48 51	C_1	100×1	294, 294, 294	2 17 20 1	0.1828	69, 72, 77, 83
100:72	1 7 9 11 23 26 30 39 44 46 50 52	C_1	100×1	294, 294, 294	3 16 19 2	0.2217	71, 73, 84

100:54

100:60

100:66

100:72

100:53

100:59

100:65

100:71

100:52

100:58

100:64

100:70

100:51

100:57

100:63

100:69

100:50

100:56

100:62

100:68

100:49

100:55

100:61

100:67

Table A.20. (*Continued*)

Isomer	Ring spiral	Point group	NMR pattern	Vibrations	Hexagon indices	Band gap	Transformations
100:73	1 7 9 11 23 26 30 40 44 46 49 52	C_1	100×1	294, 294, 294	3 15 21 1	0.2447	68, 72
100:74	1 7 9 11 23 26 32 39 41 46 48 52	C_1	100×1	294, 294, 294	2 17 20 1	0.1707	77, 81
100:75	1 7 9 11 23 26 34 39 41 43 47 51	C_1	100×1	294, 294, 294	3 16 19 2	0.0563	58, 76, 87
100:76	1 7 9 11 23 26 34 39 41 43 48 52	C_1	100×1	294, 294, 294	2 19 16 3	0.0662	57, 69, 75, 77, 88
100:77	1 7 9 11 23 26 39 41 43 45 47 49	C_1	100×1	294, 294, 294	2 18 18 2	0.1412	71, 74, 76, 89
100:78	1 7 9 11 23 27 30 34 37 43 48 52	C_1	100×1	294, 294, 294	3 16 19 2	0.2426	59, 79
100:79	1 7 9 11 23 27 30 34 37 48 50 52	C_1	100×1	294, 294, 294	2 19 16 3	0.2630	61, 78, 80, 82, 86
100:80	1 7 9 11 23 27 30 34 38 47 50 52	C_1	100×1	294, 294, 294	3 19 13 5	0.3753	60, 69, 79, 83, 88
100:81	1 7 9 11 23 27 30 36 44 47 49 51	C_1	100×1	294, 294, 294	2 18 18 2	0.1935	74, 82, 89
100:82	1 7 9 11 23 27 30 37 44 46 49 51	C_1	100×1	294, 294, 294	2 18 18 2	0.2832	79, 81, 83, 90
100:83	1 7 9 11 23 27 30 38 44 46 48 51	C_1	100×1	294, 294, 294	2 19 16 3	0.3080	71, 80, 82, 84, 89
100:84	1 7 9 11 23 27 30 38 44 46 50 52	C_1	100×1	294, 294, 294	3 16 19 2	0.3213	72, 83
100:85	1 7 9 11 23 27 34 37 40 43 50 52	C_s	6×1, 47×2	294, 294, 294	2 19 16 3	0.2325	53(2), 86(2)
100:86	1 7 9 11 23 27 34 37 41 43 49 52	C_1	100×1	294, 294, 294	2 19 16 3	0.3059	52, 79, 85, 88, 90
100:87	1 7 9 11 23 27 34 38 41 43 47 51	C_1	100×1	294, 294, 294	3 17 17 3	0.2787	51, 75, 88
100:88	1 7 9 11 23 27 34 38 41 43 48 52	C_1	100×1	294, 294, 294	2 20 14 4	0.3812	50, 76, 80, 86, 87, 89
100:89	1 7 9 11 23 27 36 41 43 45 48 50	C_1	100×1	294, 294, 294	2 19 16 3	0.3555	77, 81, 83, 88, 90
100:90	1 7 9 11 23 27 37 41 43 45 47 50	C_s	6×1, 47×2	294, 294, 294	2 20 14 4	0.3448	82(2), 86(2), 89(2)
100:91	1 7 9 11 23 30 32 36 38 40 47 50	C_1	100×1	294, 294, 294	3 17 17 3	0.1948	22, 47, 95
100:92	1 7 9 11 23 30 34 36 39 41 48 51	C_1	100×1	294, 294, 294	4 16 16 4	0.2167	55
100:93	1 7 9 11 24 26 30 32 37 44 49 52	C_s	4×1, 48×2	294, 294, 294	4 16 16 4	0.1996	64, 94
100:94	1 7 9 11 24 26 30 32 44 46 48 50	C_{2v}	6×2, 22×4	223, 294, 223	2 18 18 2	0.1742	25(2), 93(2)
100:95	1 7 9 11 24 26 30 34 40 43 46 52	C_1	100×1	294, 294, 294	2 17 20 1	0.1801	30, 91, 96, 97
100:96	1 7 9 11 24 26 30 34 40 46 50 52	C_1	100×1	294, 294, 294	4 15 18 3	0.1888	31, 95

100:78

100:84

100:90

100:96

100:77

100:83

100:89

100:95

100:76

100:82

100:88

100:94

100:75

100:81

100:87

100:93

100:74

100:80

100:86

100:92

100:73

100:79

100:85

100:91

Table A.20. (*Continued*)

Isomer	Ring spiral	Point group	NMR pattern	Vibrations	Hexagon indices	Band gap	Transformations
100:97	1 7 9 11 24 26 30 35 40 43 45 52	C_1	100×1	294, 294, 294	3 17 17 3	0.1838	32, 47, 70, 95
100:98	1 7 9 11 24 30 35 37 39 41 45 51	C_2	50×2	294, 294, 294	4 14 20 2	0.2494	
100:99	1 7 9 12 14 20 36 39 42 44 47 49	C_2	50×2	294, 294, 294	2 18 18 2	0.0164	100(2)
100:100	1 7 9 12 14 20 37 39 42 44 46 49	C_1	100×1	294, 294, 294	1 19 19 1	0.1104	99, 101, 330
100:101	1 7 9 12 14 20 37 40 42 44 46 48	C_1	100×1	294, 294, 294	2 18 18 2	0.1109	100, 102, 152
100:102	1 7 9 12 14 20 37 41 44 46 48 50	C_1	100×1	294, 294, 294	2 19 16 3	0.1438	42, 101, 116, 160
100:103	1 7 9 12 14 21 37 39 41 45 50 52	C_2	50×2	294, 294, 294	2 18 18 2	0.1275	17(2)
100:104	1 7 9 12 14 28 31 34 40 45 47 50	C_1	100×1	294, 294, 294	1 20 17 2	0.1469	107, 121, 176
100:105	1 7 9 12 14 28 31 35 40 43 46 52	C_1	100×1	294, 294, 294	1 21 15 3	0.1865	106, 108, 109, 111, 125
100:106	1 7 9 12 14 28 31 35 40 43 47 51	C_1	100×1	294, 294, 294	1 20 17 2	0.1535	105, 107, 112, 126
100:107	1 7 9 12 14 28 31 35 40 44 47 50	C_1	100×1	294, 294, 294	1 20 17 2	0.1225	104, 106, 125
100:108	1 7 9 12 14 28 31 35 43 46 48 50	C_1	100×1	294, 294, 294	2 18 18 2	0.1620	105, 110
100:109	1 7 9 12 14 28 31 36 40 43 45 52	C_1	100×1	294, 294, 294	1 21 15 3	0.0742	105, 110, 114, 121, 171
100:110	1 7 9 12 14 28 31 36 43 45 48 51	C_1	100×1	294, 294, 294	2 18 18 2	0.1590	108, 109, 147
100:111	1 7 9 12 14 28 32 35 40 42 46 52	C_1	100×1	294, 294, 294	2 20 14 4	0.1887	105, 112, 114, 123
100:112	1 7 9 12 14 28 32 35 40 42 47 51	C_1	100×1	294, 294, 294	1 20 17 2	0.2054	106, 111, 113, 122
100:113	1 7 9 12 14 28 32 35 41 47 49 51	C_1	100×1	294, 294, 294	2 17 20 1	0.1104	112
100:114	1 7 9 12 14 28 32 36 40 42 45 52	C_1	100×1	294, 294, 294	2 19 16 3	0.1983	109, 111, 118, 145
100:115	1 7 9 12 14 28 32 37 41 45 50 52	C_2	50×2	294, 294, 294	2 16 22 0	0.0380	
100:116	1 7 9 12 14 28 32 38 42 44 46 49	C_1	100×1	294, 294, 294	2 17 20 1	0.1753	54, 102
100:117	1 7 9 12 14 29 31 34 37 47 49 51	C_2	50×2	294, 294, 294	2 18 18 2	0.0751	120(2)
100:118	1 7 9 12 14 29 31 34 38 45 48 50	C_1	100×1	294, 294, 294	1 19 19 1	0.1795	114, 119, 121, 123, 173
100:119	1 7 9 12 14 29 31 34 38 45 49 52	C_1	100×1	294, 294, 294	2 18 18 2	0.1994	118, 120, 124, 175
100:120	1 7 9 12 14 29 31 34 38 46 49 51	C_1	100×1	294, 294, 294	2 17 20 1	0.2007	117, 119, 174

100:102

100:108

100:114

100:120

100:101

100:107

100:113

100:119

100:100

100:106

100:112

100:118

100:99

100:105

100:111

100:117

100:98

100:104

100:110

100:116

100:97

100:103

100:109

100:115

Table A.20. *(Continued)*

Isomer	Ring spiral	Point group	NMR pattern	Vibrations	Hexagon indices	Band gap	Transformations
100:121	1 7 9 12 14 29 31 34 39 45 47 50	C_1	100×1	294, 294, 294	1 20 17 2	0.1768	104, 109, 118, 125, 172
100:122	1 7 9 12 14 29 31 35 38 43 48 51	C_1	100×1	294, 294, 294	1 19 19 1	0.2437	112, 123, 126, 127
100:123	1 7 9 12 14 29 31 35 38 44 48 50	C_1	100×1	294, 294, 294	1 20 17 2	0.2621	111, 118, 122, 124, 125
100:124	1 7 9 12 14 29 31 35 38 44 49 52	C_1	100×1	294, 294, 294	2 17 20 1	0.2150	119, 123
100:125	1 7 9 12 14 29 31 35 39 43 46 52	C_1	100×1	294, 294, 294	1 20 17 2	0.3159	105, 107, 121, 123, 126
100:126	1 7 9 12 14 29 31 35 39 43 47 51	C_s	8×1, 46×2	294, 294, 294	1 21 15 3	0.3241	106(2), 122(2), 125(2)
100:127	1 7 9 12 14 29 32 35 38 42 48 51	C_s	8×1, 46×2	294, 294, 294	1 18 21 0	0.1647	122(2)
100:128	1 7 9 12 14 32 34 38 40 42 47 50	C_1	100×1	294, 294, 294	2 17 20 1	0.0380	46, 162
100:129	1 7 9 12 20 24 27 40 44 46 48 51	C_1	100×1	294, 294, 294	2 17 20 1	0.0903	62, 146, 168
100:130	1 7 9 12 20 24 29 37 41 44 46 48	C_2	50×2	294, 294, 294	2 18 18 2	0.1853	131(2), 137(2)
100:131	1 7 9 12 20 24 29 37 41 45 48 51	C_1	100×1	294, 294, 294	1 21 15 3	0.0300	130, 132, 138, 156, 390
100:132	1 7 9 12 20 24 29 38 41 45 47 51	C_2	50×2	294, 294, 294	2 18 18 2	0.2221	131(2), 139, 142
100:133	1 7 9 12 20 24 30 36 39 44 47 49	C_1	100×1	294, 294, 294	2 18 18 2	0.1488	134, 147, 267, 306
100:134	1 7 9 12 20 24 30 37 39 44 46 49	C_1	100×1	294, 294, 294	1 19 19 1	0.1250	133, 135, 137, 268, 391
100:135	1 7 9 12 20 24 30 37 39 45 49 51	C_1	100×1	294, 294, 294	1 20 17 2	0.0617	134, 136, 138, 269, 365
100:136	1 7 9 12 20 24 30 37 39 45 50 52	C_1	100×1	294, 294, 294	1 20 17 2	0.1718	135, 140, 270, 324
100:137	1 7 9 12 20 24 30 37 40 44 46 48	C_1	100×1	294, 294, 294	2 19 16 3	0.0803	130, 134, 138, 156, 271
100:138	1 7 9 12 20 24 30 37 40 45 48 51	C_1	100×1	294, 294, 294	1 21 15 3	0.1034	131, 135, 137, 139, 272, 392
100:139	1 7 9 12 20 24 30 38 40 45 47 51	C_2	50×2	294, 294, 294	2 18 18 2	0.1319	132, 138(2), 273
100:140	1 7 9 12 20 24 37 39 41 43 46 52	C_1	100×1	294, 294, 294	1 20 17 2	0.1369	136, 178, 274, 393
100:141	1 7 9 12 20 25 27 42 44 46 48 50	C_1	100×1	294, 294, 294	2 20 14 4	0.0971	177, 394
100:142	1 7 9 12 20 25 28 34 40 46 49 51	C_2	50×2	294, 294, 294	2 20 14 4	0.2439	132, 144(2), 156(2), 273
100:143	1 7 9 12 20 25 28 34 41 45 48 52	C_1	100×1	294, 294, 294	2 19 16 3	0.3175	144, 145, 296, 303, 306
100:144	1 7 9 12 20 25 28 34 41 46 48 51	C_1	100×1	294, 294, 294	1 20 17 2	0.2338	142, 143, 157, 297, 391, 442

100:126

100:132

100:138

100:144

100:125

100:131

100:137

100:143

100:124

100:130

100:136

100:142

100:123

100:129

100:135

100:141

100:122

100:128

100:134

100:140

100:121

100:127

100:133

100:139

Table A.20. (Continued)

Isomer	Ring spiral	Point group	NMR pattern	Vibrations	Hexagon indices	Band gap	Transformations
100:145	1 7 9 12 20 25 28 35 41 44 48 52	C_1	100×1	294, 294, 294	2 18 18 2	0.2414	114, 143, 171, 173
100:146	1 7 9 12 20 25 28 35 42 47 49 51	C_1	100×1	294, 294, 294	2 19 16 3	0.2501	60, 129, 163, 169
100:147	1 7 9 12 20 25 30 35 39 44 47 49	C_1	100×1	294, 294, 294	2 17 20 1	0.1870	110, 133, 171
100:148	1 7 9 12 20 25 34 38 40 43 47 50	C_1	100×1	294, 294, 294	2 18 18 2	0.2250	58, 164, 167, 168
100:149	1 7 9 12 20 26 29 35 43 46 48 50	C_2	50×2	294, 294, 294	2 16 22 0	0.2914	150(2)
100:150	1 7 9 12 20 26 29 36 43 45 48 50	C_s	8×1, 46×2	294, 294, 294	1 18 21 0	0.0501	149(2), 395
100:151	1 7 9 12 20 26 33 36 39 42 49 52	C_2	50×2	294, 294, 294	2 18 18 2	0.1664	152(2), 154(2)
100:152	1 7 9 12 20 26 33 36 40 42 48 52	C_1	100×1	294, 294, 294	1 19 19 1	0.1010	101, 151, 160, 330
100:153	1 7 9 12 20 27 30 35 37 44 48 52	C_1	100×1	294, 294, 294	2 17 20 1	0.1067	59, 169
100:154	1 7 9 12 20 27 33 36 38 42 49 52	C_1	100×1	294, 294, 294	2 19 16 3	0.1058	44, 151, 159, 160
100:155	1 7 9 12 20 27 34 36 40 42 48 52	C_1	100×1	294, 294, 294	1 20 17 2	0.1701	158, 179, 393
100:156	1 7 9 12 20 28 34 36 38 41 50 52	C_1	100×1	294, 294, 294	1 22 13 4	0.0608	131, 137, 142, 157, 272, 391, 392
100:157	1 7 9 12 20 28 34 36 38 42 49 52	C_1	100×1	294, 294, 294	1 21 15 3	0.1323	144, 156, 158, 298, 363, 365
100:158	1 7 9 12 20 28 34 36 39 42 48 52	C_1	100×1	294, 294, 294	1 21 15 3	0.0963	155, 157, 279, 324, 332
100:159	1 7 9 12 21 24 29 37 41 43 46 48	C_2	50×2	294, 294, 294	2 20 14 4	0.0867	40(2), 154(2)
100:160	1 7 9 12 21 24 30 37 39 43 46 49	C_1	100×1	294, 294, 294	1 20 17 2	0.1130	43, 102, 152, 154
100:161	1 7 9 12 21 25 28 40 44 46 48 51	C_1	100×1	294, 294, 294	1 21 15 3	0.0302	48, 166, 447
100:162	1 7 9 12 21 25 29 38 41 43 45 47	C_1	100×1	294, 294, 294	2 18 18 2	0.1147	49, 128, 164, 167
100:163	1 7 9 12 21 25 29 38 41 44 46 52	C_1	100×1	294, 294, 294	1 20 17 2	0.2741	50, 146, 164, 165, 168
100:164	1 7 9 12 21 25 29 38 41 44 47 51	C_1	100×1	294, 294, 294	2 19 16 3	0.1952	51, 148, 162, 163
100:165	1 7 9 12 21 25 29 38 44 46 49 51	C_1	100×1	294, 294, 294	1 20 17 2	0.1593	52, 163, 166, 169
100:166	1 7 9 12 21 25 29 39 44 46 48 51	C_1	100×1	294, 294, 294	1 20 17 2	0.1936	53, 161, 165
100:167	1 7 9 12 21 25 30 38 40 43 45 47	C_1	100×1	294, 294, 294	2 17 20 1	0.0558	56, 148, 162
100:168	1 7 9 12 21 25 30 38 40 44 46 52	C_1	100×1	294, 294, 294	1 19 19 1	0.0522	57, 129, 148, 163

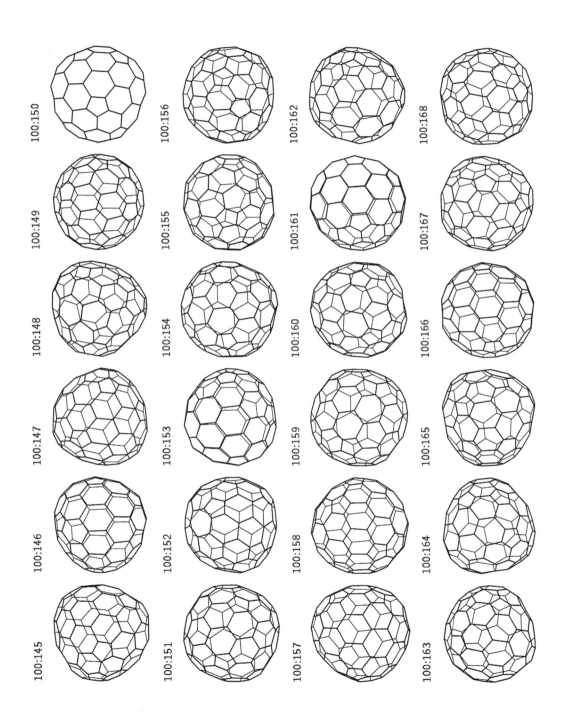

100:150

100:156

100:162

100:168

100:149

100:155

100:161

100:167

100:148

100:154

100:160

100:166

100:147

100:153

100:159

100:165

100:146

100:152

100:158

100:164

100:145

100:151

100:157

100:163

Table A.20. (*Continued*)

Isomer	Ring spiral	Point group	NMR pattern	Vibrations	Hexagon indices	Band gap	Transformations
100:169	1 7 9 12 21 26 29 37 44 46 49 51	C_1	100×1	294, 294, 294	1 20 17 2	0.1004	61, 146, 153, 165
100:170	1 7 9 12 21 26 30 36 40 45 47 50	C_1	100×1	294, 294, 294	3 15 21 1	0.1571	63
100:171	1 7 9 12 24 26 29 32 41 44 46 52	C_1	100×1	294, 294, 294	1 20 17 2	0.1268	109, 145, 147, 172, 306
100:172	1 7 9 12 24 26 30 32 40 44 46 52	C_1	100×1	294, 294, 294	1 19 19 1	0.2105	121, 171, 173, 176, 305
100:173	1 7 9 12 24 26 30 33 40 43 46 52	C_1	100×1	294, 294, 294	1 18 21 0	0.1202	118, 145, 172, 175, 303
100:174	1 7 9 12 24 26 30 33 40 45 47 50	C_2	50×2	294, 294, 294	2 16 22 0	0.0312	120(2), 175(2)
100:175	1 7 9 12 24 26 30 33 40 46 50 52	C_1	100×1	294, 294, 294	2 17 20 1	0.1677	119, 173, 174, 304
100:176	1 7 9 12 24 26 31 40 42 44 46 52	C_1	100×1	294, 294, 294	1 19 19 1	0.1519	104, 172, 302
100:177	1 7 9 12 24 27 31 37 41 43 45 52	C_1	100×1	294, 294, 294	1 19 19 1	0.0922	141, 225
100:178	1 7 9 12 24 27 32 34 40 42 48 52	C_1	100×1	294, 294, 294	1 19 19 1	0.1485	140, 195
100:179	1 7 9 12 24 28 30 32 38 44 46 52	C_1	100×1	294, 294, 294	1 20 17 2	0.0639	155, 182, 279, 375
100:180	1 7 9 12 24 28 30 33 37 45 48 50	C_1	100×1	294, 294, 294	1 20 17 2	0.1521	181, 183, 186, 196, 232, 284, 415
100:181	1 7 9 12 24 28 30 33 37 45 49 51	C_1	100×1	294, 294, 294	1 20 17 2	0.0730	180, 187, 233, 283, 368
100:182	1 7 9 12 24 28 30 33 38 43 46 52	C_1	100×1	294, 294, 294	1 19 19 1	0.1068	179, 184, 277, 386
100:183	1 7 9 12 24 28 30 33 38 45 47 50	C_1	100×1	294, 294, 294	1 19 19 1	0.1703	180, 184, 188, 197, 276, 424
100:184	1 7 9 12 24 28 30 33 38 46 50 52	C_1	100×1	294, 294, 294	1 19 19 1	0.0296	182, 183, 198, 278, 361
100:185	1 7 9 12 24 28 30 34 37 43 49 52	C_1	100×1	294, 294, 294	1 21 15 3	0.0813	187, 189, 234, 281, 389
100:186	1 7 9 12 24 28 30 34 37 44 48 50	C_1	100×1	294, 294, 294	1 21 15 3	0.2364	180, 187, 188, 199, 235, 282, 420
100:187	1 7 9 12 24 28 30 34 37 44 49 51	C_1	100×1	294, 294, 294	1 21 15 3	0.2151	181, 185, 186, 236, 280, 423
100:188	1 7 9 12 24 28 30 34 38 44 47 50	C_1	100×1	294, 294, 294	1 20 17 2	0.0657	183, 186, 200, 275, 400
100:189	1 7 9 12 24 28 30 34 43 46 48 50	C_1	100×1	294, 294, 294	1 21 15 3	0.0264	185, 190, 285, 387
100:190	1 7 9 12 24 28 30 35 43 45 48 50	C_2	50×2	294, 294, 294	2 20 14 4	0.0823	189(2)
100:191	1 7 9 12 24 28 32 34 37 41 48 51	C_1	100×1	294, 294, 294	1 21 15 3	0.1092	192, 199, 203, 237, 289, 397
100:192	1 7 9 12 24 28 32 34 38 41 47 51	C_1	100×1	294, 294, 294	1 20 17 2	0.0950	191, 193, 200, 204, 286, 427

100:174

100:180

100:186

100:192

100:173

100:179

100:185

100:191

100:172

100:178

100:184

100:190

100:171

100:177

100:183

100:189

100:170

100:176

100:182

100:188

100:169

100:175

100:181

100:187

Table A.20. (Continued)

Isomer	Ring spiral	Point group	NMR pattern	Vibrations	Hexagon indices	Band gap	Transformations
100:193	1 7 9 12 24 28 32 34 38 41 50 52	C_1	100×1	294, 294, 294	1 20 17 2	0.0772	192, 194, 205, 288, 370
100:194	1 7 9 12 24 28 32 34 38 42 49 52	C_1	100×1	294, 294, 294	1 20 17 2	0.1531	193, 195, 301, 382
100:195	1 7 9 12 24 28 32 34 39 42 48 52	C_1	100×1	294, 294, 294	1 20 17 2	0.0461	178, 194, 274, 405
100:196	1 7 9 12 24 28 33 37 41 43 46 49	C_1	100×1	294, 294, 294	1 21 15 3	0.0141	180, 197, 199, 203, 240, 289, 437
100:197	1 7 9 12 24 28 33 38 41 43 46 48	C_1	100×1	294, 294, 294	1 20 17 2	0.0788	183, 196, 198, 200, 204, 286, 448
100:198	1 7 9 12 24 28 33 38 41 43 47 52	C_1	100×1	294, 294, 294	1 20 17 2	0.0428	184, 197, 205, 288, 396
100:199	1 7 9 12 24 28 34 37 41 43 45 49	C_1	100×1	294, 294, 294	1 22 13 4	0.3305	186, 191, 196, 200, 208, 241, 290, 446
100:200	1 7 9 12 24 28 34 38 41 43 45 48	C_1	100×1	294, 294, 294	1 21 15 3	0.0320	188, 192, 197, 199, 209, 287, 437
100:201	1 7 9 12 24 29 32 34 37 39 42 52	C_1	100×1	294, 294, 294	1 19 19 1	0.1445	202, 210, 239, 242
100:202	1 7 9 12 24 29 32 34 37 39 49 51	C_1	100×1	294, 294, 294	1 19 19 1	0.1618	201, 203, 207, 240, 243
100:203	1 7 9 12 24 29 32 34 37 40 48 51	C_1	100×1	294, 294, 294	1 20 17 2	0.1047	191, 196, 202, 204, 206, 208
100:204	1 7 9 12 24 29 32 34 38 40 47 51	C_1	100×1	294, 294, 294	1 19 19 1	0.1605	192, 197, 203, 205, 209
100:205	1 7 9 12 24 29 32 34 38 40 50 52	C_1	100×1	294, 294, 294	1 19 19 1	0.0449	193, 198, 204
100:206	1 7 9 12 24 29 33 37 39 43 46 50	C_1	100×1	294, 294, 294	1 19 19 1	0.1781	203, 207, 237, 243
100:207	1 7 9 12 24 29 34 37 39 43 45 50	C_1	100×1	294, 294, 294	1 20 17 2	0.0953	202, 206, 208, 241, 244
100:208	1 7 9 12 24 29 34 37 40 43 45 49	C_s	4×1, 48×2	294, 294, 294	1 21 15 3	0.1820	199(2), 203(2), 207(2), 209
100:209	1 7 9 12 24 29 34 38 40 43 45 48	C_s	6×1, 47×2	294, 294, 294	1 20 17 2	0.0148	200(2), 204(2), 208
100:210	1 7 9 12 24 30 32 34 37 39 41 52	C_1	100×1	294, 294, 294	1 19 19 1	0.2533	201, 211, 245, 320
100:211	1 7 9 12 24 31 34 37 39 41 43 52	C_1	100×1	294, 294, 294	1 19 19 1	0.0510	210, 228, 238
100:212	1 7 9 12 24 31 35 38 40 42 44 52	C_2	50×2	294, 294, 294	2 18 18 2	0.0223	
100:213	1 7 9 12 25 27 30 32 34 45 47 49	C_1	100×1	294, 294, 294	1 18 21 0	0.0419	214, 216, 222, 263
100:214	1 7 9 12 25 27 30 32 34 45 48 52	C_1	100×1	294, 294, 294	1 18 21 0	0.1721	213, 215, 217, 223
100:215	1 7 9 12 25 27 30 32 34 46 48 51	C_1	100×1	294, 294, 294	1 18 21 0	0.1479	214, 224
100:216	1 7 9 12 25 27 30 32 35 44 47 49	C_1	100×1	294, 294, 294	1 19 19 1	0.2132	213, 217, 218, 226, 256, 264

100:198

100:197

100:196

100:195

100:194

100:193

100:204

100:203

100:202

100:201

100:200

100:199

100:210

100:209

100:208

100:207

100:206

100:205

100:216

100:215

100:214

100:213

100:212

100:211

Table A.20. (*Continued*)

Isomer	Ring spiral	Point group	NMR pattern	Vibrations	Hexagon indices	Band gap	Transformations
100:217	1 7 9 12 25 27 30 32 35 44 48 52	C_1	100×1	294, 294, 294	1 19 19 1	0.1311	214, 216, 219, 227, 254
100:218	1 7 9 12 25 27 30 33 35 43 47 49	C_1	100×1	294, 294, 294	1 18 21 0	0.1009	216, 219, 257, 265
100:219	1 7 9 12 25 27 30 33 35 43 48 52	C_1	100×1	294, 294, 294	1 18 21 0	0.1557	217, 218, 220, 255
100:220	1 7 9 12 25 27 30 33 35 48 50 52	C_1	100×1	294, 294, 294	1 18 21 0	0.0802	219, 228, 253
100:221	1 7 9 12 25 27 31 34 41 45 47 50	C_3	1×1, 33×3	194, 194, 194	1 18 21 0	0.0000	222(3)
100:222	1 7 9 12 25 27 31 34 42 45 47 49	C_1	100×1	294, 294, 294	1 19 19 1	0.0263	213, 221, 223, 226(2)
100:223	1 7 9 12 25 27 31 34 42 45 48 52	C_1	100×1	294, 294, 294	1 19 19 1	0.0251	214, 222, 224, 227
100:224	1 7 9 12 25 27 31 34 42 46 48 51	C_1	100×1	294, 294, 294	1 19 19 1	0.1216	215, 223, 225
100:225	1 7 9 12 25 27 31 34 43 46 48 50	C_1	100×1	294, 294, 294	1 19 19 1	0.0621	177, 224
100:226	1 7 9 12 25 27 31 35 42 44 47 49	C_1	100×1	294, 294, 294	1 20 17 2	0.1521	216, 222(2), 227, 263, 266
100:227	1 7 9 12 25 27 31 35 42 44 48 52	C_1	100×1	294, 294, 294	1 20 17 2	0.0882	217, 223, 226, 262
100:228	1 7 9 12 25 27 33 35 41 43 49 52	C_1	100×1	294, 294, 294	1 19 19 1	0.1246	211, 220, 245
100:229	1 7 9 12 25 28 30 33 35 46 48 50	C_1	100×1	294, 294, 294	1 19 19 1	0.1386	230, 232, 239, 308, 360
100:230	1 7 9 12 25 28 30 33 35 46 49 51	C_1	100×1	294, 294, 294	1 19 19 1	0.1338	229, 231, 233, 312, 409
100:231	1 7 9 12 25 28 30 33 35 47 49 52	C_1	100×1	294, 294, 294	1 19 19 1	0.0504	230, 307, 313, 369
100:232	1 7 9 12 25 28 30 33 36 45 48 50	C_1	100×1	294, 294, 294	1 19 19 1	0.2182	180, 229, 233, 235, 240, 426
100:233	1 7 9 12 25 28 30 33 36 45 49 51	C_1	100×1	294, 294, 294	1 19 19 1	0.0737	181, 230, 232, 236, 428
100:234	1 7 9 12 25 28 30 34 36 43 49 52	C_1	100×1	294, 294, 294	1 20 17 2	0.1241	185, 236, 349
100:235	1 7 9 12 25 28 30 34 36 44 48 50	C_1	100×1	294, 294, 294	1 20 17 2	0.1198	186, 232, 236, 241, 398
100:236	1 7 9 12 25 28 30 34 36 44 49 51	C_1	100×1	294, 294, 294	1 20 17 2	0.1333	187, 233, 234, 235, 374
100:237	1 7 9 12 25 28 32 34 36 41 48 51	C_1	100×1	294, 294, 294	1 20 17 2	0.0596	191, 206, 241, 402
100:238	1 7 9 12 25 28 33 35 39 43 50 52	C_1	100×1	294, 294, 294	1 20 17 2	0.0650	211, 246, 320, 405
100:239	1 7 9 12 25 28 33 35 41 43 47 49	C_1	100×1	294, 294, 294	1 20 17 2	0.0954	201, 229, 240, 320, 370
100:240	1 7 9 12 25 28 33 36 41 43 46 49	C_1	100×1	294, 294, 294	1 20 17 2	0.1279	196, 202, 232, 239, 241, 427

100:222

100:221

100:220

100:219

100:218

100:217

100:228

100:227

100:226

100:225

100:224

100:223

100:234

100:233

100:232

100:231

100:230

100:229

100:240

100:239

100:238

100:237

100:236

100:235

Table A.20. (*Continued*)

Isomer	Ring spiral	Point group	NMR pattern	Vibrations	Hexagon indices	Band gap	Transformations
100:241	1 7 9 12 25 28 34 36 41 43 45 49	C_1	100×1	294, 294, 294	1 21 15 3	0.1026	199, 207, 235, 237, 240, 397
100:242	1 7 9 12 25 29 32 34 36 39 42 52	C_1	100×1	294, 294, 294	1 19 19 1	0.1766	201, 243, 245, 247
100:243	1 7 9 12 25 29 32 34 36 39 49 51	C_1	100×1	294, 294, 294	1 20 17 2	0.1591	202, 206, 242, 244, 248
100:244	1 7 9 12 25 29 34 36 39 43 45 50	C_s	4×1, 48×2	294, 294, 294	1 21 15 3	0.0191	207(2), 243(2), 252
100:245	1 7 9 12 25 30 32 34 36 39 41 52	C_1	100×1	294, 294, 294	1 19 19 1	0.1311	210, 228, 242, 253
100:246	1 7 9 12 26 28 33 35 38 43 50 52	C_1	100×1	294, 294, 294	1 20 17 2	0.0335	238, 308, 315, 404
100:247	1 7 9 12 26 29 32 34 36 38 42 52	C_1	100×1	294, 294, 294	1 19 19 1	0.1897	242, 248, 249, 253
100:248	1 7 9 12 26 29 32 34 36 38 49 51	C_1	100×1	294, 294, 294	1 18 21 0	0.2136	243, 247, 252
100:249	1 7 9 12 26 29 32 34 36 42 47 49	C_1	100×1	294, 294, 294	1 18 21 0	0.1730	247, 250, 255
100:250	1 7 9 12 26 29 32 34 37 42 46 49	C_1	100×1	294, 294, 294	1 18 21 0	0.0536	249, 251, 257
100:251	1 7 9 12 26 29 32 35 37 42 45 49	C_1	100×1	294, 294, 294	1 19 19 1	0.1301	250, 251, 259, 261
100:252	1 7 9 12 26 29 34 36 38 43 45 50	C_s	6×1, 47×2	294, 294, 294	1 19 19 1	0.1078	244, 248(2)
100:253	1 7 9 12 26 30 32 34 36 38 41 52	C_1	100×1	294, 294, 294	1 18 21 0	0.2159	220, 245, 247, 255
100:254	1 7 9 12 26 30 32 34 36 40 47 50	C_1	100×1	294, 294, 294	1 18 21 0	0.1894	217, 255, 256, 262
100:255	1 7 9 12 26 30 32 34 36 41 47 49	C_1	100×1	294, 294, 294	1 19 19 1	0.1431	219, 249, 253, 254, 257
100:256	1 7 9 12 26 30 32 34 37 40 46 50	C_1	100×1	294, 294, 294	1 18 21 0	0.0261	216, 254, 257, 258, 263
100:257	1 7 9 12 26 30 32 34 37 41 46 49	C_1	100×1	294, 294, 294	1 19 19 1	0.1165	218, 250, 255, 256, 259
100:258	1 7 9 12 26 30 32 35 37 40 45 50	C_1	100×1	294, 294, 294	1 19 19 1	0.1404	256, 259, 260, 264(2), 322
100:259	1 7 9 12 26 30 32 35 37 41 45 49	C_1	100×1	294, 294, 294	1 20 17 2	0.1545	251, 257, 258, 261, 265, 323
100:260	1 7 9 12 26 30 33 35 37 40 44 50	C_1	100×1	294, 294, 294	1 18 21 0	0.1533	258, 261, 265, 323
100:261	1 7 9 12 26 30 33 35 37 41 44 49	C_1	100×1	294, 294, 294	1 18 21 0	0.0971	251, 259, 260, 323
100:262	1 7 9 12 26 31 35 37 40 42 44 52	C_1	100×1	294, 294, 294	1 19 19 1	0.0703	227, 254, 263
100:263	1 7 9 12 26 31 35 37 40 42 45 51	C_1	100×1	294, 294, 294	1 19 19 1	0.1076	213, 226, 256, 262, 264
100:264	1 7 9 12 26 31 35 37 40 43 45 50	C_1	100×1	294, 294, 294	1 20 17 2	0.1268	216, 258(2), 263, 265, 266

100:246

100:245

100:244

100:243

100:242

100:241

100:252

100:251

100:250

100:249

100:248

100:247

100:258

100:257

100:256

100:255

100:254

100:253

100:264

100:263

100:262

100:261

100:260

100:259

Table A.20. *(Continued)*

Isomer	Ring spiral	Point group	NMR pattern	Vibrations	Hexagon indices	Band gap	Transformations
100:265	1 7 9 12 26 31 35 37 41 43 45 49	C_1	100×1	294, 294, 294	1 19 19 1	0.1149	218, 259, 260, 264
100:266	1 7 9 12 27 31 35 37 39 43 45 50	C_3	1×1, 33×3	194, 194, 194	1 21 15 3	0.0561	226(3), 264(3)
100:267	1 7 9 13 20 23 30 36 39 44 47 49	C_1	100×1	294, 294, 294	2 19 16 3	0.0692	133, 268, 314, 317
100:268	1 7 9 13 20 23 30 37 39 44 46 49	C_1	100×1	294, 294, 294	1 20 17 2	0.0948	134, 267, 269, 271, 408, 439
100:269	1 7 9 13 20 23 30 37 39 45 49 51	C_1	100×1	294, 294, 294	1 21 15 3	0.1557	135, 268, 270, 272, 299, 407, 436
100:270	1 7 9 13 20 23 30 37 39 45 50 52	C_1	100×1	294, 294, 294	1 21 15 3	0.1487	136, 269, 274, 300, 325, 328
100:271	1 7 9 13 20 23 30 37 40 44 46 48	C_2	50×2	294, 294, 294	2 20 14 4	0.2082	137(2), 268(2), 272(2)
100:272	1 7 9 13 20 23 30 37 40 45 48 51	C_1	100×1	294, 294, 294	1 22 13 4	0.3165	138, 156, 269, 271, 273, 298, 406, 408
100:273	1 7 9 13 20 23 30 38 40 45 47 51	C_2	50×2	294, 294, 294	2 20 14 4	0.2229	139, 142, 272(2), 297(2)
100:274	1 7 9 13 20 23 37 39 41 43 46 52	C_1	100×1	294, 294, 294	1 22 13 4	0.0029	140, 195, 270, 301, 375, 404
100:275	1 7 9 13 20 25 30 34 38 45 48 50	C_1	100×1	294, 294, 294	1 21 15 3	0.0091	188, 276, 282, 287, 291, 367, 368
100:276	1 7 9 13 20 25 30 34 39 45 47 50	C_1	100×1	294, 294, 294	1 20 17 2	0.0375	183, 275, 278, 284, 286, 428, 429
100:277	1 7 9 13 20 25 30 34 39 46 49 51	C_1	100×1	294, 294, 294	1 20 17 2	0.1174	182, 278, 279, 299, 359, 369
100:278	1 7 9 13 20 25 30 34 39 46 50 52	C_1	100×1	294, 294, 294	1 20 17 2	0.1449	184, 276, 277, 288, 300, 409, 411
100:279	1 7 9 13 20 25 30 34 40 46 48 51	C_1	100×1	294, 294, 294	1 21 15 3	0.2132	158, 179, 277, 298, 328, 336
100:280	1 7 9 13 20 25 30 35 38 43 48 51	C_1	100×1	294, 294, 294	1 22 13 4	0.1277	187, 281, 282, 283, 292, 389, 418
100:281	1 7 9 13 20 25 30 35 38 43 49 52	C_1	100×1	294, 294, 294	1 22 13 4	0.1273	185, 280, 285, 293, 387, 388
100:282	1 7 9 13 20 25 30 35 38 44 48 50	C_1	100×1	294, 294, 294	1 22 13 4	0.3332	186, 275, 280, 284, 290, 294, 421, 423
100:283	1 7 9 13 20 25 30 35 39 43 47 51	C_1	100×1	294, 294, 294	1 21 15 3	0.0634	181, 280, 284, 349, 355
100:284	1 7 9 13 20 25 30 35 39 44 47 50	C_1	100×1	294, 294, 294	1 21 15 3	0.0882	180, 276, 282, 283, 289, 367, 374
100:285	1 7 9 13 20 25 30 38 43 45 47 50	C_s	8×1, 46×2	294, 294, 294	2 21 12 5	0.0162	189(2), 281(2)
100:286	1 7 9 13 20 25 34 38 40 43 46 50	C_1	100×1	294, 294, 294	1 21 15 3	0.2884	192, 197, 276, 287, 288, 289, 424, 426
100:287	1 7 9 13 20 25 34 38 41 43 46 49	C_s	6×1, 47×2	294, 294, 294	1 22 13 4	0.2138	200(2), 275(2), 286(2), 290, 415(2)
100:288	1 7 9 13 20 25 34 39 41 43 47 52	C_1	100×1	294, 294, 294	1 21 15 3	0.0954	193, 198, 278, 286, 301, 360, 361

100:270

100:269

100:268

100:267

100:266

100:265

100:276

100:275

100:274

100:273

100:272

100:271

100:282

100:281

100:280

100:279

100:278

100:277

100:288

100:287

100:286

100:285

100:284

100:283

Table A.20. (*Continued*)

Isomer	Ring spiral	Point group	NMR pattern	Vibrations	Hexagon indices	Band gap	Transformations
100:289	1 7 9 13 20 25 35 38 40 43 45 50	C_1	100×1	294, 294, 294	1 22 13 4	0.0496	191, 196, 284, 286, 290, 398, 400
100:290	1 7 9 13 20 25 35 38 41 43 45 49	C_s	4×1, 48×2	294, 294, 294	1 23 11 5	0.2557	199(2), 282(2), 287, 289(2), 420(2)
100:291	1 7 9 13 20 26 30 34 37 45 48 50	C_s	6×1, 47×2	294, 294, 294	1 20 17 2	0.0459	275(2), 294, 355(2)
100:292	1 7 9 13 20 26 30 35 37 43 48 51	C_1	100×1	294, 294, 294	1 21 15 3	0.1099	280, 293, 294, 295, 388, 438
100:293	1 7 9 13 20 26 30 35 37 43 49 52	C_2	50×2	294, 294, 294	2 20 14 4	0.0036	281(2), 292(2)
100:294	1 7 9 13 20 26 30 35 37 44 48 50	C_s	4×1, 48×2	294, 294, 294	1 21 15 3	0.2015	282(2), 291, 292(2), 418(2)
100:295	1 7 9 13 20 26 30 36 43 46 48 51	C_{2v}	6×2, 22×4	223, 294, 223	2 20 14 4	0.0614	292(4)
100:296	1 7 9 13 20 26 34 39 41 44 46 48	C_1	100×1	294, 294, 294	2 19 16 3	0.0770	143, 297, 309, 314
100:297	1 7 9 13 20 26 34 39 41 44 47 51	C_1	100×1	294, 294, 294	1 20 17 2	0.1959	144, 273, 296, 298, 408, 440
100:298	1 7 9 13 20 26 34 39 42 44 47 50	C_1	100×1	294, 294, 294	1 21 15 3	0.2433	157, 272, 279, 297, 299, 407, 412
100:299	1 7 9 13 20 26 34 40 42 44 47 49	C_1	100×1	294, 294, 294	1 20 17 2	0.2063	269, 277, 298, 300, 441, 443
100:300	1 7 9 13 20 30 34 36 38 40 50 52	C_1	100×1	294, 294, 294	1 20 17 2	0.1566	270, 278, 299, 301, 357, 359
100:301	1 7 9 13 20 34 36 38 40 42 44 52	C_1	100×1	294, 294, 294	1 21 15 3	0.1426	194, 274, 288, 300, 383, 386
100:302	1 7 9 13 22 25 29 32 41 46 48 51	C_1	100×1	294, 294, 294	1 20 17 2	0.0932	176, 305, 307, 332
100:303	1 7 9 13 22 25 30 32 39 46 49 51	C_1	100×1	294, 294, 294	1 19 19 1	0.1599	143, 173, 304, 305, 309, 442
100:304	1 7 9 13 22 25 30 32 39 46 50 52	C_2	50×2	294, 294, 294	2 18 18 2	0.2654	175(2), 303(2), 310
100:305	1 7 9 13 22 25 30 32 40 46 48 51	C_1	100×1	294, 294, 294	1 20 17 2	0.1424	172, 302, 303, 306, 311, 363
100:306	1 7 9 13 22 25 31 40 42 46 48 51	C_1	100×1	294, 294, 294	1 21 15 3	0.0519	133, 143, 171, 305, 314, 391
100:307	1 7 9 13 22 29 32 36 38 42 48 51	C_1	100×1	294, 294, 294	1 20 17 2	0.1323	231, 302, 311, 336
100:308	1 7 9 13 22 30 32 35 37 39 41 52	C_1	100×1	294, 294, 294	1 19 19 1	0.2414	229, 246, 312, 320, 383
100:309	1 7 9 13 22 30 32 36 38 40 49 51	C_1	100×1	294, 294, 294	1 19 19 1	0.0930	296, 303, 310, 311, 440
100:310	1 7 9 13 22 30 32 36 38 40 50 52	C_2	50×2	294, 294, 294	2 18 18 2	0.0354	304, 309(2)
100:311	1 7 9 13 22 30 32 36 38 41 48 51	C_1	100×1	294, 294, 294	1 20 17 2	0.1381	305, 307, 309, 313, 314, 412
100:312	1 7 9 13 22 30 32 36 39 41 46 52	C_1	100×1	294, 294, 294	1 19 19 1	0.1306	230, 308, 313, 315, 357

100:294

100:300

100:306

100:312

100:293

100:299

100:305

100:311

100:292

100:298

100:304

100:310

100:291

100:297

100:303

100:309

100:290

100:296

100:302

100:308

100:289

100:295

100:301

100:307

Table A.20. (*Continued*)

Isomer	Ring spiral	Point group	NMR pattern	Vibrations	Hexagon indices	Band gap	Transformations
100:313	1 7 9 13 22 30 32 36 39 41 47 51	C_1	100×1	294, 294, 294	1 19 19 1	0.1677	231, 311, 312, 316, 441
100:314	1 7 9 13 22 31 36 38 41 43 48 51	C_1	100×1	294, 294, 294	1 21 15 3	0.1739	267, 296, 306, 311, 316, 408
100:315	1 7 9 13 22 31 36 39 41 43 46 52	C_1	100×1	294, 294, 294	1 20 17 2	0.1067	246, 312, 316, 325
100:316	1 7 9 13 22 31 36 39 41 43 47 51	C_1	100×1	294, 294, 294	1 20 17 2	0.0742	313, 314, 315, 317, 436
100:317	1 7 9 13 22 31 36 39 41 44 47 50	C_1	100×1	294, 294, 294	1 19 19 1	0.0555	267, 316, 318, 439
100:318	1 7 9 13 22 31 36 39 42 44 47 49	C_2	50×2	294, 294, 294	2 18 18 2	0.0187	317(2)
100:319	1 7 9 13 23 25 29 33 42 47 49 51	C_s	4×1, 48×2	294, 294, 294	2 19 16 3	0.0021	447
100:320	1 7 9 13 23 30 32 34 37 39 41 52	C_1	100×1	294, 294, 294	1 20 17 2	0.2493	210, 238, 239, 308, 382
100:321	1 7 9 22 24 27 31 33 36 39 41 51	T	1×4, 8×12	73, 123, 73	4 12 24 0	0.0000	258(3), 323(3)
100:322	1 7 9 22 24 27 31 33 38 41 46 51	C_3	1×1, 33×3	194, 194, 194	1 18 21 0	0.0560	259, 260, 261, 322, 323(2)
100:323	1 7 9 22 24 27 31 33 38 42 46 50	C_1	100×1	294, 294, 294	1 19 19 1	0.1666	136, 158, 328, 331, 365, 393
100:324	1 7 10 12 14 18 34 39 41 45 47 50	C_1	100×1	294, 294, 294	0 23 14 3	0.1250	270, 315, 327, 357, 404, 436
100:325	1 7 10 12 14 18 35 38 41 43 48 51	C_1	100×1	294, 294, 294	0 21 18 1	0.1032	
100:326	1 7 10 12 14 18 35 39 41 43 46 52	C_2	50×2	294, 294, 294	0 22 16 2	0.0599	327(2), 329(2), 333(2)
100:327	1 7 10 12 14 18 35 39 41 43 47 51	C_1	100×1	294, 294, 294	0 22 16 2	0.0523	325, 326, 328, 334, 358, 403, 435
100:328	1 7 10 12 14 18 35 39 41 44 47 50	C_1	100×1	294, 294, 294	0 23 14 3	0.1022	270, 279, 324, 327, 335, 359, 375, 407
100:329	1 7 10 12 14 18 36 39 41 43 45 52	C_1	100×1	294, 294, 294	0 21 18 1	0.0703	326, 337, 401, 403
100:330	1 7 10 12 14 19 37 39 42 44 46 49	C_2	50×2	294, 294, 294	0 20 20 0	0.0875	100(2), 152(2)
100:331	1 7 10 12 14 28 30 35 42 46 48 51	C_1	100×1	294, 294, 294	0 22 16 2	0.1330	324, 332, 335, 338, 364
100:332	1 7 10 12 14 28 30 35 43 46 48 50	C_1	100×1	294, 294, 294	0 21 18 1	0.0879	158, 302, 331, 336, 363
100:333	1 7 10 12 14 28 30 36 42 44 47 49	C_1	100×1	294, 294, 294	0 21 18 1	0.0940	326, 334, 337, 341, 358, 381
100:334	1 7 10 12 14 28 30 36 42 44 48 52	C_1	100×1	294, 294, 294	0 21 18 1	0.0585	327, 333, 335, 342, 380, 434
100:335	1 7 10 12 14 28 30 36 42 45 48 51	C_1	100×1	294, 294, 294	0 22 16 2	0.1292	328, 331, 334, 336, 343, 385, 413
100:336	1 7 10 12 14 28 30 36 43 45 48 50	C_1	100×1	294, 294, 294	0 21 18 1	0.1584	279, 307, 332, 335, 369, 412

100:318

100:324

100:330

100:336

100:317

100:323

100:329

100:335

100:316

100:322

100:328

100:334

100:315

100:321

100:327

100:333

100:314

100:320

100:326

100:332

100:313

100:319

100:325

100:331

Table A.20. (*Continued*)

Isomer	Ring spiral	Point group	NMR pattern	Vibrations	Hexagon indices	Band gap	Transformations
100:337	1 7 10 12 14 28 30 37 42 44 46 49	C_1	100×1	294, 294, 294	0 20 20 0	0.1320	329, 333, 346, 384
100:338	1 7 10 12 14 28 31 35 41 46 48 51	C_2	50×2	294, 294, 294	0 22 16 2	0.1221	331(2), 343
100:339	1 7 10 12 14 28 31 36 40 43 47 51	C_1	100×1	294, 294, 294	0 22 16 2	0.0802	340, 345, 347, 351, 366, 372
100:340	1 7 10 12 14 28 31 36 40 44 47 50	C_1	100×1	294, 294, 294	0 21 18 1	0.1464	339, 341, 344, 352, 377, 416
100:341	1 7 10 12 14 28 31 36 41 44 47 49	C_1	100×1	294, 294, 294	0 21 18 1	0.0463	333, 340, 342, 346, 379, 410
100:342	1 7 10 12 14 28 31 36 41 44 48 52	C_1	100×1	294, 294, 294	0 21 18 1	0.0342	334, 341, 343, 378, 385
100:343	1 7 10 12 14 28 31 36 41 45 48 51	C_2	50×2	294, 294, 294	0 22 16 2	0.1221	335(2), 338, 342(2)
100:344	1 7 10 12 14 28 31 37 40 43 45 52	C_1	100×1	294, 294, 294	0 20 20 0	0.1667	340, 345, 346, 353, 425
100:345	1 7 10 12 14 28 31 37 40 43 46 51	C_s	8×1, 46×2	294, 294, 294	0 21 18 1	0.1098	339(2), 344(2), 354, 414
100:346	1 7 10 12 14 28 31 37 41 44 46 49	C_1	100×1	294, 294, 294	0 20 20 0	0.1367	337, 341, 344, 362
100:347	1 7 10 12 14 28 32 36 40 42 47 51	C_{2v}	8×2, 21×4	224, 294, 224	0 23 14 3	0.0356	339(4), 356(2)
100:348	1 7 10 12 14 29 31 36 38 43 48 51	C_1	100×1	294, 294, 294	0 21 18 1	0.1849	349, 350, 351, 355, 373, 422
100:349	1 7 10 12 14 29 31 36 38 43 49 52	C_1	100×1	294, 294, 294	0 21 18 1	0.1490	234, 283, 348, 374, 389
100:350	1 7 10 12 14 29 31 36 38 44 48 50	C_2	50×2	294, 294, 294	0 20 20 0	0.1947	348(2), 352(2)
100:351	1 7 10 12 14 29 31 36 39 43 47 51	C_1	100×1	294, 294, 294	0 22 16 2	0.1891	339, 348, 352, 354, 356, 371, 419
100:352	1 7 10 12 14 29 31 36 39 44 47 50	C_1	100×1	294, 294, 294	0 21 18 1	0.0931	340, 350, 351, 353, 373, 376
100:353	1 7 10 12 14 29 31 37 39 43 45 52	C_1	100×1	294, 294, 294	0 20 20 0	0.0600	344, 352, 354, 399
100:354	1 7 10 12 14 29 31 37 39 43 46 51	C_s	6×1, 47×2	294, 294, 294	0 21 18 1	0.0806	345, 351(2), 353(2), 444
100:355	1 7 10 12 14 29 32 36 38 42 48 51	C_1	100×1	294, 294, 294	0 22 16 2	0.1400	283, 291, 348, 356, 367, 368, 418
100:356	1 7 10 12 14 29 32 36 39 42 47 51	C_s	6×1, 47×2	294, 294, 294	0 23 14 3	0.1798	347, 351(2), 355(2), 366(2), 417
100:357	1 7 10 12 18 24 26 39 42 44 46 49	C_1	100×1	294, 294, 294	0 20 20 0	0.0774	300, 312, 325, 358, 383, 409, 441
100:358	1 7 10 12 18 24 26 40 42 44 46 48	C_1	100×1	294, 294, 294	0 21 18 1	0.1429	327, 333, 357, 359, 380, 384, 410, 434
100:359	1 7 10 12 18 24 26 40 42 45 48 51	C_1	100×1	294, 294, 294	0 22 16 2	0.1120	277, 300, 328, 358, 385, 386, 411, 443
100:360	1 7 10 12 18 24 27 36 41 43 48 51	C_1	100×1	294, 294, 294	0 21 18 1	0.1186	229, 288, 362, 370, 383, 409, 426

357

100:342

100:341

100:340

100:339

100:338

100:337

100:348

100:347

100:346

100:345

100:344

100:343

100:354

100:353

100:352

100:351

100:350

100:349

100:360

100:359

100:358

100:357

100:356

100:355

Table A.20. (*Continued*)

Isomer	Ring spiral	Point group	NMR pattern	Vibrations	Hexagon indices	Band gap	Transformations
100:361	1 7 10 12 18 24 27 40 43 45 47 50	C_1	100×1	294, 294, 294	0 21 18 1	0.0594	184, 288, 362, 386, 396, 411, 424
100:362	1 7 10 12 18 24 27 41 43 45 47 49	C_1	100×1	294, 294, 294	0 20 20 0	0.1205	346, 360, 361, 384, 410, 425
100:363	1 7 10 12 18 24 31 37 39 44 46 49	C_1	100×1	294, 294, 294	0 21 18 1	0.1672	157, 305, 332, 364, 391, 412, 442
100:364	1 7 10 12 18 24 31 37 39 45 49 51	C_2	50×2	294, 294, 294	0 22 16 2	0.0843	331(2), 363(2), 365(2), 413
100:365	1 7 10 12 18 24 31 37 40 45 48 51	C_1	100×1	294, 294, 294	0 23 14 3	0.0403	135, 157, 324, 364, 391, 392, 407
100:366	1 7 10 12 18 24 32 36 38 40 49 51	C_1	100×1	294, 294, 294	0 22 16 2	0.0662	339, 356, 367, 368, 371, 414, 416, 419
100:367	1 7 10 12 18 24 32 36 38 41 48 51	C_1	100×1	294, 294, 294	0 23 14 3	0.0381	275, 284, 355, 366, 373, 400, 415, 421, 429
100:368	1 7 10 12 18 24 32 38 40 45 47 50	C_1	100×1	294, 294, 294	0 21 18 1	0.0528	181, 275, 355, 366, 415, 423, 428
100:369	1 7 10 12 18 25 27 32 44 46 48 50	C_1	100×1	294, 294, 294	0 20 20 0	0.0928	231, 277, 336, 385, 409, 441
100:370	1 7 10 12 18 25 27 35 41 43 48 51	C_1	100×1	294, 294, 294	0 20 20 0	0.0732	193, 239, 360, 382, 427
100:371	1 7 10 12 18 25 32 35 38 40 49 51	C_1	100×1	294, 294, 294	0 21 18 1	0.0968	351, 366, 372, 373, 376, 430, 445
100:372	1 7 10 12 18 25 32 35 38 40 50 52	C_2	50×2	294, 294, 294	0 21 18 1	0.0015	339(2), 371(2), 377(2)
100:373	1 7 10 12 18 25 32 35 38 41 48 51	C_1	100×1	294, 294, 294	0 22 16 2	0.1275	348, 352, 367, 371, 374, 399, 416, 419
100:374	1 7 10 12 18 25 32 35 39 41 47 51	C_1	100×1	294, 294, 294	0 21 18 1	0.1389	236, 284, 349, 373, 398, 423, 428
100:375	1 7 10 12 18 25 33 35 39 41 47 50	C_1	100×1	294, 294, 294	0 22 16 2	0.0067	179, 274, 328, 386, 393, 403
100:376	1 7 10 12 18 26 32 35 37 40 49 51	C_2	50×2	294, 294, 294	0 20 20 0	0.0046	352(2), 371(2), 377(2)
100:377	1 7 10 12 18 26 32 35 37 40 50 52	C_1	100×1	294, 294, 294	0 20 20 0	0.0955	340, 372, 376, 379, 430, 432
100:378	1 7 10 12 18 26 32 35 39 46 49 51	C_2	50×2	294, 294, 294	0 20 20 0	0.1067	342(2), 379(2), 380(2)
100:379	1 7 10 12 18 26 32 35 40 46 48 51	C_1	100×1	294, 294, 294	0 20 20 0	0.0623	341, 377, 378, 381, 431, 433
100:380	1 7 10 12 18 26 32 36 39 45 49 51	C_1	100×1	294, 294, 294	0 20 20 0	0.1274	334, 358, 378, 381, 385, 433, 449
100:381	1 7 10 12 18 26 32 36 40 45 48 51	C_2	50×2	294, 294, 294	0 20 20 0	0.0029	333(2), 379(2), 380(2)
100:382	1 7 10 12 18 26 33 36 39 42 47 51	C_1	100×1	294, 294, 294	0 20 20 0	0.2131	194, 320, 370, 383, 405
100:383	1 7 10 12 18 26 33 36 39 43 47 50	C_1	100×1	294, 294, 294	0 21 18 1	0.2152	301, 308, 357, 360, 382, 384, 404
100:384	1 7 10 12 18 26 33 36 39 43 49 52	C_1	100×1	294, 294, 294	0 20 20 0	0.2206	337, 358, 362, 383, 386, 403

100:366

100:365

100:364

100:363

100:362

100:361

100:372

100:371

100:370

100:369

100:368

100:367

100:378

100:377

100:376

100:375

100:374

100:373

100:384

100:383

100:382

100:381

100:380

100:379

Table A.20. (*Continued*)

Isomer	Ring spiral	Point group	NMR pattern	Vibrations	Hexagon indices	Band gap	Transformations
100:385	1 7 10 12 18 27 32 36 38 45 49 51	C_1	100×1	294, 294, 294	0 21 18 1	0.0580	335, 342, 359, 369, 380, 410, 434
100:386	1 7 10 12 18 27 33 36 38 43 49 52	C_1	100×1	294, 294, 294	0 21 18 1	0.1381	182, 301, 359, 361, 375, 384
100:387	1 7 10 12 18 27 33 36 40 45 48 51	C_2	50×2	294, 294, 294	0 22 16 2	0.1561	189(2), 281(2), 389(2)
100:388	1 7 10 12 18 27 34 36 40 44 47 50	C_s	8×1, 46×2	294, 294, 294	0 23 14 3	0.2006	281(2), 292(2), 389(2), 418(2)
100:389	1 7 10 12 18 27 34 36 40 44 48 51	C_1	100×1	294, 294, 294	0 22 16 2	0.1756	185, 280, 349, 387, 388, 422, 423
100:390	1 7 10 12 19 24 29 37 41 45 48 51	D_2	25×4	219, 294, 219	0 24 12 4	0.0805	131(4), 392(2)
100:391	1 7 10 12 19 24 30 37 39 44 46 49	C_1	100×1	294, 294, 294	0 22 16 2	0.0496	134, 144, 156, 306, 363, 365, 408
100:392	1 7 10 12 19 24 30 37 40 45 48 51	C_2	50×2	294, 294, 294	0 24 12 4	0.1012	138(2), 156(2), 365(2), 390, 406
100:393	1 7 10 12 19 24 37 39 41 43 46 52	C_1	100×1	294, 294, 294	0 22 16 2	0.1478	140, 155, 324, 375
100:394	1 7 10 12 19 25 27 42 44 46 48 50	C_s	10×1, 45×2	294, 294, 294	0 21 18 1	0.1185	141(2)
100:395	1 7 10 12 19 26 29 36 43 45 48 50	D_{2d}	3×4, 11×8	111, 186, 111	0 20 20 0	0.0000	150(4)
100:396	1 7 10 12 24 27 29 31 44 46 48 51	C_2	50×2	294, 294, 294	0 20 20 0	0.0420	198(2), 361(2), 448
100:397	1 7 10 12 24 27 30 32 37 44 48 50	C_1	100×1	294, 294, 294	0 21 18 1	0.0669	191, 241, 398, 402, 427, 446
100:398	1 7 10 12 24 27 30 32 38 44 47 50	C_1	100×1	294, 294, 294	0 22 16 2	0.1106	235, 289, 374, 397, 399, 420, 426
100:399	1 7 10 12 24 27 30 32 38 44 49 52	C_1	100×1	294, 294, 294	0 21 18 1	0.0973	353, 373, 398, 400, 425, 444
100:400	1 7 10 12 24 27 30 33 38 43 49 52	C_1	100×1	294, 294, 294	0 22 16 2	0.0695	188, 289, 367, 399, 420, 424, 437
100:401	1 7 10 12 25 27 30 32 34 46 48 50	C_2	50×2	294, 294, 294	0 20 20 0	0.0345	329(2)
100:402	1 7 10 12 25 27 30 32 36 44 48 50	C_2	50×2	294, 294, 294	0 20 20 0	0.0689	237(2), 397(2)
100:403	1 7 10 12 25 27 32 34 40 43 47 50	C_1	100×1	294, 294, 294	0 21 18 1	0.0288	327, 329, 375, 384, 404
100:404	1 7 10 12 25 28 32 34 39 43 47 50	C_1	100×1	294, 294, 294	0 22 16 2	0.0485	246, 274, 325, 383, 403, 405
100:405	1 7 10 12 26 29 31 34 36 38 49 51	C_1	100×1	294, 294, 294	0 20 20 0	0.0160	195, 238, 382, 404
100:406	1 7 10 13 18 22 34 36 39 41 47 50	D_2	25×4	219, 294, 219	0 24 12 4	0.2699	272(4), 392(2), 407(4)
100:407	1 7 10 13 18 22 34 37 39 41 46 50	C_1	100×1	294, 294, 294	0 23 14 3	0.1906	269, 298, 328, 365, 406, 408, 413, 435, 443
100:408	1 7 10 13 18 22 34 37 39 42 46 49	C_1	100×1	294, 294, 294	0 22 16 2	0.1453	268, 272, 297, 314, 391, 407, 412, 436

361

Table A.20. (*Continued*)

Isomer	Ring spiral	Point group	NMR pattern	Vibrations	Hexagon indices	Band gap	Transformations
100:409	1 7 10 13 18 23 26 39 42 44 46 49	C_1	100×1	294, 294, 294	0 20 20 0	0.1128	230, 278, 357, 360, 369, 410, 428
100:410	1 7 10 13 18 23 26 40 42 44 46 48	C_1	100×1	294, 294, 294	0 21 18 1	0.0709	341, 358, 362, 385, 409, 411, 416, 433
100:411	1 7 10 13 18 23 26 40 42 45 48 51	C_2	50×2	294, 294, 294	0 22 16 2	0.0274	278(2), 359(2), 361(2), 410(2), 429
100:412	1 7 10 13 18 23 31 37 39 44 46 49	C_1	100×1	294, 294, 294	0 21 18 1	0.0672	298, 311, 336, 363, 408, 413, 440, 441
100:413	1 7 10 13 18 23 31 37 39 45 49 51	C_2	50×2	294, 294, 294	0 22 16 2	0.1644	335(2), 364, 407(2), 412(2), 434(2)
100:414	1 7 10 13 18 23 32 36 38 40 49 51	C_s	6×1, 47×2	294, 294, 294	0 21 18 1	0.1544	345, 366(2), 415(2), 425(2), 444
100:415	1 7 10 13 18 23 32 36 38 41 48 51	C_1	100×1	294, 294, 294	0 22 16 2	0.0958	180, 287, 367, 368, 414, 420, 424, 426, 437
100:416	1 7 10 13 18 25 31 34 39 45 47 50	C_1	100×1	294, 294, 294	0 21 18 1	0.0062	340, 366, 373, 410, 425, 428, 429, 430
100:417	1 7 10 13 18 25 31 35 38 43 48 51	C_{2v}	6×2, 22×4	223, 294, 223	0 24 12 4	0.3649	356(2), 418(4), 419(4)
100:418	1 7 10 13 18 25 31 35 38 43 49 52	C_1	100×1	294, 294, 294	0 23 14 3	0.2738	280, 294, 355, 388, 417, 421, 422, 423, 438
100:419	1 7 10 13 18 25 31 35 38 44 48 50	C_1	100×1	294, 294, 294	0 23 14 3	0.3435	351, 366, 373, 417, 421, 422, 423, 444, 445
100:420	1 7 10 13 18 25 32 35 38 41 50 52	C_1	100×1	294, 294, 294	0 23 14 3	0.2666	186, 290, 398, 400, 415, 421, 423, 444, 446
100:421	1 7 10 13 18 25 32 35 38 42 49 52	C_2	50×2	294, 294, 294	0 24 12 4	0.3827	282(2), 367(2), 418(2), 419(2), 420(2)
100:422	1 7 10 13 18 26 31 35 37 44 48 50	C_2	50×2	294, 294, 294	0 22 16 2	0.2471	348(2), 389(2), 418(2), 419(2)
100:423	1 7 10 13 18 26 32 35 37 42 48 51	C_1	100×1	294, 294, 294	0 22 16 2	0.2919	187, 282, 368, 374, 389, 418, 419, 420
100:424	1 7 10 13 18 26 32 35 39 44 47 50	C_1	100×1	294, 294, 294	0 21 18 1	0.1722	183, 286, 361, 400, 415, 425, 429, 448
100:425	1 7 10 13 18 26 33 35 39 43 47 50	C_1	100×1	294, 294, 294	0 20 20 0	0.2111	344, 362, 399, 414, 416, 424, 426
100:426	1 7 10 13 18 26 33 35 39 43 49 52	C_1	100×1	294, 294, 294	0 21 18 1	0.1747	232, 286, 360, 398, 415, 425, 427, 428
100:427	1 7 10 13 18 26 33 35 40 43 48 52	C_1	100×1	294, 294, 294	0 20 20 0	0.1151	192, 240, 370, 397, 426, 437
100:428	1 7 10 13 18 27 32 34 37 45 48 50	C_1	100×1	294, 294, 294	0 20 20 0	0.0479	233, 276, 368, 374, 409, 416, 426
100:429	1 7 10 13 18 27 32 35 38 44 47 50	C_2	50×2	294, 294, 294	0 22 16 2	0.0840	276(2), 367(2), 411, 416(2), 424(2)
100:430	1 7 10 13 18 27 33 35 37 43 48 50	C_2	50×2	294, 294, 294	0 20 20 0	0.0438	371(2), 377(2), 416(2), 433
100:431	1 7 10 13 18 27 33 36 42 45 48 52	D_2	25×4	219, 294, 219	0 20 20 0	0.2569	379(4), 432(2)
100:432	1 7 10 13 18 27 33 36 42 46 48 51	D_2	25×4	219, 294, 219	0 20 20 0	0.2569	377(4), 431(2)

100:414
100:413
100:412
100:411
100:410
100:409
100:420
100:419
100:418
100:417
100:416
100:415
100:426
100:425
100:424
100:423
100:422
100:421
100:432
100:431
100:430
100:429
100:428
100:427

Table A.20. (*Continued*)

Isomer	Ring spiral	Point group	NMR pattern	Vibrations	Hexagon indices	Band gap	Transformations
100:433	1 7 10 13 18 27 34 36 42 44 47 49	C_2	50×2	294, 294, 294	0 20 20 0	0.0405	379(2), 380(2), 410(2), 430
100:434	1 7 10 13 18 31 34 36 38 40 49 51	C_1	100×1	294, 294, 294	0 21 18 1	0.0927	334, 358, 385, 413, 435, 441, 443, 449
100:435	1 7 10 13 18 31 34 36 38 41 48 51	C_2	50×2	294, 294, 294	0 22 16 2	0.0563	327(2), 407(2), 434(2), 436(2)
100:436	1 7 10 13 18 31 34 36 39 41 47 51	C_1	100×1	294, 294, 294	0 21 18 1	0.0421	269, 316, 325, 408, 435, 439, 441
100:437	1 7 10 13 18 32 34 36 39 41 47 50	C_1	100×1	294, 294, 294	0 21 18 1	0.0342	196, 200, 400, 415, 427, 446, 448
100:438	1 7 10 13 19 26 30 35 37 43 48 51	D_2	25×4	219, 294, 219	0 22 16 2	0.1767	292(4), 418(4)
100:439	1 7 10 13 19 26 33 35 40 42 48 52	C_2	50×2	294, 294, 294	0 20 20 0	0.0048	268(2), 317(2), 436(2)
100:440	1 7 10 13 19 26 34 39 41 44 47 51	C_2	50×2	294, 294, 294	0 20 20 0	0.0584	297(2), 309(2), 412(2), 442
100:441	1 7 10 13 19 26 34 40 42 44 47 49	C_1	100×1	294, 294, 294	0 20 20 0	0.0641	299, 313, 357, 369, 412, 434, 436
100:442	1 7 10 13 19 27 34 38 41 44 47 51	C_2	50×2	294, 294, 294	0 20 20 0	0.0803	144(2), 303(2), 363(2), 440
100:443	1 7 10 13 19 30 34 36 38 40 49 51	C_2	50×2	294, 294, 294	0 22 16 2	0.1296	299(2), 359(2), 407(2), 434(2)
100:444	1 7 10 14 18 24 31 35 38 44 48 50	C_s	4×1, 48×2	294, 294, 294	0 22 16 2	0.1792	354, 399(2), 414, 419(2), 420(2)
100:445	1 7 10 14 18 24 32 35 38 42 48 51	D_2	25×4	219, 294, 219	0 22 16 2	0.2364	371(4), 419(4)
100:446	1 7 10 14 18 24 32 38 41 44 47 52	C_2	50×2	294, 294, 294	0 22 16 2	0.2157	199(2), 397(2), 420(2), 437(2)
100:447	1 7 10 14 19 23 27 42 44 46 48 50	C_s	6×1, 47×2	294, 294, 294	0 21 18 1	0.0698	161(2), 319
100:448	1 7 10 14 19 24 34 39 41 43 46 48	C_2	50×2	294, 294, 294	0 20 20 0	0.1141	197(2), 396, 424(2), 437(2)
100:449	1 7 10 18 23 25 27 33 40 43 46 48	D_2	25×4	219, 294, 219	0 20 20 0	0.1562	380(4), 434(4)
100:450	1 8 10 12 14 16 37 39 41 43 45 52	D_5	10×10	86, 148, 58	0 20 20 0	0.0000	

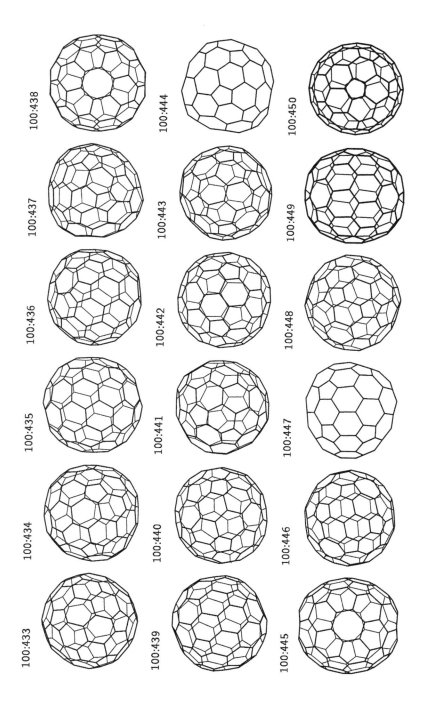

100:438

100:444

100:450

100:437

100:443

100:449

100:436

100:442

100:448

100:435

100:441

100:447

100:434

100:440

100:446

100:433

100:439

100:445

Table A.21 π electronic parameters of general fullerene isomers: C_{20} to C_{50}

Isomer	HOMO	LUMO	E_{res}	Isomer	HOMO	LUMO	E_{res}	Isomer	HOMO	LUMO	E_{res}	Isomer	HOMO	LUMO	E_{res}
20:1	.0000	.0000	.4708	34:5	.3046	.3026	.5168	38:6	.4069	.2001	.5208	40:12	.4280	.2420	.5229
24:1	.3011	.3011	.4758	34:6	.3007	.3007	.5136	38:7	.3692	.3040	.5196	40:13	.4605	.2173	.5234
								38:8	.4009	.3027	.5187	40:14	.4202	.4104	.5190
26:1	.4142	.3473	.4751	36:1	.3524	.3473	.5083	38:9	.3465	.1492	.5173	40:15	.3591	.3063	.5194
				36:2	.3473	.2103	.5144	38:10	.4246	.4069	.5186	40:16	.3656	.2692	.5197
				36:3	.3668	.3285	.5114	38:11	.3958	.2859	.5189	40:17	.4264	.2460	.5212
28:1	.4707	.3773	.4970	36:4	.3721	.3564	.5110	38:12	.4261	.4215	.5180	40:18	.4330	.2275	.5242
28:2	.4142	.4142	.4803	36:5	.3998	.2521	.5194	38:13	.4534	.3029	.5206	40:19	.3765	.3309	.5178
				36:6	.3473	.2103	.5176	38:14	.3949	.2913	.5200	40:20	.4142	.3316	.5170
30:1	.1487	.1487	.4960	36:7	.3695	.3116	.5168	38:15	.4142	.3831	.5127	40:21	.4746	.3339	.5239
30:2	.5531	.1487	.5114	36:8	.3904	.2943	.5168	38:16	.3404	.2950	.5184	40:22	.4415	.3209	.5214
30:3	.4727	.4142	.5044	36:9	.3561	.2362	.5185	38:17	.4438	.2362	.5236	40:23	.4983	.3339	.5260
				36:10	.3672	.3259	.5157					40:24	.4222	.3474	.5222
32:1	.2541	.2424	.5072	36:11	.3877	.3264	.5185	40:1	.2256	.2256	.5100	40:25	.4582	.2970	.5239
32:2	.2541	.2541	.5072	36:12	.3264	.3232	.5167	40:2	.4219	.2978	.5200	40:26	.4276	.3250	.5228
32:3	.2541	.2541	.5054	36:13	.3561	.3561	.5130	40:3	.2513	.1891	.5166	40:27	.4342	.3528	.5206
32:4	.4715	.2188	.5164	36:14	.4142	.1140	.5219	40:4	.4309	.2694	.5230	40:28	.4470	.3415	.5247
32:5	.4938	.0995	.5144	36:15	.4142	.4142	.5115	40:5	.4240	.3807	.5208	40:29	.4519	.3656	.5223
32:6	.6180	.2541	.5201					40:6	.4281	.2812	.5213	40:30	.3692	.3692	.5196
				38:1	.4115	.3803	.5127	40:7	.4285	.2400	.5212	40:31	.4617	.3004	.5232
34:1	.3230	.3003	.5075	38:2	.3473	.3473	.5155	40:8	.3482	.2040	.5187	40:32	.4276	.4234	.5166
34:2	.3304	.2907	.5111	38:3	.4278	.3588	.5157	40:9	.4196	.4079	.5202	40:33	.4450	.4142	.5192
34:3	.3664	.2447	.5176	38:4	.3990	.2563	.5182	40:10	.4535	.2745	.5239	40:34	.4313	.2997	.5227
34:4	.3347	.2210	.5162	38:5	.4269	.2867	.5199	40:11	.3339	.2191	.5223	40:35	.3568	.3385	.5202

Table A.21 (*Continued*)

Isomer	HOMO	LUMO	E_{res}	Isomer	HOMO	LUMO	E_{res}	Isomer	HOMO	LUMO	E_{res}	Isomer	HOMO	LUMO	E_{res}
40:36	.3385	.3319	.5195	42:19	.4382	.2559	.5248	42:43	.4750	.3036	.5264	44:21	.3476	.2786	.5258
40:37	.4471	.2040	.5248	42:20	.3454	.2717	.5205	42:44	.3998	.3352	.5242	44:22	.4143	.2211	.5283
40:38	.4610	.1475	.5255	42:21	.4841	.2242	.5251	42:45	.5394	.2594	.5284	44:23	.3885	.3135	.5260
40:39	.4865	.1134	.5282	42:22	.4530	.4226	.5231					44:24	.3017	.2960	.5239
40:40	.3004	.3004	.5156	42:23	.4494	.3204	.5278	44:1	.3633	.2653	.5241	44:25	.3807	.2692	.5287
				42:24	.4603	.4051	.5257	44:2	.3660	.2870	.5215	44:26	.3420	.3012	.5265
42:1	.3792	.2710	.5232	42:25	.4246	.3595	.5245	44:3	.3627	.3627	.5191	44:27	.3738	.2784	.5255
42:2	.4208	.2727	.5246	42:26	.4260	.2909	.5253	44:4	.3351	.2768	.5262	44:28	.3532	.3322	.5245
42:3	.3627	.2661	.5225	42:27	.4142	.2932	.5264	44:5	.3202	.2653	.5259	44:29	.3942	.2805	.5274
42:4	.4063	.2740	.5248	42:28	.4133	.3098	.5258	44:6	.3380	.2996	.5250	44:30	.4404	.3112	.5276
42:5	.3818	.2528	.5229	42:29	.3945	.2963	.5224	44:7	.3184	.2653	.5261	44:31	.3433	.3121	.5244
42:6	.4142	.3586	.5238	42:30	.4504	.2628	.5273	44:8	.3613	.2635	.5266	44:32	.4769	.2353	.5309
42:7	.3138	.2230	.5241	42:31	.4672	.3718	.5249	44:9	.3223	.2637	.5236	44:33	.3638	.3018	.5259
42:8	.3850	.2565	.5262	42:32	.4781	.2972	.5256	44:10	.4094	.3030	.5268	44:34	.3669	.2752	.5247
42:9	.4566	.3078	.5262	42:33	.4703	.2958	.5257	44:11	.4106	.3045	.5264	44:35	.3669	.2198	.5214
42:10	.4779	.2546	.5284	42:34	.4390	.3167	.5230	44:12	.4744	.2227	.5300	44:36	.3286	.3244	.5235
42:11	.4384	.2176	.5263	42:35	.4343	.3847	.5246	44:13	.4627	.1789	.5308	44:37	.5675	.3608	.5317
42:12	.4920	.3195	.5264	42:36	.4629	.3021	.5255	44:14	.3368	.2997	.5258	44:38	.5339	.3608	.5296
42:13	.3738	.2897	.5250	42:37	.4455	.2906	.5271	44:15	.3875	.3019	.5298	44:39	.5675	.3086	.5344
42:14	.4514	.3131	.5247	42:38	.4307	.2268	.5273	44:16	.4793	.2084	.5322	44:40	.4887	.2758	.5319
42:15	.3770	.2693	.5238	42:39	.4414	.2949	.5256	44:17	.3202	.2927	.5246	44:41	.4325	.2647	.5298
42:16	.3738	.2897	.5231	42:40	.3364	.3006	.5201	44:18	.4401	.2227	.5299	44:42	.4304	.3127	.5282
42:17	.3894	.2849	.5226	42:41	.4133	.3687	.5237	44:19	.4010	.2491	.5281	44:43	.3665	.3116	.5271
42:18	.4294	.2545	.5243	42:42	.3729	.3581	.5247	44:20	.2878	.2378	.5258	44:44	.3720	.3210	.5282

Table A.21 (*Continued*)

Isomer	HOMO	LUMO	E_{res}	Isomer	HOMO	LUMO	E_{res}	Isomer	HOMO	LUMO	E_{res}	Isomer	HOMO	LUMO	E_{res}
44:45	.3720	.3337	.5275	44:69	.4632	.3164	.5291	46:3	.3761	.2699	.5264	46:27	.3773	.2718	.5288
44:46	.4142	.3049	.5272	44:70	.4012	.3029	.5249	46:4	.3758	.3358	.5263	46:28	.4530	.2370	.5304
44:47	.4419	.3412	.5285	44:71	.3684	.3432	.5265	46:5	.3686	.2885	.5285	46:29	.3764	.3103	.5284
44:48	.4398	.3445	.5281	44:72	.6180	.1939	.5327	46:6	.3556	.3093	.5282	46:30	.3370	.2518	.5298
44:49	.3811	.3557	.5270	44:73	.5397	.1338	.5269	46:7	.3537	.1840	.5278	46:31	.3180	.3163	.5245
44:50	.4211	.3112	.5289	44:74	.4424	.3853	.5264	46:8	.3358	.1961	.5297	46:32	.4052	.2976	.5295
44:51	.4784	.3251	.5294	44:75	.5314	.2854	.5316	46:9	.3655	.2990	.5235	46:33	.3291	.3066	.5231
44:52	.4896	.3143	.5309	44:76	.5160	.1340	.5319	46:10	.3790	.3197	.5271	46:34	.4049	.3222	.5272
44:53	.4385	.2947	.5276	44:77	.4435	.3340	.5278	46:11	.4337	.2030	.5337	46:35	.3785	.2404	.5288
44:54	.4825	.3655	.5286	44:78	.4100	.3391	.5273	46:12	.3835	.2067	.5314	46:36	.3581	.2317	.5297
44:55	.5686	.2579	.5342	44:79	.4424	.3476	.5276	46:13	.3813	.1733	.5340	46:37	.3741	.3338	.5252
44:56	.4641	.3141	.5307	44:80	.3820	.3669	.5269	46:14	.3375	.1711	.5311	46:38	.4086	.3288	.5261
44:57	.4401	.3940	.5272	44:81	.4411	.2824	.5297	46:15	.3463	.2783	.5291	46:39	.4042	.3081	.5261
44:58	.4284	.2746	.5268	44:82	.3405	.3405	.5206	46:16	.3146	.2304	.5274	46:40	.3889	.2425	.5296
44:59	.4747	.3326	.5274	44:83	.3540	.3244	.5260	46:17	.3554	.3057	.5295	46:41	.3549	.2324	.5295
44:60	.4628	.3142	.5272	44:84	.4740	.2085	.5299	46:18	.3943	.2659	.5298	46:42	.3614	.2822	.5265
44:61	.3669	.3448	.5273	44:85	.4849	.2465	.5303	46:19	.3997	.2534	.5303	46:43	.3708	.3195	.5315
44:62	.4225	.3063	.5263	44:86	.4350	.2501	.5276	46:20	.3786	.2489	.5268	46:44	.3938	.3268	.5307
44:63	.4344	.3413	.5261	44:87	.4206	.3126	.5278	46:21	.3587	.2604	.5271	46:45	.4015	.3198	.5311
44:64	.4044	.2750	.5280	44:88	.4379	.3238	.5273	46:22	.2890	.2863	.5251	46:46	.3789	.3285	.5319
44:65	.4705	.2808	.5295	44:89	.5397	.2854	.5305	46:23	.3506	.2695	.5270	46:47	.3998	.3694	.5289
44:66	.3446	.2824	.5280	46:1	.3580	.3381	.5248	46:24	.4163	.2357	.5301	46:48	.3941	.3121	.5292
44:67	.3535	.3328	.5230	46:2	.4017	.3344	.5281	46:25	.3686	.2309	.5300	46:49	.4021	.3715	.5277
44:68	.4542	.3342	.5266					46:26	.3495	.2419	.5282	46:50	.4176	.2689	.5308

Table A.21 (*Continued*)

Isomer	HOMO	LUMO	E_{res}	Isomer	HOMO	LUMO	E_{res}	Isomer	HOMO	LUMO	E_{res}	Isomer	HOMO	LUMO	E_{res}
46:51	.4187	.3183	.5299	46:75	.3923	.2072	.5311	46:99	.4396	.3111	.5330	48:6	.4054	.2837	.5287
46:52	.4443	.3112	.5293	46:76	.4458	.2292	.5328	46:100	.4815	.2984	.5330	48:7	.3730	.2880	.5298
46:53	.4244	.3679	.5288	46:77	.4244	.1974	.5327	46:101	.4842	.2818	.5331	48:8	.3529	.2211	.5299
46:54	.4313	.3083	.5316	46:78	.3407	.3203	.5294	46:102	.4849	.1549	.5344	48:9	.3485	.2490	.5271
46:55	.3948	.3090	.5315	46:79	.4528	.2612	.5318	46:103	.4587	.3346	.5324	48:10	.3628	.2062	.5308
46:56	.4316	.3005	.5317	46:80	.3761	.2804	.5281	46:104	.4299	.3468	.5295	48:11	.3736	.3115	.5293
46:57	.3970	.3132	.5293	46:81	.3823	.3282	.5301	46:105	.3748	.3574	.5290	48:12	.4063	.2427	.5296
46:58	.4454	.3163	.5315	46:82	.3894	.2847	.5287	46:106	.4505	.3277	.5304	48:13	.4045	.3598	.5285
46:59	.4866	.2416	.5340	46:83	.3885	.2977	.5288	46:107	.5535	.2190	.5347	48:14	.3787	.3359	.5246
46:60	.4614	.2842	.5340	46:84	.3343	.3339	.5264	46:108	.5324	.3382	.5334	48:15	.3390	.2950	.5328
46:61	.4173	.3022	.5301	46:85	.3985	.3156	.5297	46:109	.5051	.3403	.5329	48:16	.2722	.2025	.5312
46:62	.3743	.3121	.5297	46:86	.4901	.3141	.5318	46:110	.4555	.3414	.5306	48:17	.3390	.2950	.5324
46:63	.4399	.2477	.5316	46:87	.4416	.3618	.5298	46:111	.4482	.1996	.5329	48:18	.3969	.2134	.5330
46:64	.3854	.2520	.5299	46:88	.4787	.2188	.5333	46:112	.3680	.3146	.5286	48:19	.4172	.2211	.5332
46:65	.4389	.3425	.5304	46:89	.3302	.3123	.5251	46:113	.3864	.1401	.5318	48:20	.4025	.1846	.5328
46:66	.4532	.3373	.5312	46:90	.4515	.3349	.5300	46:114	.4655	.3346	.5311	48:21	.3794	.3111	.5300
46:67	.4925	.2586	.5333	46:91	.3467	.3048	.5274	46:115	.3438	.3344	.5245	48:22	.3530	.1996	.5321
46:68	.3893	.3329	.5283	46:92	.5007	.3467	.5305	46:116	.4659	.3595	.5305	48:23	.4239	.2733	.5316
46:69	.3728	.3099	.5316	46:93	.4246	.2857	.5287	48:1	.3796	.3079	.5263	48:24	.3870	.3114	.5312
46:70	.4042	.3610	.5284	46:94	.3669	.3669	.5259	48:2	.3005	.2870	.5251	48:25	.3625	.3257	.5295
46:71	.4061	.2512	.5297	46:95	.3786	.3359	.5246	48:3	.3910	.3149	.5289	48:26	.4242	.3195	.5294
46:72	.4346	.2523	.5312	46:96	.3595	.3234	.5270	48:4	.3661	.3567	.5283	48:27	.3870	.3113	.5290
46:73	.3570	.3126	.5280	46:97	.4505	.2309	.5325	48:5	.3768	.2820	.5275	48:28	.3086	.2254	.5292
46:74	.4517	.2255	.5325	46:98	.4294	.3197	.5314					48:29	.3960	.2618	.5306

Table A.21 (*Continued*)

Isomer	HOMO	LUMO	E_{res}	Isomer	HOMO	LUMO	E_{res}	Isomer	HOMO	LUMO	E_{res}	Isomer	HOMO	LUMO	E_{res}
48:30	.3641	.2603	.5296	48:54	.4091	.3245	.5300	48:78	.3785	.3404	.5327	48:102	.4323	.2616	.5335
48:31	.3571	.2489	.5286	48:55	.4481	.3255	.5309	48:79	.3793	.2625	.5324	48:103	.4172	.3030	.5334
48:32	.2945	.2275	.5267	48:56	.4715	.2489	.5305	48:80	.4298	.3878	.5315	48:104	.3721	.3355	.5331
48:33	.3006	.2445	.5287	48:57	.3349	.3282	.5270	48:81	.4241	.2946	.5347	48:105	.4041	.3282	.5302
48:34	.3935	.2260	.5317	48:58	.3867	.2571	.5313	48:82	.3871	.3267	.5345	48:106	.4139	.3087	.5335
48:35	.3550	.3293	.5285	48:59	.4044	.3867	.5266	48:83	.3785	.3439	.5355	48:107	.3829	.3483	.5286
48:36	.3781	.3050	.5293	48:60	.4066	.2652	.5308	48:84	.3808	.3200	.5316	48:108	.3944	.2276	.5350
48:37	.3847	.2259	.5288	48:61	.3557	.2816	.5274	48:85	.3659	.2508	.5334	48:109	.3933	.3406	.5306
48:38	.3911	.2686	.5316	48:62	.2943	.2434	.5282	48:86	.3769	.3177	.5319	48:110	.3802	.2857	.5329
48:39	.3015	.2761	.5294	48:63	.3436	.2960	.5280	48:87	.3638	.2863	.5333	48:111	.3883	.2185	.5340
48:40	.2945	.2573	.5273	48:64	.4026	.3871	.5320	48:88	.3886	.3142	.5335	48:112	.3690	.2743	.5319
48:41	.3390	.2950	.5294	48:65	.4027	.3575	.5336	48:89	.4119	.2912	.5308	48:113	.3446	.2245	.5326
48:42	.3970	.2647	.5332	48:66	.3956	.2630	.5337	48:90	.3843	.3321	.5312	48:114	.3707	.2876	.5283
48:43	.3420	.3161	.5288	48:67	.3956	.2581	.5333	48:91	.4280	.2398	.5362	48:115	.4022	.3325	.5307
48:44	.3912	.3160	.5313	48:68	.4054	.3962	.5311	48:92	.4452	.3748	.5327	48:116	.3605	.2846	.5321
48:45	.3273	.3021	.5300	48:69	.4059	.3400	.5308	48:93	.4402	.3077	.5339	48:117	.4003	.2832	.5326
48:46	.3602	.3317	.5292	48:70	.4043	.3951	.5277	48:94	.4349	.3134	.5332	48:118	.3610	.2707	.5335
48:47	.3729	.2416	.5314	48:71	.4112	.2041	.5332	48:95	.4654	.3856	.5322	48:119	.4033	.3154	.5312
48:48	.3603	.3100	.5298	48:72	.3967	.3227	.5308	48:96	.4550	.3013	.5318	48:120	.3889	.3209	.5320
48:49	.3407	.3147	.5258	48:73	.4112	.2603	.5341	48:97	.4404	.3373	.5325	48:121	.3670	.2869	.5336
48:50	.3498	.2777	.5283	48:74	.4139	.3386	.5316	48:98	.4010	.3045	.5318	48:122	.3901	.2813	.5334
48:51	.3697	.2660	.5311	48:75	.3595	.2732	.5307	48:99	.3692	.3239	.5321	48:123	.3965	.3132	.5304
48:52	.3935	.3543	.5285	48:76	.4393	.3201	.5321	48:100	.3590	.2895	.5324	48:124	.3590	.3075	.5304
48:53	.3563	.2535	.5298	48:77	.3818	.2503	.5325	48:101	.3981	.3200	.5314	48:125	.4688	.2103	.5330

Table A.21 (Continued)

Isomer	HOMO	LUMO	E_{res}	Isomer	HOMO	LUMO	E_{res}	Isomer	HOMO	LUMO	E_{res}	Isomer	HOMO	LUMO	E_{res}
48:126	.4443	.2194	.5318	48:150	.4137	.3271	.5325	48:174	.3782	.3529	.5296	48:198	.4170	.1718	.5351
48:127	.4278	.2417	.5341	48:151	.3683	.3261	.5286	48:175	.4708	.2420	.5353	48:199	.4681	.3848	.5338
48:128	.4151	.2433	.5320	48:152	.3715	.3242	.5315	48:176	.3957	.2430	.5343	50:1	.2711	.2711	.5197
48:129	.3840	.3613	.5286	48:153	.3454	.2679	.5309	48:177	.2619	.2563	.5301	50:2	.3858	.3532	.5290
48:130	.3882	.2632	.5325	48:154	.3828	.2793	.5302	48:178	.4613	.2584	.5331	50:3	.3709	.3709	.5273
48:131	.3692	.2498	.5317	48:155	.3809	.3623	.5294	48:179	.3714	.3223	.5308	50:4	.4395	.3165	.5328
48:132	.3742	.3220	.5270	48:156	.3906	.3785	.5307	48:180	.5152	.2555	.5359	50:5	.3921	.3143	.5314
48:133	.3893	.3241	.5325	48:157	.4365	.2203	.5351	48:181	.3608	.3115	.5318	50:6	.3236	.2546	.5276
48:134	.4170	.2383	.5321	48:158	.4431	.2325	.5345	48:182	.3935	.2672	.5342	50:7	.3894	.2757	.5322
48:135	.3690	.3265	.5322	48:159	.4056	.2592	.5338	48:183	.4163	.2451	.5343	50:8	.4008	.2280	.5331
48:136	.3853	.2235	.5313	48:160	.5358	.2547	.5381	48:184	.3764	.2871	.5342	50:9	.3716	.2178	.5316
48:137	.3674	.2493	.5314	48:161	.5517	.1767	.5379	48:185	.5106	.2679	.5362	50:10	.3425	.2786	.5302
48:138	.4715	.3504	.5333	48:162	.4857	.2756	.5364	48:186	.4142	.2008	.5317	50:11	.3521	.2278	.5289
48:139	.4496	.2426	.5354	48:163	.5610	.1767	.5378	48:187	.4394	.3567	.5333	50:12	.3624	.2763	.5302
48:140	.5001	.2579	.5349	48:164	.3643	.2916	.5324	48:188	.5259	.3764	.5340	50:13	.3924	.2604	.5312
48:141	.4474	.2385	.5349	48:165	.4955	.2463	.5356	48:189	.4142	.4142	.5295	50:14	.3945	.3172	.5316
48:142	.3690	.3518	.5293	48:166	.5098	.1905	.5360	48:190	.4101	.2619	.5323	50:15	.3493	.3264	.5295
48:143	.3952	.3500	.5315	48:167	.3953	.3488	.5326	48:191	.3543	.2555	.5321	50:16	.4355	.3124	.5305
48:144	.4170	.2180	.5321	48:168	.4641	.2032	.5362	48:192	.4651	.2952	.5346	50:17	.4397	.3080	.5315
48:145	.4256	.3313	.5304	48:169	.3758	.2050	.5338	48:193	.4206	.2445	.5339	50:18	.2548	.2171	.5298
48:146	.3658	.2446	.5307	48:170	.4129	.1883	.5343	48:194	.4506	.2223	.5338	50:19	.4381	.1914	.5352
48:147	.3822	.2989	.5291	48:171	.4708	.3242	.5359	48:195	.4145	.3433	.5316	50:20	.3933	.2183	.5341
48:148	.4374	.3243	.5312	48:172	.3481	.3220	.5326	48:196	.4663	.3677	.5343	50:21	.3657	.2217	.5335
48:149	.4386	.2959	.5339	48:173	.5191	.2461	.5367	48:197	.4498	.2030	.5356				

Table A.21 (*Continued*)

Isomer	HOMO	LUMO	E_{res}	Isomer	HOMO	LUMO	E_{res}	Isomer	HOMO	LUMO	E_{res}	Isomer	HOMO	LUMO	E_{res}
50:22	.4158	.2934	.5318	50:46	.3908	.2846	.5318	50:70	.3371	.2732	.5303	50:94	.4104	.3134	.5331
50:23	.3763	.2694	.5339	50:47	.4067	.2546	.5325	50:71	.3612	.2612	.5299	50:95	.3651	.3643	.5313
50:24	.2878	.2809	.5301	50:48	.3876	.2646	.5305	50:72	.3519	.2785	.5307	50:96	.3853	.2694	.5352
50:25	.4228	.2896	.5338	50:49	.3853	.3133	.5311	50:73	.4210	.2650	.5319	50:97	.4044	.3120	.5352
50:26	.4607	.2230	.5342	50:50	.3960	.3382	.5311	50:74	.4422	.3438	.5354	50:98	.4053	.3364	.5357
50:27	.4229	.2664	.5312	50:51	.3345	.2924	.5302	50:75	.3887	.3552	.5327	50:99	.3804	.3083	.5353
50:28	.4398	.3581	.5313	50:52	.3865	.2818	.5334	50:76	.3756	.3130	.5337	50:100	.3847	.3318	.5361
50:29	.2963	.2646	.5311	50:53	.3540	.2750	.5314	50:77	.4218	.2897	.5354	50:101	.3646	.3062	.5323
50:30	.4221	.2883	.5323	50:54	.3914	.3099	.5313	50:78	.4422	.3832	.5331	50:102	.3953	.3030	.5340
50:31	.2968	.2298	.5307	50:55	.3666	.2944	.5335	50:79	.4259	.3015	.5328	50:103	.4107	.3815	.5319
50:32	.3605	.2160	.5308	50:56	.3730	.3310	.5294	50:80	.4453	.4023	.5312	50:104	.4217	.3299	.5359
50:33	.3874	.2792	.5311	50:57	.3084	.3016	.5297	50:81	.4129	.2798	.5334	50:105	.4229	.2883	.5357
50:34	.3138	.2717	.5296	50:58	.3599	.2993	.5326	50:82	.4484	.2833	.5357	50:106	.3638	.2920	.5331
50:35	.3645	.2734	.5302	50:59	.3617	.3283	.5312	50:83	.3975	.2223	.5352	50:107	.4174	.2175	.5372
50:36	.3838	.3001	.5322	50:60	.3661	.2914	.5313	50:84	.4162	.3594	.5311	50:108	.4172	.2794	.5355
50:37	.3812	.2330	.5317	50:61	.2968	.2489	.5289	50:85	.4202	.1924	.5354	50:109	.4475	.3440	.5341
50:38	.3522	.2817	.5295	50:62	.3855	.2803	.5312	50:86	.4228	.2950	.5335	50:110	.4477	.2476	.5357
50:39	.3621	.2473	.5321	50:63	.4057	.2544	.5327	50:87	.4008	.3228	.5340	50:111	.4211	.2003	.5371
50:40	.3740	.2973	.5327	50:64	.4358	.3145	.5323	50:88	.4485	.1718	.5359	50:112	.3911	.3382	.5325
50:41	.4158	.2744	.5303	50:65	.4054	.3044	.5303	50:89	.4062	.3369	.5328	50:113	.3454	.3001	.5324
50:42	.4088	.2917	.5332	50:66	.3521	.2865	.5321	50:90	.4060	.3490	.5299	50:114	.4042	.1986	.5365
50:43	.3463	.2419	.5294	50:67	.3565	.3210	.5294	50:91	.4212	.2488	.5336	50:115	.4087	.2959	.5348
50:44	.3623	.2981	.5321	50:68	.3893	.2852	.5306	50:92	.4137	.3447	.5322	50:116	.4452	.2853	.5355
50:45	.3527	.2712	.5290	50:69	.3190	.2858	.5296	50:93	.3873	.3607	.5319	50:117	.3774	.3018	.5343

Table A.21 (*Continued*)

Isomer	HOMO	LUMO	E_{res}	Isomer	HOMO	LUMO	E_{res}	Isomer	HOMO	LUMO	E_{res}	Isomer	HOMO	LUMO	E_{res}
50:118	.3858	.3373	.5321	50:142	.3487	.3090	.5321	50:166	.4655	.2653	.5364	50:190	.4287	.2449	.5348
50:119	.3637	.3430	.5329	50:143	.3521	.2055	.5334	50:167	.3766	.2693	.5345	50:191	.3975	.3366	.5334
50:120	.3942	.2944	.5344	50:144	.3987	.2772	.5345	50:168	.3579	.3117	.5335	50:192	.3924	.3052	.5314
50:121	.3667	.2686	.5345	50:145	.4073	.2570	.5346	50:169	.3596	.3063	.5310	50:193	.4074	.2947	.5332
50:122	.4738	.2628	.5360	50:146	.3651	.2636	.5331	50:170	.3923	.3075	.5348	50:194	.3973	.3371	.5323
50:123	.3783	.3141	.5339	50:147	.4089	.2918	.5325	50:171	.3563	.2661	.5341	50:195	.3592	.3307	.5327
50:124	.4081	.3427	.5318	50:148	.4070	.3410	.5299	50:172	.4373	.2682	.5362	50:196	.3513	.3249	.5300
50:125	.4142	.3464	.5312	50:149	.4057	.3681	.5315	50:173	.3411	.2786	.5344	50:197	.4112	.3009	.5333
50:126	.3767	.3633	.5307	50:150	.3711	.3680	.5328	50:174	.3081	.2677	.5323	50:198	.4050	.3390	.5330
50:127	.4332	.3254	.5337	50:151	.3815	.2830	.5330	50:175	.4566	.2630	.5330	50:199	.4307	.2371	.5370
50:128	.3941	.3134	.5316	50:152	.4087	.3286	.5324	50:176	.4187	.2483	.5335	50:200	.4227	.3172	.5356
50:129	.4350	.2649	.5341	50:153	.4271	.2549	.5356	50:177	.4073	.2798	.5353	50:201	.4250	.3496	.5344
50:130	.3677	.2829	.5330	50:154	.4225	.2813	.5342	50:178	.3788	.2704	.5321	50:202	.4226	.2424	.5372
50:131	.4080	.3277	.5313	50:155	.4074	.2936	.5320	50:179	.3346	.3030	.5325	50:203	.4128	.3842	.5332
50:132	.4105	.2169	.5343	50:156	.3926	.3601	.5323	50:180	.4741	.2292	.5357	50:204	.3933	.3803	.5315
50:133	.3472	.3085	.5327	50:157	.4327	.1932	.5333	50:181	.3928	.2696	.5315	50:205	.2791	.2584	.5318
50:134	.3376	.2916	.5324	50:158	.4337	.2216	.5357	50:182	.3454	.3342	.5321	50:206	.3952	.2807	.5321
50:135	.3670	.3156	.5335	50:159	.3737	.3690	.5288	50:183	.3416	.3360	.5318	50:207	.4023	.1477	.5364
50:136	.4361	.3049	.5326	50:160	.3917	.3585	.5301	50:184	.3523	.3377	.5334	50:208	.4751	.2397	.5374
50:137	.3872	.3163	.5333	50:161	.4038	.3610	.5311	50:185	.4128	.2704	.5310	50:209	.3860	.3733	.5319
50:138	.4582	.3572	.5336	50:162	.4272	.3344	.5321	50:186	.3906	.2700	.5337	50:210	.3117	.2877	.5322
50:139	.3936	.2513	.5347	50:163	.3649	.3550	.5320	50:187	.4184	.2314	.5345	50:211	.3580	.3156	.5333
50:140	.3642	.3103	.5322	50:164	.4188	.3238	.5331	50:188	.3547	.3281	.5342	50:212	.4066	.2907	.5358
50:141	.4146	.3469	.5329	50:165	.4252	.3329	.5336	50:189	.3890	.3447	.5348	50:213	.4116	.2829	.5366

Table A.21 (*Continued*)

Isomer	HOMO	LUMO	E_{res}	Isomer	HOMO	LUMO	E_{res}	Isomer	HOMO	LUMO	E_{res}	Isomer	HOMO	LUMO	E_{res}
50:214	.3964	.1442	.5372	50:229	.3600	.2886	.5351	50:244	.3718	.2905	.5332	50:259	.3711	.3509	.5326
50:215	.4274	.2106	.5375	50:230	.3960	.3091	.5343	50:245	.3946	.3026	.5320	50:260	.4109	.2708	.5380
50:216	.3923	.2894	.5367	50:231	.3154	.2720	.5335	50:246	.4078	.3327	.5362	50:261	.4497	.2121	.5376
50:217	.3637	.2411	.5339	50:232	.3588	.2617	.5341	50:247	.4261	.3368	.5353	50:262	.4434	.2339	.5391
50:218	.3892	.2075	.5353	50:233	.3980	.2957	.5356	50:248	.4132	.2846	.5369	50:263	.6180	.2539	.5407
50:219	.3515	.2918	.5353	50:234	.3976	.2344	.5364	50:249	.3743	.3304	.5338	50:264	.4934	.3344	.5390
50:220	.3551	.2850	.5352	50:235	.4106	.3585	.5327	50:250	.4228	.3364	.5327	50:265	.4246	.2320	.5367
50:221	.3896	.3365	.5368	50:236	.3531	.2440	.5351	50:251	.4075	.3165	.5347	50:266	.4943	.2461	.5399
50:222	.4270	.2426	.5376	50:237	.4116	.2446	.5370	50:252	.3983	.3178	.5354	50:267	.4809	.2521	.5388
50:223	.3463	.2685	.5337	50:238	.4559	.2277	.5360	50:253	.3184	.3151	.5319	50:268	.4312	.2343	.5379
50:224	.3769	.2669	.5349	50:239	.4268	.2086	.5366	50:254	.3919	.3104	.5321	50:269	.2950	.2791	.5332
50:225	.3795	.2753	.5366	50:240	.4485	.2712	.5370	50:255	.4109	.2677	.5344	50:270	.6180	.1501	.5414
50:226	.4716	.1441	.5393	50:241	.4090	.3263	.5358	50:256	.4542	.2146	.5358	50:271	.4142	.3111	.5391
50:227	.3776	.3344	.5317	50:242	.3943	.2886	.5305	50:257	.4228	.2707	.5324				
50:228	.3556	.3079	.5343	50:243	.3931	.3004	.5334	50:258	.3753	.3482	.5268				

Table A.22 π-electronic parameters of isolated-pentagon fullerene isomers: C_{60} to C_{100}

Isomer	HOMO	LUMO	E_{res}
60:1	.6180	$\overline{.1386}$	.5527
70:1	.5293	.0000	.5545
72:1	.5637	$\overline{.1386}$	.5565
74:1	.2950	.1919	.5517
76:1	.4139	.0704	.5551
76:2	.2153	.2153	.5482
78:1	.3683	.1152	.5552
78:2	.4546	.1065	.5557
78:3	.3581	.1778	.5528
78:4	.5157	$\overline{.1176}$	.5578
78:5	.3581	.2851	.5490
80:1	.2470	.1742	.5547
80:2	.3246	.1497	.5552
80:3	.3474	.3136	.5521
80:4	.2999	.1648	.5538
80:5	.3955	.2968	.5500
80:6	.2968	.2968	.5473
80:7	.2739	.2739	.5438
82:1	.3355	.1860	.5552
82:2	.3963	.0650	.5564
82:3	.3994	.1426	.5552
82:4	.4248	.1798	.5541
82:5	.4148	.2848	.5522
82:6	.3819	.3136	.5505
82:7	.3571	.3571	.5519
82:8	.3851	.3383	.5484
82:9	.3795	.3635	.5489
84:1	.4975	$\overline{.1168}$	.5590
84:2	.4647	.1125	.5570
84:3	.2905	.2714	.5530
84:4	.4381	.0862	.5564
84:5	.3861	.1458	.5546
84:6	.3696	.1804	.5555
84:7	.3696	.1804	.5527
84:8	.3690	.1914	.5541
84:9	.3690	.3133	.5531
84:10	.3667	.2751	.5523
84:11	.4359	.1818	.5553
84:12	.4247	.2083	.5540
84:13	.4140	.3152	.5526
84:14	.4713	.0659	.5566
84:15	.3964	.1772	.5540
84:16	.4651	.1282	.5559
84:17	.4466	.2721	.5546
84:18	.4563	.1278	.5567
84:19	.4142	.2281	.5552
84:20	.5720	$\overline{.1242}$	.5588
84:21	.3393	.2012	.5522
84:22	.4600	.1151	.5549
84:23	.4600	.1151	.5546
84:24	.5293	.0000	.5565
86:1	.2759	.2470	.5544
86:2	.4327	.0563	.5576
86:3	.3497	.2791	.5545
86:4	.3323	.2607	.5546
86:5	.3765	.2146	.5541
86:6	.3323	.1655	.5553
86:7	.3582	.3225	.5529
86:8	.3355	.2856	.5510
86:9	.3645	.3594	.5490
86:10	.3140	.2374	.5544
86:11	.3700	.1996	.5546
86:12	.3372	.2093	.5543
86:13	.3376	.2094	.5527
86:14	.2581	.2178	.5523
86:15	.3647	.2133	.5522
86:16	.4597	.0510	.5571
86:17	.3979	.1641	.5557
86:18	.3932	.1451	.5555
86:19	.3694	.1727	.5512
88:1	.3731	.0612	.5582
88:2	.3880	.0416	.5579
88:3	.3742	.2156	.5562
88:4	.3770	.1867	.5554
88:5	.3731	.1215	.5574
88:6	.2822	.2383	.5545
88:7	.4106	.1650	.5561
88:8	.2697	.2290	.5548
88:9	.2642	.2295	.5525
88:10	.3699	.1319	.5573
88:11	.4120	.0881	.5569
88:12	.3818	.2797	.5542
88:13	.3752	.2741	.5552
88:14	.3815	.3505	.5538
88:15	.3657	.2275	.5547
88:16	.3439	.3151	.5541
88:17	.4232	.1789	.5558

Table A.22 (*Continued*)

Isomer	HOMO	LUMO	E_{res}	Isomer	HOMO	LUMO	E_{res}	Isomer	HOMO	LUMO	E_{res}	Isomer	HOMO	LUMO	E_{res}
88:18	.3791	.3387	.5534	90:6	.3595	.2559	.5557	90:30	.4165	.1372	.5562	92:7	.3617	.1074	.5578
88:19	.3492	.3374	.5536	90:7	.3552	.2819	.5549	90:31	.4356	.0614	.5576	92:8	.3372	.2155	.5559
88:20	.3932	.1212	.5566	90:8	.3279	.1923	.5558	90:32	.4423	.1415	.5565	92:9	.3430	.1371	.5567
88:21	.3340	.2199	.5552	90:9	.3901	.3019	.5545	90:33	.4698	.0538	.5581	92:10	.3649	.2608	.5545
88:22	.3180	.2261	.5547	90:10	.4080	.3333	.5525	90:34	.4590	.1188	.5580	92:11	.3738	.2855	.5554
88:23	.4274	.0511	.5579	90:11	.3284	.2204	.5557	90:35	.4678	.1545	.5561	92:12	.3343	.1744	.5564
88:24	.4080	.2480	.5550	90:12	.3458	.1890	.5556	90:36	.4277	.1822	.5554	92:13	.2870	.2120	.5562
88:25	.3724	.3093	.5536	90:13	.4277	.2581	.5560	90:37	.4591	.2632	.5560	92:14	.4296	.2249	.5561
88:26	.3746	.2922	.5518	90:14	.4153	.1172	.5580	90:38	.4218	.2874	.5542	92:15	.3971	.3465	.5538
88:27	.3627	.3162	.5534	90:15	.4491	.0516	.5581	90:39	.4336	.0167	.5580	92:16	.4045	.2506	.5533
88:28	.3133	.2511	.5527	90:16	.5289	.1210	.5599	90:40	.4591	.2670	.5543	92:17	.3835	.2759	.5556
88:29	.3850	.2097	.5547	90:17	.4356	.0967	.5575	90:41	.4463	.3649	.5520	92:18	.3510	.2217	.5556
88:30	.3859	.2812	.5531	90:18	.3978	.0987	.5574	90:42	.4512	.3414	.5531	92:19	.3460	.2664	.5540
88:31	.4063	.3480	.5519	90:19	.4144	.2970	.5539	90:43	.3918	.3847	.5509	92:20	.3541	.2798	.5557
88:32	.3675	.3515	.5515	90:20	.3931	.1888	.5558	90:44	.4463	.2646	.5530	92:21	.4444	.0281	.5585
88:33	.3884	.1155	.5559	90:21	.4055	.2889	.5527	90:45	.4564	.2045	.5556	92:22	.3020	.2415	.5558
88:34	.1798	.1798	.5527	90:22	.3943	.3325	.5538	90:46	.4336	.1969	.5560	92:23	.4304	.3023	.5554
88:35	.3788	.3235	.5496	90:23	.3646	.3009	.5532					92:24	.4152	.3662	.5536
				90:24	.3767	.1771	.5559	92:1	.2795	.1179	.5582	92:25	.4309	.2756	.5553
90:1	.4142	.0846	.5598	90:25	.2861	.1766	.5540	92:2	.2530	.2215	.5561	92:26	.4170	.0749	.5578
90:2	.4904	.1084	.5599	90:26	.4259	.2036	.5555	92:3	.3998	.0532	.5589	92:27	.3477	.2729	.5551
90:3	.4193	.1107	.5581	90:27	.4605	.1070	.5575	92:4	.4542	.1681	.5579	92:28	.4675	.0080	.5588
90:4	.3853	.1906	.5571	90:28	.3658	.0848	.5560	92:5	.4191	.1898	.5574	92:29	.4094	.2750	.5552
90:5	.3176	.1836	.5557	90:29	.4503	.0962	.5574	92:6	.3476	.3236	.5549	92:30	.3543	.1738	.5560

Table A.22 (Continued)

Isomer	HOMO	LUMO	E_{res}	Isomer	HOMO	LUMO	E_{res}	Isomer	HOMO	LUMO	E_{res}	Isomer	HOMO	LUMO	E_{res}
92:31	.3919	.3092	.5541	92:55	.4391	.2372	.5552	92:79	.3107	.2964	.5547	94:16	.3516	.1548	.5571
92:32	.3300	.2960	.5532	92:56	.3736	.2125	.5556	92:80	.3154	.2964	.5546	94:17	.3869	.0395	.5590
92:33	.3578	.2608	.5556	92:57	.4500	.2119	.5561	92:81	.3582	.1497	.5559	94:18	.3471	.1681	.5562
92:34	.3541	.2859	.5543	92:58	.3834	.2162	.5556	92:82	.3582	.0950	.5566	94:19	.3840	.2727	.5562
92:35	.3456	.2415	.5538	92:59	.3133	.2957	.5533	92:83	.2964	.2211	.5549	94:20	.2840	.2718	.5551
92:36	.3921	.3176	.5529	92:60	.3757	.2793	.5541	92:84	.4160	.1014	.5562	94:21	.2855	.2799	.5557
92:37	.4516	.0433	.5584	92:61	.4133	.2809	.5539	92:85	.3844	.3449	.5490	94:22	.2855	.1922	.5563
92:38	.4137	.1153	.5565	92:62	.3133	.3118	.5536	92:86	.3996	.1497	.5530	94:23	.2615	.2377	.5561
92:39	.4149	.1279	.5571	92:63	.3464	.2346	.5548					94:24	.3571	.3210	.5549
92:40	.3416	.2061	.5564	92:64	.3844	.3449	.5516	94:1	.2942	.1530	.5582	94:25	.3439	.3289	.5553
92:41	.4115	.0970	.5578	92:65	.3331	.2968	.5527	94:2	.3792	.0456	.5589	94:26	.3811	.1286	.5582
92:42	.4162	.2914	.5548	92:66	.3888	.2208	.5530	94:3	.3659	.1305	.5586	94:27	.3990	.2448	.5563
92:43	.4156	.1743	.5564	92:67	.3594	.2727	.5528	94:4	.4259	.1304	.5587	94:28	.3986	.1230	.5576
92:44	.4383	.2179	.5541	92:68	.3136	.2345	.5533	94:5	.4319	.1633	.5572	94:29	.3743	.2741	.5546
92:45	.3362	.1116	.5565	92:69	.3589	.3045	.5543	94:6	.3588	.1436	.5581	94:30	.4225	.0323	.5591
92:46	.3029	.1408	.5558	92:70	.3751	.2172	.5553	94:7	.4188	.1570	.5577	94:31	.4159	.1511	.5571
92:47	.3473	.2090	.5546	92:71	.4133	.1547	.5560	94:8	.3271	.2479	.5557	94:32	.3588	.1776	.5569
92:48	.3260	.2346	.5560	92:72	.3119	.2145	.5550	94:9	.4110	.1169	.5577	94:33	.3910	.2236	.5548
92:49	.4155	.1063	.5579	92:73	.3463	.2096	.5555	94:10	.2928	.2021	.5566	94:34	.3912	.1314	.5569
92:50	.4648	.1038	.5572	92:74	.2529	.2193	.5536	94:11	.2668	.2362	.5543	94:35	.3580	.1537	.5577
92:51	.4537	.1650	.5578	92:75	.3015	.2251	.5537	94:12	.3030	.2628	.5553	94:36	.3948	.2477	.5558
92:52	.3051	.2429	.5555	92:76	.3266	.2375	.5547	94:13	.2935	.1601	.5568	94:37	.4126	.0818	.5582
92:53	.4132	.2458	.5556	92:77	.3358	.2346	.5540	94:14	.2998	.2837	.5559	94:38	.3412	.1482	.5571
92:54	.4137	.2392	.5564	92:78	.2964	.2341	.5552	94:15	.3714	.1292	.5574	94:39	.4161	.1160	.5572

Table A.22 (*Continued*)

Isomer	HOMO	LUMO	E_{res}	Isomer	HOMO	LUMO	E_{res}	Isomer	HOMO	LUMO	E_{res}	Isomer	HOMO	LUMO	E_{res}
94:40	.4146	.0359	.5591	94:64	.3270	.2595	.5561	94:88	.3412	.2178	.5558	94:112	.3795	.3508	.5536
94:41	.3610	.1240	.5571	94:65	.3661	.1840	.5566	94:89	.3156	.1114	.5562	94:113	.2934	.2392	.5559
94:42	.4456	.0517	.5584	94:66	.3002	.2131	.5566	94:90	.3400	.2702	.5541	94:114	.3627	.3627	.5548
94:43	.4370	.0702	.5580	94:67	.3641	.1970	.5561	94:91	.4266	.1485	.5569	94:115	.3420	.2366	.5544
94:44	.3730	.1252	.5580	94:68	.3697	.2330	.5567	94:92	.3437	.1485	.5563	94:116	.4005	.2571	.5548
94:45	.3470	.2682	.5556	94:69	.2760	.2376	.5556	94:93	.3696	.1356	.5574	94:117	.4132	.3831	.5523
94:46	.3914	.2125	.5569	94:70	.2678	.2425	.5558	94:94	.2664	.1500	.5559	94:118	.4038	.3479	.5536
94:47	.3892	.2747	.5553	94:71	.3962	.2552	.5560	94:95	.4214	.2943	.5547	94:119	.3821	.3066	.5544
94:48	.3829	.2411	.5561	94:72	.4153	.2213	.5554	94:96	.2016	.1959	.5539	94:120	.4114	.3378	.5528
94:49	.3454	.2593	.5547	94:73	.3697	.2600	.5553	94:97	.3395	.1961	.5560	94:121	.3399	.2516	.5533
94:50	.3740	.2604	.5565	94:74	.3314	.3037	.5542	94:98	.2963	.2262	.5554	94:122	.3786	.2489	.5554
94:51	.3948	.2153	.5556	94:75	.3738	.3061	.5552	94:99	.2853	.2584	.5539	94:123	.3772	.3323	.5546
94:52	.4199	.1856	.5551	94:76	.3535	.2643	.5545	94:100	.2965	.2168	.5549	94:124	.3814	.3628	.5539
94:53	.3472	.1835	.5567	94:77	.3555	.1874	.5558	94:101	.3224	.3129	.5543	94:125	.3445	.3310	.5536
94:54	.4078	.3049	.5530	94:78	.3910	.1371	.5572	94:102	.3228	.2471	.5542	94:126	.3399	.2154	.5548
94:55	.3751	.2009	.5563	94:79	.3550	.1391	.5566	94:103	.3466	.2292	.5560	94:127	.3541	.2393	.5558
94:56	.3569	.2716	.5548	94:80	.4147	.1684	.5567	94:104	.2730	.2361	.5540	94:128	.3523	.2408	.5551
94:57	.4098	.2629	.5548	94:81	.3697	.2115	.5560	94:105	.3723	.1572	.5568	94:129	.2879	.2344	.5543
94:58	.4059	.2170	.5569	94:82	.3722	.2804	.5557	94:106	.2887	.2097	.5561	94:130	.3640	.2771	.5546
94:59	.3994	.2400	.5549	94:83	.3956	.0980	.5579	94:107	.3734	.0658	.5581	94:131	.3923	.3380	.5542
94:60	.3728	.2718	.5550	94:84	.3652	.1279	.5585	94:108	.3140	.2776	.5557	94:132	.4246	.3589	.5538
94:61	.3950	.1013	.5579	94:85	.3803	.2696	.5565	94:109	.3140	.1529	.5567	94:133	.3816	.1287	.5565
94:62	.3359	.2084	.5564	94:86	.3556	.1391	.5579	94:110	.3210	.3210	.5526	94:134	.4238	.4238	.5533
94:63	.3955	.0712	.5582	94:87	.3171	.2502	.5557	94:111	.3765	.2961	.5551				

Table A.22 (Continued)

Isomer	HOMO	LUMO	E_{res}	Isomer	HOMO	LUMO	E_{res}	Isomer	HOMO	LUMO	E_{res}	Isomer	HOMO	LUMO	E_{res}
96:1	.4467	.0906	.5608	96:25	.3723	.3010	.5561	96:49	.3691	.1106	.5582	96:73	.3022	.1849	.5569
96:2	.2977	.0872	.5593	96:26	.4034	.2696	.5559	96:50	.3403	.2455	.5563	96:74	.3949	.0735	.5588
96:3	.4328	.0844	.5609	96:27	.4599	.1134	.5585	96:51	.3902	.2133	.5567	96:75	.4070	.2549	.5561
96:4	.4769	.0969	.5608	96:28	.4158	.1224	.5590	96:52	.3805	.2537	.5564	96:76	.3837	.1905	.5570
96:5	.4145	.1092	.5592	96:29	.4638	.2548	.5573	96:53	.3704	.2703	.5549	96:77	.2892	.2081	.5569
96:6	.3858	.1229	.5590	96:30	.5072	.1151	.5608	96:54	.3956	.2623	.5568	96:78	.3114	.1937	.5568
96:7	.2789	.2402	.5565	96:31	.4653	.1260	.5589	96:55	.2835	.2220	.5550	96:79	.4063	.1228	.5577
96:8	.4287	.0835	.5592	96:32	.4243	.0318	.5593	96:56	.3579	.2572	.5546	96:80	.3834	.1115	.5590
96:9	.2789	.2447	.5568	96:33	.5054	.1265	.5609	96:57	.3731	.2880	.5555	96:81	.3493	.3293	.5541
96:10	.3114	.1893	.5559	96:34	.2889	.2606	.5560	96:58	.3673	.2850	.5547	96:82	.3281	.2015	.5564
96:11	.3125	.1700	.5582	96:35	.3348	.2632	.5560	96:59	.3686	.2839	.5553	96:83	.4343	.1893	.5573
96:12	.3471	.2003	.5564	96:36	.3829	.2611	.5566	96:60	.3158	.2330	.5548	96:84	.4236	.2128	.5559
96:13	.3825	.1831	.5581	96:37	.4017	.2149	.5563	96:61	.3677	.2053	.5564	96:85	.3101	.2148	.5559
96:14	.2859	.2318	.5569	96:38	.4414	.1729	.5572	96:62	.3685	.2535	.5568	96:86	.3601	.2410	.5569
96:15	.3033	.2702	.5556	96:39	.3423	.1897	.5569	96:63	.3404	.1872	.5569	96:87	.4358	.1105	.5582
96:16	.3098	.2895	.5561	96:40	.3825	.1483	.5572	96:64	.3732	.2181	.5566	96:88	.4210	.0877	.5584
96:17	.3496	.2862	.5546	96:41	.4444	.1483	.5578	96:65	.3741	.3127	.5553	96:89	.3406	.3166	.5542
96:18	.3104	.2675	.5559	96:42	.3965	.2725	.5564	96:66	.3826	.1891	.5568	96:90	.3480	.1543	.5569
96:19	.4026	.2624	.5570	96:43	.3743	.1148	.5579	96:67	.3432	.2260	.5554	96:91	.3488	.2558	.5560
96:20	.3014	.2956	.5552	96:44	.3481	.2481	.5560	96:68	.3817	.2606	.5559	96:92	.3947	.1867	.5560
96:21	.3494	.1621	.5584	96:45	.3479	.2135	.5573	96:69	.4505	.2267	.5568	96:93	.3098	.1812	.5574
96:22	.4270	.0503	.5591	96:46	.3658	.1765	.5571	96:70	.4441	.2132	.5564	96:94	.3886	.1181	.5579
96:23	.3987	.2119	.5567	96:47	.4141	.1022	.5580	96:71	.2575	.2257	.5563	96:95	.2793	.2397	.5555
96:24	.3793	.1720	.5582	96:48	.3303	.1152	.5579	96:72	.2816	.2146	.5551	96:96	.3110	.1532	.5580

Table A.22 (*Continued*)

Isomer	HOMO	LUMO	E_{res}	Isomer	HOMO	LUMO	E_{res}	Isomer	HOMO	LUMO	E_{res}	Isomer	HOMO	LUMO	E_{res}
96:97	.3107	.2202	.5566	96:121	.3022	.2718	.5543	96:145	.4007	.1294	.5577	96:169	.2979	.2397	.5554
96:98	.2811	.2332	.5552	96:122	.3584	.2633	.5552	96:146	.4443	.1545	.5583	96:170	.3355	.2948	.5525
96:99	.4161	.0661	.5588	96:123	.3451	.2758	.5550	96:147	.3818	.2123	.5565	96:171	.3137	.2304	.5570
96:100	.4257	.2076	.5566	96:124	.3990	.3356	.5548	96:148	.3834	.0368	.5592	96:172	.3462	.1907	.5569
96:101	.4207	.2790	.5548	96:125	.3930	.1619	.5574	96:149	.3834	.1828	.5572	96:173	.4206	.1204	.5573
96:102	.4228	.1131	.5570	96:126	.4471	.1674	.5584	96:150	.3709	.0290	.5593	96:174	.3561	.1718	.5572
96:103	.3855	.1153	.5573	96:127	.4066	.1687	.5573	96:151	.3709	.1759	.5573	96:175	.3614	.1778	.5572
96:104	.4203	.2232	.5562	96:128	.4172	.1117	.5577	96:152	.3611	.2462	.5544	96:176	.3561	.1560	.5562
96:105	.4284	.1353	.5579	96:129	.4502	.0561	.5591	96:153	.3717	.2653	.5554	96:177	.3710	.1548	.5567
96:106	.4053	.1114	.5591	96:130	.4192	.1614	.5570	96:154	.3831	.3770	.5531	96:178	.3133	.1901	.5572
96:107	.4153	.1462	.5575	96:131	.4074	.1638	.5575	96:155	.3918	.2759	.5552	96:179	.3514	.1765	.5560
96:108	.3219	.1485	.5572	96:132	.4471	.1463	.5576	96:156	.3401	.3001	.5552	96:180	.3056	.1747	.5571
96:109	.3300	.0772	.5581	96:133	.4432	.2509	.5573	96:157	.4046	.3770	.5521	96:181	.4002	.1092	.5577
96:110	.4179	.1161	.5591	96:134	.4482	.0913	.5587	96:158	.3865	.2393	.5542	96:182	.3828	.1367	.5569
96:111	.2390	.1304	.5562	96:135	.4221	.1130	.5590	96:159	.3905	.2739	.5546	96:183	.3709	.1777	.5560
96:112	.4427	.2574	.5556	96:136	.5294	<u>.1124</u>	.5607	96:160	.4007	.2726	.5540	96:184	.3834	.2541	.5550
96:113	.3586	.1946	.5570	96:137	.4570	.0369	.5593	96:161	.4022	.2575	.5553	96:185	.2470	.1742	.5575
96:114	.4495	.1471	.5574	96:138	.4184	.2043	.5574	96:162	.3158	.2674	.5554	96:186	.4046	.3770	.5515
96:115	.2139	.1697	.5557	96:139	.2906	.2237	.5566	96:163	.4301	.2467	.5536	96:187	.3776	.3776	.5507
96:116	.3458	.1599	.5566	96:140	.4106	.2716	.5557	96:164	.3917	.1931	.5562				
96:117	.3670	.1404	.5575	96:141	.4126	.2498	.5560	96:165	.4002	.2598	.5557	98:1	.3826	.0523	.5600
96:118	.3990	.1697	.5572	96:142	.3998	.0985	.5578	96:166	.2980	.2584	.5544	98:2	.2450	.2162	.5574
96:119	.3564	.2105	.5556	96:143	.3663	.1600	.5577	96:167	.2735	.2546	.5537	98:3	.4116	.1796	.5587
96:120	.3421	.1826	.5569	96:144	.4446	.1104	.5580	96:168	.2973	.2345	.5545	98:4	.3361	.1949	.5572

Table A.22 (Continued)

Isomer	HOMO	LUMO	E_{res}	Isomer	HOMO	LUMO	E_{res}	Isomer	HOMO	LUMO	E_{res}	Isomer	HOMO	LUMO	E_{res}
98:5	.3115	.2382	.5575	98:29	.3547	.1829	.5576	98:53	.2774	.2038	.5574	98:77	.3376	.2797	.5557
98:6	.3322	.2716	.5567	98:30	.3466	.2159	.5567	98:54	.2863	.2243	.5559	98:78	.3710	.3049	.5551
98:7	.2791	.2530	.5548	98:31	.3452	.2558	.5569	98:55	.3383	.1458	.5571	98:79	.3619	.3583	.5539
98:8	.3521	.1378	.5589	98:32	.4191	.1679	.5588	98:56	.3267	.1620	.5581	98:80	.3772	.2098	.5568
98:9	.4656	.1617	.5589	98:33	.4152	.0343	.5595	98:57	.3827	.2197	.5562	98:81	.3411	.2554	.5545
98:10	.3142	.2713	.5566	98:34	.3249	.2115	.5574	98:58	.3375	.2809	.5558	98:82	.3815	.2666	.5571
98:11	.3100	.1500	.5576	98:35	.4084	.0415	.5595	98:59	.3611	.2044	.5572	98:83	.3586	.2967	.5551
98:12	.3240	.2453	.5573	98:36	.3660	.0997	.5591	98:60	.3579	.1160	.5583	98:84	.3816	.2782	.5561
98:13	.3268	.2145	.5569	98:37	.4189	.1054	.5584	98:61	.2670	.2251	.5572	98:85	.3987	.1964	.5574
98:14	.3402	.2774	.5551	98:38	.4126	.2148	.5574	98:62	.4162	.0899	.5582	98:86	.2740	.2309	.5555
98:15	.3858	.2336	.5571	98:39	.4143	.2311	.5566	98:63	.3687	.3218	.5548	98:87	.3191	.1400	.5576
98:16	.3709	.2790	.5569	98:40	.2752	.2154	.5569	98:64	.3893	.0707	.5592	98:88	.4258	.1193	.5583
98:17	.3847	.0907	.5586	98:41	.3049	.1967	.5577	98:65	.3642	.0688	.5592	98:89	.3377	.1881	.5562
98:18	.3346	.2591	.5560	98:42	.3728	.2115	.5571	98:66	.3396	.0883	.5589	98:90	.3283	.1633	.5576
98:19	.3264	.2190	.5569	98:43	.3895	.2133	.5578	98:67	.4099	.1491	.5578	98:91	.3814	.0993	.5591
98:20	.3370	.2916	.5559	98:44	.3450	.2003	.5569	98:68	.3270	.1774	.5573	98:92	.2844	.2440	.5567
98:21	.3610	.2881	.5562	98:45	.3109	.2259	.5568	98:69	.3170	.2995	.5546	98:93	.2920	.1990	.5574
98:22	.3053	.2373	.5567	98:46	.2849	.2133	.5570	98:70	.3347	.2905	.5563	98:94	.3872	.1956	.5570
98:23	.4159	.2088	.5578	98:47	.3454	.2297	.5574	98:71	.3989	.2910	.5565	98:95	.4471	.1219	.5582
98:24	.3053	.2953	.5566	98:48	.3811	.1651	.5578	98:72	.3352	.2560	.5555	98:96	.3980	.1431	.5580
98:25	.3210	.2215	.5572	98:49	.3700	.2177	.5568	98:73	.3567	.2718	.5559	98:97	.4461	.0438	.5593
98:26	.3297	.1758	.5562	98:50	.3139	.2351	.5557	98:74	.3653	.3028	.5550	98:98	.3567	.1663	.5578
98:27	.3770	.1733	.5584	98:51	.3352	.1660	.5582	98:75	.3971	.3490	.5530	98:99	.3728	.1162	.5575
98:28	.3172	.2993	.5563	98:52	.3362	.2132	.5571	98:76	.3455	.1886	.5572	98:100	.4267	.1507	.5578

Table A.22 (*Continued*)

Isomer	HOMO	LUMO	E_{res}	Isomer	HOMO	LUMO	E_{res}	Isomer	HOMO	LUMO	E_{res}	Isomer	HOMO	LUMO	E_{res}
98:101	.4008	.2093	.5565	98:125	.3843	.2221	.5563	98:149	.3131	.2430	.5569	98:173	.3877	.3121	.5545
98:102	.4050	.2493	.5568	98:126	.4001	.0930	.5588	98:150	.4351	.2393	.5565	98:174	.4153	.3054	.5545
98:103	.4442	.1593	.5590	98:127	.3835	.2009	.5580	98:151	.3761	.2802	.5542	98:175	.3874	.3108	.5542
98:104	.4395	.1099	.5582	98:128	.3735	.1370	.5582	98:152	.3366	.2703	.5567	98:176	.3973	.2919	.5554
98:105	.4043	.2373	.5575	98:129	.3896	.1583	.5568	98:153	.3966	.2494	.5567	98:177	.3086	.3031	.5550
98:106	.3942	.1167	.5583	98:130	.3423	.0859	.5579	98:154	.4030	.3208	.5555	98:178	.3354	.2913	.5558
98:107	.4057	.0693	.5585	98:131	.2950	.2572	.5551	98:155	.4095	.1624	.5573	98:179	.3532	.2356	.5562
98:108	.3673	.1446	.5577	98:132	.3168	.2367	.5559	98:156	.4190	.1358	.5584	98:180	.3619	.2865	.5545
98:109	.4078	.1393	.5580	98:133	.3437	.1723	.5576	98:157	.3369	.2706	.5555	98:181	.3520	.3170	.5557
98:110	.4545	.1339	.5578	98:134	.3162	.1569	.5565	98:158	.2953	.2686	.5551	98:182	.3926	.3170	.5557
98:111	.4595	.1665	.5587	98:135	.3081	.2309	.5561	98:159	.4124	.2631	.5548	98:183	.4314	.2784	.5544
98:112	.2826	.2396	.5562	98:136	.3190	.2626	.5547	98:160	.2951	.2451	.5547	98:184	.4112	.2229	.5565
98:113	.2981	.1846	.5571	98:137	.3022	.2096	.5563	98:161	.4120	.2419	.5558	98:185	.3479	.2867	.5552
98:114	.3520	.2620	.5565	98:138	.3415	.1929	.5572	98:162	.4065	.3099	.5544	98:186	.3104	.2797	.5549
98:115	.3922	.1457	.5574	98:139	.3176	.1984	.5558	98:163	.4013	.2919	.5553	98:187	.3426	.2436	.5554
98:116	.4207	.1304	.5576	98:140	.3137	.2304	.5562	98:164	.3831	.2612	.5557	98:188	.3858	.1165	.5568
98:117	.4331	.0493	.5592	98:141	.2773	.1869	.5562	98:165	.4265	.2855	.5546	98:189	.3702	.2813	.5545
98:118	.3999	.1659	.5577	98:142	.3248	.1457	.5578	98:166	.3542	.2079	.5547	98:190	.3655	.2840	.5551
98:119	.4334	.2463	.5562	98:143	.3971	.1614	.5573	98:167	.3639	.2356	.5546	98:191	.3280	.2883	.5534
98:120	.4355	.1427	.5577	98:144	.2931	.1943	.5564	98:168	.3932	.2850	.5545	98:192	.3273	.3217	.5533
98:121	.3428	.1751	.5569	98:145	.3595	.2802	.5552	98:169	.3371	.2800	.5542	98:193	.3527	.3129	.5544
98:122	.3748	.1548	.5577	98:146	.3496	.2283	.5565	98:170	.3918	.2412	.5546	98:194	.3606	.3175	.5544
98:123	.3443	.2040	.5577	98:147	.1931	.1579	.5561	98:171	.3302	.2538	.5543	98:195	.3212	.2482	.5538
98:124	.3838	.2039	.5566	98:148	.4265	.1320	.5581	98:172	.3407	.2685	.5550	98:196	.3591	.2709	.5557

Table A.22 (Continued)

Isomer	HOMO	LUMO	E_{res}	Isomer	HOMO	LUMO	E_{res}	Isomer	HOMO	LUMO	E_{res}	Isomer	HOMO	LUMO	E_{res}
98:197	.3187	.2118	.5553	98:221	.3678	.2794	.5533	98:245	.3069	.2508	.5561	100:9	.3504	.2369	.5580
98:198	.3871	.1688	.5571	98:222	.4132	.2722	.5558	98:246	.3529	.2592	.5547	100:10	.3493	.2468	.5571
98:199	.2592	.2575	.5543	98:223	.4025	.3165	.5556	98:247	.4026	.2320	.5553	100:11	.3639	.0748	.5591
98:200	.3184	.2389	.5562	98:224	.3990	.2936	.5552	98:248	.4141	.1127	.5569	100:12	.3354	.1393	.5593
98:201	.3322	.2695	.5558	98:225	.4028	.3354	.5548	98:249	.3206	.2278	.5568	100:13	.3815	.0993	.5598
98:202	.3419	.2677	.5560	98:226	.4287	.2449	.5561	98:250	.2762	.2422	.5558	100:14	.3557	.2747	.5550
98:203	.2873	.2727	.5557	98:227	.3903	.2074	.5565	98:251	.4028	.2104	.5569	100:15	.3527	.3175	.5567
98:204	.3533	.2347	.5564	98:228	.4824	.2519	.5568	98:252	.4130	.2885	.5550	100:16	.2976	.2528	.5574
98:205	.3620	.2782	.5558	98:229	.4220	.2348	.5572	98:253	.4710	.1799	.5574	100:17	.3632	.2102	.5577
98:206	.3616	.3185	.5563	98:230	.4603	.2743	.5570	98:254	.4132	.1687	.5575	100:18	.4263	.0571	.5595
98:207	.3107	.2838	.5553	98:231	.4603	.2314	.5565	98:255	.3895	.1747	.5569	100:19	.2945	.2630	.5572
98:208	.3675	.2607	.5558	98:232	.4338	.2844	.5559	98:256	.3591	.1510	.5571	100:20	.3285	.2446	.5562
98:209	.3064	.1967	.5553	98:233	.3529	.3294	.5541	98:257	.3185	.1802	.5578	100:21	.3699	.3022	.5565
98:210	.3049	.2254	.5565	98:234	.4240	.2861	.5554	98:258	.3122	.2174	.5553	100:22	.2992	.1396	.5582
98:211	.2744	.2416	.5549	98:235	.3111	.2043	.5570	98:259	.4260	.2327	.5536	100:23	.4339	.2013	.5581
98:212	.3056	.2525	.5548	98:236	.2620	.2486	.5566	100:1	.3501	.0000	.5609	100:24	.2578	.2183	.5576
98:213	.2942	.2167	.5566	98:237	.2917	.2178	.5566	100:2	.3227	.0488	.5603	100:25	.3794	.1378	.5582
98:214	.3007	.2102	.5557	98:238	.2593	.2563	.5561	100:3	.3739	.0490	.5599	100:26	.3261	.2881	.5550
98:215	.3351	.1733	.5548	98:239	.3653	.0891	.5588	100:4	.3273	.2388	.5580	100:27	.3670	.1721	.5576
98:216	.3482	.2661	.5567	98:240	.3991	.2188	.5574	100:5	.3020	.2411	.5577	100:28	.3709	.2863	.5565
98:217	.3990	.2449	.5563	98:241	.3111	.3037	.5564	100:6	.2708	.2218	.5573	100:29	.3809	.1787	.5584
98:218	.3673	.3063	.5542	98:242	.3096	.1940	.5560	100:7	.3760	.1126	.5597	100:30	.3611	.1457	.5580
98:219	.3656	.2679	.5560	98:243	.3603	.1397	.5579	100:8	.3179	.1436	.5592	100:31	.3808	.0277	.5600
98:220	.2876	.2252	.5566	98:244	.3536	.1655	.5573					100:32	.3399	.1430	.5581

384

Table A.22 (*Continued*)

Isomer	HOMO	LUMO	E_{res}	Isomer	HOMO	LUMO	E_{res}	Isomer	HOMO	LUMO	E_{res}	Isomer	HOMO	LUMO	E_{res}
100:33	.3795	.0812	.5592	100:57	.2737	.2465	.5569	100:81	.4054	.2119	.5574	100:105	.3823	.1958	.5571
100:34	.3055	.1694	.5578	100:58	.3352	.1299	.5593	100:82	.4058	.1225	.5583	100:106	.3714	.2179	.5567
100:35	.3016	.2422	.5571	100:59	.2979	.2432	.5573	100:83	.4204	.1124	.5588	100:107	.3704	.2480	.5557
100:36	.2848	.1041	.5586	100:60	.3429	.1213	.5593	100:84	.4381	.1167	.5586	100:108	.3938	.2318	.5568
100:37	.2752	.1410	.5593	100:61	.2988	.2383	.5570	100:85	.4215	.1890	.5578	100:109	.2807	.2065	.5559
100:38	.3645	.2694	.5558	100:62	.2647	.2143	.5576	100:86	.4251	.1192	.5589	100:110	.3645	.2055	.5573
100:39	.3090	.2956	.5568	100:63	.3186	.1692	.5580	100:87	.3981	.1193	.5589	100:111	.4047	.2161	.5580
100:40	.3693	.1629	.5579	100:64	.3793	.0109	.5602	100:88	.4308	.0495	.5594	100:112	.3972	.1918	.5569
100:41	.3723	.2279	.5583	100:65	.3907	.2450	.5569	100:89	.4474	.0919	.5592	100:113	.3994	.2889	.5562
100:42	.3312	.2907	.5567	100:66	.3754	.1721	.5585	100:90	.4674	.1227	.5596	100:114	.3717	.1735	.5579
100:43	.3384	.2681	.5555	100:67	.3011	.2807	.5563	100:91	.3497	.1549	.5583	100:115	.3781	.3401	.5550
100:44	.3383	.2678	.5572	100:68	.3543	.2537	.5567	100:92	.3452	.1285	.5586	100:116	.3523	.1770	.5568
100:45	.2998	.2057	.5579	100:69	.3270	.1669	.5585	100:93	.3321	.1326	.5590	100:117	.3465	.2714	.5564
100:46	.3742	.2648	.5575	100:70	.3253	.1664	.5571	100:94	.3143	.1401	.5571	100:118	.3516	.1721	.5574
100:47	.4037	.0333	.5601	100:71	.3442	.1614	.5577	100:95	.3259	.1458	.5571	100:119	.3780	.1786	.5583
100:48	.3728	.3554	.5561	100:72	.3963	.1746	.5587	100:96	.3213	.1325	.5590	100:120	.3328	.1321	.5577
100:49	.4011	.0663	.5594	100:73	.3965	.1518	.5582	100:97	.3312	.1474	.5587	100:121	.3819	.2051	.5575
100:50	.4038	.0842	.5592	100:74	.3999	.2292	.5567	100:98	.3957	.1463	.5586	100:122	.4085	.1648	.5575
100:51	.3652	.1135	.5596	100:75	.2714	.2151	.5574	100:99	.3359	.3194	.5561	100:123	.4175	.1554	.5576
100:52	.3883	.2474	.5579	100:76	.2884	.2222	.5572	100:100	.3599	.2495	.5556	100:124	.4076	.1926	.5574
100:53	.4081	.2466	.5573	100:77	.3672	.2260	.5578	100:101	.3554	.2445	.5571	100:125	.4231	.1073	.5581
100:54	.3795	.2024	.5580	100:78	.3804	.1377	.5580	100:102	.3397	.1960	.5577	100:126	.4481	.1240	.5587
100:55	.3999	.0244	.5602	100:79	.4102	.1473	.5584	100:103	.3708	.2433	.5567	100:127	.3871	.2224	.5559
100:56	.2507	.2413	.5569	100:80	.4181	.0427	.5600	100:104	.3785	.2316	.5557	100:128	.3137	.2757	.5559

Table A.22 (*Continued*)

Isomer	HOMO	LUMO	E_{res}	Isomer	HOMO	LUMO	E_{res}	Isomer	HOMO	LUMO	E_{res}	Isomer	HOMO	LUMO	E_{res}
100:129	.2968	.2065	.5571	100:153	.3329	.2262	.5566	100:177	.3601	.2679	.5550	100:201	.3593	.2148	.5570
100:130	.3263	.1409	.5573	100:154	.3083	.2025	.5572	100:178	.3687	.2202	.5556	100:202	.3590	.1972	.5575
100:131	.2931	.2630	.5553	100:155	.3777	.2077	.5558	100:179	.3408	.2769	.5559	100:203	.3205	.2158	.5573
100:132	.3589	.1368	.5580	100:156	.2878	.2269	.5573	100:180	.3271	.1750	.5579	100:204	.3436	.1832	.5568
100:133	.3502	.2014	.5581	100:157	.3457	.2135	.5574	100:181	.2960	.2229	.5560	100:205	.3465	.3017	.5550
100:134	.3095	.1845	.5568	100:158	.3671	.2709	.5566	100:182	.3618	.2550	.5560	100:206	.3856	.2075	.5570
100:135	.3300	.2682	.5564	100:159	.2525	.1658	.5564	100:183	.3350	.1647	.5568	100:207	.3280	.2327	.5569
100:136	.3772	.2054	.5566	100:160	.3221	.2091	.5557	100:184	.3307	.3011	.5556	100:208	.3431	.1612	.5582
100:137	.2831	.2028	.5577	100:161	.3757	.3455	.5552	100:185	.3720	.2908	.5562	100:209	.2751	.2603	.5554
100:138	.3030	.1997	.5570	100:162	.2978	.1831	.5574	100:186	.3893	.1529	.5583	100:210	.4017	.1484	.5575
100:139	.3088	.1768	.5577	100:163	.4123	.1382	.5578	100:187	.3804	.1653	.5577	100:211	.3295	.2785	.5556
100:140	.3807	.2438	.5564	100:164	.3294	.1342	.5577	100:188	.2903	.2246	.5558	100:212	.3473	.3250	.5550
100:141	.3964	.2993	.5550	100:165	.3991	.2398	.5568	100:189	.3475	.3211	.5550	100:213	.3008	.2590	.5556
100:142	.3691	.1253	.5595	100:166	.4101	.2165	.5565	100:190	.3797	.2974	.5561	100:214	.3917	.2196	.5568
100:143	.4069	.0894	.5593	100:167	.2819	.2261	.5563	100:191	.3391	.2300	.5572	100:215	.4120	.2641	.5556
100:144	.3744	.1406	.5579	100:168	.3115	.2594	.5559	100:192	.3203	.2253	.5571	100:216	.3461	.1329	.5578
100:145	.3736	.1323	.5585	100:169	.3471	.2467	.5564	100:193	.3656	.2884	.5565	100:217	.3383	.2073	.5570
100:146	.3804	.1302	.5584	100:170	.2942	.1371	.5577	100:194	.3624	.2093	.5565	100:218	.3225	.2216	.5560
100:147	.3655	.1786	.5579	100:171	.3430	.2162	.5571	100:195	.3452	.2991	.5559	100:219	.3586	.2029	.5564
100:148	.3464	.1214	.5580	100:172	.3797	.1693	.5578	100:196	.2608	.2467	.5568	100:220	.3605	.2803	.5557
100:149	.4381	.1466	.5577	100:173	.2942	.1740	.5569	100:197	.2826	.2039	.5573	100:221	.2888	.2888	.5541
100:150	.3298	.2797	.5554	100:174	.1891	.1579	.5565	100:198	.3037	.2609	.5559	100:222	.3144	.2881	.5558
100:151	.3765	.2100	.5579	100:175	.3195	.1518	.5580	100:199	.4151	.0846	.5592	100:223	.3212	.2961	.5552
100:152	.3117	.2107	.5557	100:176	.3816	.2297	.5563	100:200	.2676	.2357	.5570	100:224	.4042	.2825	.5558

Table A.22 (*Continued*)

Isomer	HOMO	LUMO	E_{res}	Isomer	HOMO	LUMO	E_{res}	Isomer	HOMO	LUMO	E_{res}	Isomer	HOMO	LUMO	E_{res}
100:225	.3678	.3057	.5547	100:249	.3868	.2138	.5567	100:273	.3573	.1344	.5594	100:297	.3411	.1452	.5577
100:226	.3377	.1856	.5571	100:250	.3257	.2721	.5558	100:274	.3342	.3312	.5564	100:298	.3904	.1470	.5585
100:227	.3164	.2282	.5554	100:251	.3416	.2115	.5569	100:275	.2564	.2473	.5568	100:299	.3743	.1680	.5577
100:228	.3797	.2552	.5559	100:252	.4042	.2964	.5563	100:276	.2672	.2296	.5570	100:300	.3584	.2019	.5574
100:229	.3631	.2244	.5571	100:253	.3886	.1727	.5577	100:277	.3711	.2537	.5572	100:301	.3753	.2327	.5574
100:230	.3748	.2410	.5571	100:254	.3697	.1803	.5571	100:278	.3703	.2254	.5580	100:302	.3578	.2646	.5561
100:231	.3326	.2822	.5559	100:255	.3644	.2213	.5575	100:279	.3920	.1788	.5578	100:303	.3337	.1738	.5577
100:232	.3583	.1402	.5581	100:256	.2431	.2170	.5562	100:280	.3772	.2495	.5578	100:304	.4290	.1636	.5590
100:233	.3193	.2457	.5566	100:257	.3162	.1997	.5572	100:281	.3809	.2535	.5571	100:305	.3556	.2132	.5577
100:234	.3867	.2626	.5556	100:258	.3074	.1671	.5577	100:282	.4101	.0769	.5592	100:306	.2929	.2409	.5569
100:235	.3533	.2335	.5569	100:259	.3328	.1784	.5579	100:283	.3127	.2494	.5561	100:307	.3353	.2029	.5561
100:236	.3919	.2586	.5570	100:260	.3277	.1744	.5568	100:284	.3026	.2144	.5574	100:308	.4104	.1690	.5578
100:237	.3789	.3193	.5560	100:261	.2811	.1840	.5563	100:285	.3460	.3298	.5565	100:309	.3063	.2133	.5561
100:238	.3564	.2914	.5563	100:262	.3697	.2994	.5554	100:286	.3728	.0844	.5589	100:310	.3234	.2880	.5562
100:239	.3739	.2785	.5569	100:263	.3456	.2380	.5568	100:287	.3433	.1295	.5593	100:311	.3550	.2169	.5576
100:240	.3628	.2349	.5577	100:264	.3200	.1932	.5573	100:288	.3576	.2622	.5575	100:312	.3469	.2164	.5571
100:241	.3760	.2734	.5574	100:265	.3195	.2046	.5562	100:289	.2832	.2336	.5571	100:313	.3629	.1952	.5574
100:242	.4022	.2256	.5573	100:266	.2407	.1845	.5555	100:290	.3749	.1192	.5595	100:314	.3534	.1795	.5578
100:243	.4045	.2454	.5576	100:267	.3171	.2479	.5575	100:291	.2656	.2197	.5559	100:315	.3559	.2493	.5564
100:244	.3350	.3159	.5563	100:268	.3046	.2098	.5573	100:292	.3430	.2331	.5572	100:316	.3581	.2838	.5568
100:245	.3766	.2455	.5569	100:269	.3818	.2261	.5582	100:293	.3226	.3190	.5565	100:317	.3285	.2730	.5563
100:246	.3473	.3138	.5559	100:270	.3731	.2244	.5575	100:294	.4003	.1988	.5583	100:318	.3361	.3173	.5563
100:247	.4243	.2346	.5574	100:271	.3387	.1304	.5591	100:295	.3584	.2971	.5570	100:319	.4033	.4011	.5548
100:248	.4223	.2087	.5572	100:272	.3893	.0728	.5592	100:296	.3035	.2265	.5574	100:320	.4161	.1668	.5580

Table A.22 (*Continued*)

Isomer	HOMO	LUMO	E_{res}	Isomer	HOMO	LUMO	E_{res}	Isomer	HOMO	LUMO	E_{res}	Isomer	HOMO	LUMO	E_{res}
100:321	.1309	.1309	.5565	100:345	.3551	.2452	.5562	100:369	.3468	.2540	.5555	100:393	.3887	.2409	.5546
100:322	.2076	.1516	.5565	100:346	.3870	.2502	.5547	100:370	.3514	.2782	.5554	100:394	.4125	.2941	.5539
100:323	.3182	.1516	.5579	100:347	.2911	.2556	.5551	100:371	.3225	.2257	.5568	100:395	.3262	.3262	.5534
100:324	.3553	.2303	.5552	100:348	.3710	.1861	.5564	100:372	.3007	.2992	.5553	100:396	.3060	.2640	.5547
100:325	.3305	.2273	.5560	100:349	.3696	.2206	.5556	100:373	.3623	.2349	.5575	100:397	.3419	.2750	.5560
100:326	.3674	.3075	.5552	100:350	.3678	.1731	.5554	100:374	.3517	.2128	.5569	100:398	.3493	.2387	.5566
100:327	.3313	.2791	.5558	100:351	.3735	.1844	.5573	100:375	.3357	.3290	.5547	100:399	.3410	.2437	.5562
100:328	.3736	.2714	.5564	100:352	.3324	.2394	.5561	100:376	.2461	.2414	.5549	100:400	.2952	.2257	.5560
100:329	.3692	.2989	.5543	100:353	.3030	.2430	.5546	100:377	.3382	.2427	.5562	100:401	.3695	.3350	.5535
100:330	.3166	.2291	.5542	100:354	.3437	.2631	.5562	100:378	.3456	.2390	.5560	100:402	.3829	.3139	.5551
100:331	.3865	.2535	.5550	100:355	.3113	.1713	.5564	100:379	.3257	.2635	.5557	100:403	.3400	.3111	.5549
100:332	.3543	.2664	.5550	100:356	.3553	.1755	.5575	100:380	.3034	.1760	.5565	100:404	.3446	.2961	.5556
100:333	.3324	.2384	.5554	100:357	.2851	.2077	.5561	100:381	.3035	.3007	.5545	100:405	.3072	.2912	.5547
100:334	.3000	.2415	.5552	100:358	.3511	.2083	.5572	100:382	.3764	.1633	.5564	100:406	.4140	.1441	.5584
100:335	.3768	.2476	.5560	100:359	.3527	.2407	.5565	100:383	.3877	.1725	.5571	100:407	.3788	.1882	.5576
100:336	.3586	.2002	.5559	100:360	.3697	.2511	.5563	100:384	.4025	.1818	.5567	100:408	.3271	.1818	.5571
100:337	.3909	.2589	.5549	100:361	.3340	.2747	.5557	100:385	.3332	.2751	.5556	100:409	.3473	.2345	.5564
100:338	.4098	.2877	.5536	100:362	.3715	.2510	.5561	100:386	.3586	.2204	.5557	100:410	.3431	.2722	.5567
100:339	.3034	.2232	.5555	100:363	.3653	.1981	.5564	100:387	.3896	.2335	.5559	100:411	.3108	.2834	.5567
100:340	.3531	.2068	.5563	100:364	.3747	.2904	.5557	100:388	.3932	.1926	.5575	100:412	.2994	.2323	.5565
100:341	.3448	.2985	.5553	100:365	.3038	.2635	.5552	100:389	.3912	.2156	.5568	100:413	.3911	.2267	.5575
100:342	.3335	.2994	.5545	100:366	.2970	.2308	.5568	100:390	.3333	.2528	.5542	100:414	.3439	.1895	.5576
100:343	.4098	.2877	.5549	100:367	.2730	.2349	.5569	100:391	.2905	.2409	.5556	100:415	.3053	.2095	.5574
100:344	.4046	.2379	.5561	100:368	.2886	.2358	.5557	100:392	.3055	.2043	.5560	100:416	.2790	.2728	.5563

Table A.22 (Continued)

Isomer	HOMO	LUMO	E_{res}	Isomer	HOMO	LUMO	E_{res}	Isomer	HOMO	LUMO	E_{res}	Isomer	HOMO	LUMO	E_{res}
100:417	.5016	.1366	.5594	100:426	.3690	.1943	.5575	100:435	.3364	.2801	.5563	100:444	.3868	.2076	.5579
100:418	.4130	.1392	.5585	100:427	.3443	.2291	.5563	100:436	.3190	.2769	.5561	100:445	.3883	.1519	.5582
100:419	.4496	.1061	.5589	100:428	.2801	.2322	.5558	100:437	.2806	.2463	.5558	100:446	.4017	.1860	.5578
100:420	.4054	.1388	.5584	100:429	.2834	.1995	.5571	100:438	.3300	.1533	.5575	100:447	.4026	.3328	.5544
100:421	.4316	.0490	.5593	100:430	.2845	.2407	.5564	100:439	.2774	.2726	.5555	100:448	.3060	.1919	.5563
100:422	.4435	.1963	.5580	100:431	.4467	.1898	.5568	100:440	.3016	.2432	.5554	100:449	.2737	.1175	.5569
100:423	.4027	.1108	.5580	100:432	.4467	.1898	.5572	100:441	.2963	.2321	.5561	100:450	.3042	.3042	.5518
100:424	.3373	.1651	.5571	100:433	.2845	.2440	.5561	100:442	.3016	.2213	.5560				
100:425	.3708	.1598	.5573	100:434	.3081	.2154	.5566	100:443	.2930	.1634	.5562				

INDEX

adjacency matrix 44, 51, 101–2, 165–67
aromaticity 46
aromatic sextet 56–57
'accidental' closed shell 63–66

band gap, *see* HOMO-LUMO gap
henzene 45–46
benzenoid hydrocarbon 56–57
benzenoid ring 53, 56
bond angles, idealized 70, 80
bonding connectivity 15, 23
borane 17, 47, 131
Brester tables 113

C_2 insertion/extrusion 150–63
   selection rules 153–57
C_2 patch 153–60
   site symmetry 153–56
$C_2B_{10}H_{12}$ 140
C_4 insertion 151–52
C_6 insertion 151–52
C_{20} 12, 66, 71–73, 79, 180, 366
C_{24} 79, 162, 180, 366
C_{26} 79, 180, 366
C_{28} 79, 121–24, 180, 366
C_{30} 79, 136, 162, 180, 366
C_{32} 79, 136, 162, 180, 366
C_{34} 79, 136, 180, 366
C_{36} 137, 162, 165, 176, 182–83, 366
C_{38} 79, 162, 184–85, 366
C_{40} 74, 113, 162, 186–89, 366–67
C_{42} 79, 190–93, 367
C_{44} 194–201, 367–68
C_{46} 202–11, 368–69
C_{48} 212–29, 369–71
C_{50} 79, 230–53, 371–74

C_{52} 79
C_{54} 79
C_{58} 79, 157, 162–63
C_{60} 1, 5, 47, 74, 123–24, 140, 157,
   162–63, 254-55, 375
   IR spectrum 5
   NMR spectrum 5, 112
   X-ray structure 46
   addition chemistry 11, 46
   electronic structure 44–47
   Stone-Wales map 141–44
   vibrational modes 117
   1812 isomers 31–32, 142–44, 165
C_{60}^{3-} 46
C_{62} 162
C_{70} 5, 60, 79, 254–55, 375
   NMR spectrum 6
   electronic structure 6 1–62
C_{72} 54, 57, 83, 254–55, 375
C_{74} 254–55, 375

C_{76} 6–8, 88–89, 254–55, 375
   NMR spectrum 6, 89, 113
   enantiomers 6, 89, 130
C_{78} 6–8, 54, 89–90, 138, 140–41, 254–55, 375
   NMR spectrum 6, 90
   isomer mixture 6, 90, 146
C_{80} 29, 138, 254–55, 375
C_{82} 8, 113, 138, 256–57, 375
C_{84} 6–8, 54, 60, 91–92, 139–40, 258–59,
   NMR spectrum 6, 92, 146
   isomer mixture 6, 92, 146–47
C_{86} 165, 176, 260–61, 375
C_{88} 262–65, 375–76
C_{90} 8, 57, 266–69, 376
C_{92} 270–77, 376–77
C_{94} 8, 278–89, 377–78

C_{96} 8, 57, 290–305, 379–80
C_{98} 306–27,380–83
C_{100} 328–65,383–84
C_{112} 63
C_{120} 57–59,62–63
C_{180} 56, 85
C_{380} 36–37,40
C_{800} 37
canonical spiral 30–31, 165, 170, 177
carbon:
    chain 2–5
    nanotube 11, 150
    ring 3–5, 12
    toroid 4
carbon cylinder:
    fullerene 59–62, 86–88, 179
    2p(7 + 3m) rule 48, 60
carbon-arc synthesis 1, 6, 8
carborane 131, 140
chemical potential 49–50
chirality 6, 98, 105, 129–30, 145–46,
    155–56,158—59
Clar sextet 56–57
Coulomb integral 44
Coxeter construction 10, 18–22, 35–38,
    50, 59
Coxeter net 19, 60, 98
Coxeter parameters 20–21, 37–38
cube 17–18, 55, 97, 125

dangling bonds 4, 149, 163
deltahedron 17, 19, 52
diamond 3—4
diamond-square-diamond (DSD)
    rearrangement 131, 140
dihedral symmetry 59–62
dodecahedron 16–18, 24, 50, 52, 97
dual operation 17–18, 52, 57
    see also fullerene, dual

electron counting rules 47
electronic structure:
    closed shell 47–65, 179

electronic structure (contd.)
    localized 46
    open shell 47–49, 51
    pseudo-closed shell 47-49
endohedral metallofullerene 11, 110
epikernel principle 105, 110
equivalent positions 97
ethylene 45, 69
Euler's theorem 4, 16–17, 29, 41, 79

face spiral, see ring spiral
failing spiral 27, 37
'first-order' results 10–12
Fries structure 56–57
fullerene:
    definition 1, 9, 16
    drawing 101, 103, 179
    dual 17–18, 28–30, 165–66, 169–70
    enumeration 3 1–34
    formation 149, 162
    fragmentation 150
    graph 22–29
    hypothesis 2–5
    isomer problem 10, 16-17, 41
    point groups 96–100, 106–11, 126–27
    polyhedron 16
    road 149, 161

Goldberg polyhedron:
    medial 16, 41
    multi-symmetric 20, 41, 51, 55
graph, see fullerene, graph
graph theory 42, 55
graphite 3–4, 45, 71–73
laser vaporization 1–5
great rhombicosidodecahedron 50

HOMO-LUMO gap 45, 47, 50, 61, 178
Hückel molecular orbital (HMO) theory
    43–51,68–69, 101–4, 122–23
Hückel 4n rule 51, 74
Hückel 4n+2 rule 3, 47, 51

hardness 49–50
heptagonal ring 4, 55, 125, 163
hexagon neighbour indices 80–92, 179
higher fullereries 6–9, 80–84, 88–93,
    146-47
    see also isolated-pentagon isomers
hybridization 44, 69–71, 122

icosahedral symmetry 19–21, 50–51, 82,
    128, 156
icosahedron 17–18, 50
idealized symmetry, *see* point group,
    maximal
infra-red (IR) spectroscopy 113–17
isolated-pentagon isomers 30, 33, 63–65,
    80-84, 88-93, 108-9, 121, 133-35,
    138-41, 165, 179
isolated-pentagon rule (IPR) 73–75,
    80—81
isomer distribution:
    enthalpic 146
    entropic 145-46, 157
isomorphic graphs 22–23

Jahn-Teller distortion 48, 76, 89, 177
    first order 105, 110
    second order 110

Kekulé structures 46-47, 53, 56
    kinetic factors 6, 8, 147

laser vaporization, see graphite, laser
    vaporization
leapfrog:
    fullerene 52-59, 84-86, 128, 156, 179
    operation 35, 51-57, 67, 85-86
    60+6k rule 48,53
low symmetry 20-21, 23, 105
lower fullerenes 75-79

magic numbers 3, 73
mass spectrometry 2 3
maximal point group, *see* point group,
    maximal
microscopic reversibility 146
molecular graph, *see* fullerene. graph
mutual exclusion rule 114

naphthalene, pyrolysis of 8, 149
nuclear magnetic resonance (NMR)
    spectroscopy 8, 111-13
NMR signature 111-13, 117, 165,
    170-71, 177-78

octahedron 17-18
optical activity 130
orbit, *see* point group, orbit
orbital energy mismatch 72-73, 80
order, *see* point group, order
osmylation 6, 89, 130

π-electronic stability 69, 88-92
π -orbital axis vector 69, 72
π -orbital misalignment 71-73, 80
pentagon:
    isolation criterion 4-5, 74-79
    neighbour indices 75-79, 178
    road 149
    start 32-33, 167
Platonic solids 17-18
point group:
    assignment 105-10
    orbit 97, 111-17, 126-30, 153-55
    order 25-26, 97, 99, 126, 145-46, 155,
        157, 160, 171
    maximal 76, 99, 105, 165, 170, 177
polarity 98
polarized Raman modes 114-17, 178
principle of maximum hardness 49-50
pyracylene rearrangement, *see*
Stone-Wales rearrangement

quadrilateral ring, *see* square ring

Raman spectroscopy 113–17
regular orbit 97
rehybridization 70–73, 80
resonance energy 45, 47, 74, 178
resonance integral 44
ring spiral 24–31, 35–41, 166–76
see also canonical spiral; spiral
    algorithm; spiral computer
    program; spiral conjecture; spiral
    proof; unspirallable fullerene
rotational (microwave) spectroscopy
98–99

σ orbital, bond directed 69, 72
σ–π orthogonality 69
σ–π separability 44, 68
Schönflies symbol 99
Schlegel diagram 22
separating triangle 17, 41–42, 167
site symmetry 26, 95–97, 114-17,
    125-29, 153-56, 171
snub dodecahedron 50
softness 49-50
solubility 8
spiral algorithm 27-31, 177
    generalized 39-40
spiral computer program 165-76
    example output 176
spiral conjecture 23-27, 35-38, 169
    modified 40
spiral proof 35, 37
sporadic closed shell 63-66
square ring 42, 79-80, 125, 167
steric inhibition of resonance 71
    see also π -orbital misalignment
steric strain 68-93, 178-79
Stone-Wales bond 120-2 1
    site symmetry 125-29
Stone-Wales patch 120–21, 126–28, 151,
157

Stone-Wales rearrangement 35, 74,
    120-47, 157, 159, 161-63, 178-79
    activation energy 123-24, 145
    generalized 124-25
    selection rules 126-30
    topological model 122-24, 145
sublimation 8
successful spiral 24–27, 40
superconductivity 11, 46
symmetry group, *see* point group
symmetry number 145
systematic theory 1, 9–10, 15

tensor surface harmonic (TSH) theory
    101
tetrahedral symmetry 20–22, 35–38, 128
tetrahedron 17–18, 55, 97
thermodynamic stability 4, 6, 147
topological coordinates 101–5, 165, 170,
    177
topological radius 111
triangular ring 42, 167
trigonal prism 42, 125, 167
truncated dodecahedron 112–13
truncated icosahedron 22–23, 26, 52
truncated tetrahedron 55

uniform curvature rule 80
unspirallable fullerene 36-38, 40

valence bond (VB) theory 55–56
'valence state' 71
vertex spiral 28–29
vibrational mode counts 113–17, 165,
178

Wade's n + 1 rule 47
Walsh's rules 110
Woodward-Hoffman rules 122

Chemistry

THE SCEPTICAL CHYMIST: THE CLASSIC 1661 TEXT, Robert Boyle. Boyle defines the term "element," asserting that all natural phenomena can be explained by the motion and organization of primary particles. 1911 ed. viii+232pp. 5⅜ x 8½.
0-486-42825-7

RADIOACTIVE SUBSTANCES, Marie Curie. Here is the celebrated scientist's doctoral thesis, the prelude to her receipt of the 1903 Nobel Prize. Curie discusses establishing atomic character of radioactivity found in compounds of uranium and thorium; extraction from pitchblende of polonium and radium; isolation of pure radium chloride; determination of atomic weight of radium; plus electric, photographic, luminous, heat, color effects of radioactivity. ii+94pp. 5⅜ x 8½.
0-486-42550-9

CHEMICAL MAGIC, Leonard A. Ford. Second Edition, Revised by E. Winston Grundmeier. Over 100 unusual stunts demonstrating cold fire, dust explosions, much more. Text explains scientific principles and stresses safety precautions. 128pp. 5⅜ x 8½.
0-486-67628-5

THE DEVELOPMENT OF MODERN CHEMISTRY, Aaron J. Ihde. Authoritative history of chemistry from ancient Greek theory to 20th-century innovation. Covers major chemists and their discoveries. 209 illustrations. 14 tables. Bibliographies. Indices. Appendices. 851pp. 5⅜ x 8½.
0-486-64235-6

CATALYSIS IN CHEMISTRY AND ENZYMOLOGY, William P. Jencks. Exceptionally clear coverage of mechanisms for catalysis, forces in aqueous solution, carbonyl- and acyl-group reactions, practical kinetics, more. 864pp. 5⅜ x 8½.
0-486-65460-5

ELEMENTS OF CHEMISTRY, Antoine Lavoisier. Monumental classic by founder of modern chemistry in remarkable reprint of rare 1790 Kerr translation. A must for every student of chemistry or the history of science. 539pp. 5⅜ x 8½. 0-486-64624-6

THE HISTORICAL BACKGROUND OF CHEMISTRY, Henry M. Leicester. Evolution of ideas, not individual biography. Concentrates on formulation of a coherent set of chemical laws. 260pp. 5⅜ x 8½.
0-486-61053-5

A SHORT HISTORY OF CHEMISTRY, J. R. Partington. Classic exposition explores origins of chemistry, alchemy, early medical chemistry, nature of atmosphere, theory of valency, laws and structure of atomic theory, much more. 428pp. 5⅜ x 8½. (Available in U.S. only.)
0-486-65977-1

GENERAL CHEMISTRY, Linus Pauling. Revised 3rd edition of classic first-year text by Nobel laureate. Atomic and molecular structure, quantum mechanics, statistical mechanics, thermodynamics correlated with descriptive chemistry. Problems. 992pp. 5⅜ x 8½.
0-486-65622-5

FROM ALCHEMY TO CHEMISTRY, John Read. Broad, humanistic treatment focuses on great figures of chemistry and ideas that revolutionized the science. 50 illustrations. 240pp. 5⅜ x 8½.
0-486-28690-8

Engineering

DE RE METALLICA, Georgius Agricola. The famous Hoover translation of greatest treatise on technological chemistry, engineering, geology, mining of early modern times (1556). All 289 original woodcuts. 638pp. 6¾ x 11. 0-486-60006-8

FUNDAMENTALS OF ASTRODYNAMICS, Roger Bate et al. Modern approach developed by U.S. Air Force Academy. Designed as a first course. Problems, exercises. Numerous illustrations. 455pp. 5⅜ x 8½. 0-486-60061-0

DYNAMICS OF FLUIDS IN POROUS MEDIA, Jacob Bear. For advanced students of ground water hydrology, soil mechanics and physics, drainage and irrigation engineering and more. 335 illustrations. Exercises, with answers. 784pp. 6⅛ x 9¼. 0-486-65675-6

THEORY OF VISCOELASTICITY (Second Edition), Richard M. Christensen. Complete consistent description of the linear theory of the viscoelastic behavior of materials. Problem-solving techniques discussed. 1982 edition. 29 figures. xiv+364pp. 6⅛ x 9¼. 0-486-42880-X

MECHANICS, J. P. Den Hartog. A classic introductory text or refresher. Hundreds of applications and design problems illuminate fundamentals of trusses, loaded beams and cables, etc. 334 answered problems. 462pp. 5⅜ x 8½. 0-486-60754-2

MECHANICAL VIBRATIONS, J. P. Den Hartog. Classic textbook offers lucid explanations and illustrative models, applying theories of vibrations to a variety of practical industrial engineering problems. Numerous figures. 233 problems, solutions. Appendix. Index. Preface. 436pp. 5⅜ x 8½. 0-486-64785-4

STRENGTH OF MATERIALS, J. P. Den Hartog. Full, clear treatment of basic material (tension, torsion, bending, etc.) plus advanced material on engineering methods, applications. 350 answered problems. 323pp. 5⅜ x 8½. 0-486-60755-0

A HISTORY OF MECHANICS, René Dugas. Monumental study of mechanical principles from antiquity to quantum mechanics. Contributions of ancient Greeks, Galileo, Leonardo, Kepler, Lagrange, many others. 671pp. 5⅜ x 8½. 0-486-65632-2

STABILITY THEORY AND ITS APPLICATIONS TO STRUCTURAL MECHANICS, Clive L. Dym. Self-contained text focuses on Koiter postbuckling analyses, with mathematical notions of stability of motion. Basing minimum energy principles for static stability upon dynamic concepts of stability of motion, it develops asymptotic buckling and postbuckling analyses from potential energy considerations, with applications to columns, plates, and arches. 1974 ed. 208pp. 5⅜ x 8½. 0-486-42541-X

METAL FATIGUE, N. E. Frost, K. J. Marsh, and L. P. Pook. Definitive, clearly written, and well-illustrated volume addresses all aspects of the subject, from the historical development of understanding metal fatigue to vital concepts of the cyclic stress that causes a crack to grow. Includes 7 appendixes. 544pp. 5⅜ x 8½. 0-486-40927-9

ROCKETS, Robert Goddard. Two of the most significant publications in the history of rocketry and jet propulsion: "A Method of Reaching Extreme Altitudes" (1919) and "Liquid Propellant Rocket Development" (1936). 128pp. 5⅜ x 8½.   0-486-42537-1

STATISTICAL MECHANICS: PRINCIPLES AND APPLICATIONS, Terrell L. Hill. Standard text covers fundamentals of statistical mechanics, applications to fluctuation theory, imperfect gases, distribution functions, more. 448pp. 5⅜ x 8½.

0-486-65390-0

ENGINEERING AND TECHNOLOGY 1650–1750: ILLUSTRATIONS AND TEXTS FROM ORIGINAL SOURCES, Martin Jensen. Highly readable text with more than 200 contemporary drawings and detailed engravings of engineering projects dealing with surveying, leveling, materials, hand tools, lifting equipment, transport and erection, piling, bailing, water supply, hydraulic engineering, and more. Among the specific projects outlined-transporting a 50-ton stone to the Louvre, erecting an obelisk, building timber locks, and dredging canals. 207pp. 8⅜ x 11¼.

0-486-42232-1

THE VARIATIONAL PRINCIPLES OF MECHANICS, Cornelius Lanczos. Graduate level coverage of calculus of variations, equations of motion, relativistic mechanics, more. First inexpensive paperbound edition of classic treatise. Index. Bibliography. 418pp. 5⅜ x 8½.   0-486-65067-7

PROTECTION OF ELECTRONIC CIRCUITS FROM OVERVOLTAGES, Ronald B. Standler. Five-part treatment presents practical rules and strategies for circuits designed to protect electronic systems from damage by transient overvoltages. 1989 ed. xxiv+434pp. 6⅛ x 9¼.   0-486-42552-5

ROTARY WING AERODYNAMICS, W. Z. Stepniewski. Clear, concise text covers aerodynamic phenomena of the rotor and offers guidelines for helicopter performance evaluation. Originally prepared for NASA. 537 figures. 640pp. 6⅛ x 9¼.

0-486-64647-5

INTRODUCTION TO SPACE DYNAMICS, William Tyrrell Thomson. Comprehensive, classic introduction to space-flight engineering for advanced undergraduate and graduate students. Includes vector algebra, kinematics, transformation of coordinates. Bibliography. Index. 352pp. 5⅜ x 8½.   0-486-65113-4

HISTORY OF STRENGTH OF MATERIALS, Stephen P. Timoshenko. Excellent historical survey of the strength of materials with many references to the theories of elasticity and structure. 245 figures. 452pp. 5⅜ x 8½.   0-486-61187-6

ANALYTICAL FRACTURE MECHANICS, David J. Unger. Self-contained text supplements standard fracture mechanics texts by focusing on analytical methods for determining crack-tip stress and strain fields. 336pp. 6⅛ x 9¼.   0-486-41737-9

STATISTICAL MECHANICS OF ELASTICITY, J. H. Weiner. Advanced, self-contained treatment illustrates general principles and elastic behavior of solids. Part 1, based on classical mechanics, studies thermoelastic behavior of crystalline and polymeric solids. Part 2, based on quantum mechanics, focuses on interatomic force laws, behavior of solids, and thermally activated processes. For students of physics and chemistry and for polymer physicists. 1983 ed. 96 figures. 496pp. 5⅜ x 8½.

0-486-42260-7

Mathematics

FUNCTIONAL ANALYSIS (Second Corrected Edition), George Bachman and Lawrence Narici. Excellent treatment of subject geared toward students with background in linear algebra, advanced calculus, physics and engineering. Text covers introduction to inner-product spaces, normed, metric spaces, and topological spaces; complete orthonormal sets, the Hahn-Banach Theorem and its consequences, and many other related subjects. 1966 ed. 544pp. 6⅛ x 9¼. 0-486-40251-7

ASYMPTOTIC EXPANSIONS OF INTEGRALS, Norman Bleistein & Richard A. Handelsman. Best introduction to important field with applications in a variety of scientific disciplines. New preface. Problems. Diagrams. Tables. Bibliography. Index. 448pp. 5⅜ x 8½. 0-486-65082-0

VECTOR AND TENSOR ANALYSIS WITH APPLICATIONS, A. I. Borisenko and I. E. Tarapov. Concise introduction. Worked-out problems, solutions, exercises. 257pp. 5⅝ x 8¼. 0-486-63833-2

AN INTRODUCTION TO ORDINARY DIFFERENTIAL EQUATIONS, Earl A. Coddington. A thorough and systematic first course in elementary differential equations for undergraduates in mathematics and science, with many exercises and problems (with answers). Index. 304pp. 5⅜ x 8½. 0-486-65942-9

FOURIER SERIES AND ORTHOGONAL FUNCTIONS, Harry F. Davis. An incisive text combining theory and practical example to introduce Fourier series, orthogonal functions and applications of the Fourier method to boundary-value problems. 570 exercises. Answers and notes. 416pp. 5⅜ x 8½. 0-486-65973-9

COMPUTABILITY AND UNSOLVABILITY, Martin Davis. Classic graduate-level introduction to theory of computability, usually referred to as theory of recurrent functions. New preface and appendix. 288pp. 5⅜ x 8½. 0-486-61471-9

ASYMPTOTIC METHODS IN ANALYSIS, N. G. de Bruijn. An inexpensive, comprehensive guide to asymptotic methods–the pioneering work that teaches by explaining worked examples in detail. Index. 224pp. 5⅜ x 8½ 0-486-64221-6

APPLIED COMPLEX VARIABLES, John W. Dettman. Step-by-step coverage of fundamentals of analytic function theory–plus lucid exposition of five important applications: Potential Theory; Ordinary Differential Equations; Fourier Transforms; Laplace Transforms; Asymptotic Expansions. 66 figures. Exercises at chapter ends. 512pp. 5⅜ x 8½. 0-486-64670-X

INTRODUCTION TO LINEAR ALGEBRA AND DIFFERENTIAL EQUATIONS, John W. Dettman. Excellent text covers complex numbers, determinants, orthonormal bases, Laplace transforms, much more. Exercises with solutions. Undergraduate level. 416pp. 5⅜ x 8½. 0-486-65191-6

RIEMANN'S ZETA FUNCTION, H. M. Edwards. Superb, high-level study of landmark 1859 publication entitled "On the Number of Primes Less Than a Given Magnitude" traces developments in mathematical theory that it inspired. xiv+315pp. 5⅜ x 8½. 0-486-41740-9

CALCULUS OF VARIATIONS WITH APPLICATIONS, George M. Ewing. Applications-oriented introduction to variational theory develops insight and promotes understanding of specialized books, research papers. Suitable for advanced undergraduate/graduate students as primary, supplementary text. 352pp. 5⅜ x 8½.
0-486-64856-7

COMPLEX VARIABLES, Francis J. Flanigan. Unusual approach, delaying complex algebra till harmonic functions have been analyzed from real variable viewpoint. Includes problems with answers. 364pp. 5⅜ x 8½.   0-486-61388-7

AN INTRODUCTION TO THE CALCULUS OF VARIATIONS, Charles Fox. Graduate-level text covers variations of an integral, isoperimetrical problems, least action, special relativity, approximations, more. References. 279pp. 5⅜ x 8½.
0-486-65499-0

COUNTEREXAMPLES IN ANALYSIS, Bernard R. Gelbaum and John M. H. Olmsted. These counterexamples deal mostly with the part of analysis known as "real variables." The first half covers the real number system, and the second half encompasses higher dimensions. 1962 edition. xxiv+198pp. 5⅜ x 8½. 0-486-42875-3

CATASTROPHE THEORY FOR SCIENTISTS AND ENGINEERS, Robert Gilmore. Advanced-level treatment describes mathematics of theory grounded in the work of Poincaré, R. Thom, other mathematicians. Also important applications to problems in mathematics, physics, chemistry and engineering. 1981 edition. References. 28 tables. 397 black-and-white illustrations. xvii + 666pp. 6⅛ x 9¼.
0-486-67539-4

INTRODUCTION TO DIFFERENCE EQUATIONS, Samuel Goldberg. Exceptionally clear exposition of important discipline with applications to sociology, psychology, economics. Many illustrative examples; over 250 problems. 260pp. 5⅜ x 8½.
0-486-65084-7

NUMERICAL METHODS FOR SCIENTISTS AND ENGINEERS, Richard Hamming. Classic text stresses frequency approach in coverage of algorithms, polynomial approximation, Fourier approximation, exponential approximation, other topics. Revised and enlarged 2nd edition. 721pp. 5⅜ x 8½.   0-486-65241-6

INTRODUCTION TO NUMERICAL ANALYSIS (2nd Edition), F. B. Hildebrand. Classic, fundamental treatment covers computation, approximation, interpolation, numerical differentiation and integration, other topics. 150 new problems. 669pp. 5⅜ x 8½.   0-486-65363-3

THREE PEARLS OF NUMBER THEORY, A. Y. Khinchin. Three compelling puzzles require proof of a basic law governing the world of numbers. Challenges concern van der Waerden's theorem, the Landau-Schnirelmann hypothesis and Mann's theorem, and a solution to Waring's problem. Solutions included. 64pp. 5⅜ x 8½.
0-486-40026-3

THE PHILOSOPHY OF MATHEMATICS: AN INTRODUCTORY ESSAY, Stephan Körner. Surveys the views of Plato, Aristotle, Leibniz & Kant concerning propositions and theories of applied and pure mathematics. Introduction. Two appendices. Index. 198pp. 5⅜ x 8½.   0-486-25048-2

INTRODUCTORY REAL ANALYSIS, A.N. Kolmogorov, S. V. Fomin. Translated by Richard A. Silverman. Self-contained, evenly paced introduction to real and functional analysis. Some 350 problems. 403pp. 5⅜ x 8½.   0-486-61226-0

APPLIED ANALYSIS, Cornelius Lanczos. Classic work on analysis and design of finite processes for approximating solution of analytical problems. Algebraic equations, matrices, harmonic analysis, quadrature methods, much more. 559pp. 5⅜ x 8½.
0-486-65656-X

AN INTRODUCTION TO ALGEBRAIC STRUCTURES, Joseph Landin. Superb self-contained text covers "abstract algebra": sets and numbers, theory of groups, theory of rings, much more. Numerous well-chosen examples, exercises. 247pp. 5⅜ x 8½.
0-486-65940-2

QUALITATIVE THEORY OF DIFFERENTIAL EQUATIONS, V. V. Nemytskii and V.V. Stepanov. Classic graduate-level text by two prominent Soviet mathematicians covers classical differential equations as well as topological dynamics and ergodic theory. Bibliographies. 523pp. 5⅜ x 8½.   0-486-65954-2

THEORY OF MATRICES, Sam Perlis. Outstanding text covering rank, nonsingularity and inverses in connection with the development of canonical matrices under the relation of equivalence, and without the intervention of determinants. Includes exercises. 237pp. 5⅜ x 8½.   0-486-66810-X

INTRODUCTION TO ANALYSIS, Maxwell Rosenlicht. Unusually clear, accessible coverage of set theory, real number system, metric spaces, continuous functions, Riemann integration, multiple integrals, more. Wide range of problems. Undergraduate level. Bibliography. 254pp. 5⅜ x 8½.   0-486-65038-3

MODERN NONLINEAR EQUATIONS, Thomas L. Saaty. Emphasizes practical solution of problems; covers seven types of equations. ". . . a welcome contribution to the existing literature...."–*Math Reviews.* 490pp. 5⅜ x 8½.   0-486-64232-1

MATRICES AND LINEAR ALGEBRA, Hans Schneider and George Phillip Barker. Basic textbook covers theory of matrices and its applications to systems of linear equations and related topics such as determinants, eigenvalues and differential equations. Numerous exercises. 432pp. 5⅜ x 8½.   0-486-66014-1

LINEAR ALGEBRA, Georgi E. Shilov. Determinants, linear spaces, matrix algebras, similar topics. For advanced undergraduates, graduates. Silverman translation. 387pp. 5⅜ x 8½.   0-486-63518-X

ELEMENTS OF REAL ANALYSIS, David A. Sprecher. Classic text covers fundamental concepts, real number system, point sets, functions of a real variable, Fourier series, much more. Over 500 exercises. 352pp. 5⅜ x 8½.   0-486-65385-4

SET THEORY AND LOGIC, Robert R. Stoll. Lucid introduction to unified theory of mathematical concepts. Set theory and logic seen as tools for conceptual understanding of real number system. 496pp. 5⅜ x 8¼.   0-486-63829-4

Math–Decision Theory, Statistics, Probability

ELEMENTARY DECISION THEORY, Herman Chernoff and Lincoln E. Moses. Clear introduction to statistics and statistical theory covers data processing, probability and random variables, testing hypotheses, much more. Exercises. 364pp. 5⅜ x 8½. 0-486-65218-1

STATISTICS MANUAL, Edwin L. Crow et al. Comprehensive, practical collection of classical and modern methods prepared by U.S. Naval Ordnance Test Station. Stress on use. Basics of statistics assumed. 288pp. 5⅜ x 8½. 0-486-60599-X

SOME THEORY OF SAMPLING, William Edwards Deming. Analysis of the problems, theory and design of sampling techniques for social scientists, industrial managers and others who find statistics important at work. 61 tables. 90 figures. xvii +602pp. 5⅜ x 8½. 0-486-64684-X

LINEAR PROGRAMMING AND ECONOMIC ANALYSIS, Robert Dorfman, Paul A. Samuelson and Robert M. Solow. First comprehensive treatment of linear programming in standard economic analysis. Game theory, modern welfare economics, Leontief input-output, more. 525pp. 5⅜ x 8½. 0-486-65491-5

PROBABILITY: AN INTRODUCTION, Samuel Goldberg. Excellent basic text covers set theory, probability theory for finite sample spaces, binomial theorem, much more. 360 problems. Bibliographies. 322pp. 5⅜ x 8½. 0-486-65252-1

GAMES AND DECISIONS: INTRODUCTION AND CRITICAL SURVEY, R. Duncan Luce and Howard Raiffa. Superb nontechnical introduction to game theory, primarily applied to social sciences. Utility theory, zero-sum games, n-person games, decision-making, much more. Bibliography. 509pp. 5⅜ x 8½. 0-486-65943-7

INTRODUCTION TO THE THEORY OF GAMES, J. C. C. McKinsey. This comprehensive overview of the mathematical theory of games illustrates applications to situations involving conflicts of interest, including economic, social, political, and military contexts. Appropriate for advanced undergraduate and graduate courses; advanced calculus a prerequisite. 1952 ed. x+372pp. 5⅜ x 8½. 0-486-42811-7

FIFTY CHALLENGING PROBLEMS IN PROBABILITY WITH SOLUTIONS, Frederick Mosteller. Remarkable puzzlers, graded in difficulty, illustrate elementary and advanced aspects of probability. Detailed solutions. 88pp. 5⅜ x 8½. 65355-2

PROBABILITY THEORY: A CONCISE COURSE, Y. A. Rozanov. Highly readable, self-contained introduction covers combination of events, dependent events, Bernoulli trials, etc. 148pp. 5⅜ x 8¼. 0-486-63544-9

STATISTICAL METHOD FROM THE VIEWPOINT OF QUALITY CONTROL, Walter A. Shewhart. Important text explains regulation of variables, uses of statistical control to achieve quality control in industry, agriculture, other areas. 192pp. 5⅜ x 8½. 0-486-65232-7

TENSOR CALCULUS, J.L. Synge and A. Schild. Widely used introductory text covers spaces and tensors, basic operations in Riemannian space, non-Riemannian spaces, etc. 324pp. 5⅜ x 8¼. 0-486-63612-7

ORDINARY DIFFERENTIAL EQUATIONS, Morris Tenenbaum and Harry Pollard. Exhaustive survey of ordinary differential equations for undergraduates in mathematics, engineering, science. Thorough analysis of theorems. Diagrams. Bibliography. Index. 818pp. 5⅜ x 8½. 0-486-64940-7

INTEGRAL EQUATIONS, F. G. Tricomi. Authoritative, well-written treatment of extremely useful mathematical tool with wide applications. Volterra Equations, Fredholm Equations, much more. Advanced undergraduate to graduate level. Exercises. Bibliography. 238pp. 5⅜ x 8½. 0-486-64828-1

FOURIER SERIES, Georgi P. Tolstov. Translated by Richard A. Silverman. A valuable addition to the literature on the subject, moving clearly from subject to subject and theorem to theorem. 107 problems, answers. 336pp. 5⅜ x 8½. 0-486-63317-9

INTRODUCTION TO MATHEMATICAL THINKING, Friedrich Waismann. Examinations of arithmetic, geometry, and theory of integers; rational and natural numbers; complete induction; limit and point of accumulation; remarkable curves; complex and hypercomplex numbers, more. 1959 ed. 27 figures. xii+260pp. 5⅜ x 8½. 0-486-63317-9

POPULAR LECTURES ON MATHEMATICAL LOGIC, Hao Wang. Noted logician's lucid treatment of historical developments, set theory, model theory, recursion theory and constructivism, proof theory, more. 3 appendixes. Bibliography. 1981 edition. ix + 283pp. 5⅜ x 8½. 0-486-67632-3

CALCULUS OF VARIATIONS, Robert Weinstock. Basic introduction covering isoperimetric problems, theory of elasticity, quantum mechanics, electrostatics, etc. Exercises throughout. 326pp. 5⅜ x 8½. 0-486-63069-2

THE CONTINUUM: A CRITICAL EXAMINATION OF THE FOUNDATION OF ANALYSIS, Hermann Weyl. Classic of 20th-century foundational research deals with the conceptual problem posed by the continuum. 156pp. 5⅜ x 8½. 0-486-67982-9

CHALLENGING MATHEMATICAL PROBLEMS WITH ELEMENTARY SOLUTIONS, A. M. Yaglom and I. M. Yaglom. Over 170 challenging problems on probability theory, combinatorial analysis, points and lines, topology, convex polygons, many other topics. Solutions. Total of 445pp. 5⅜ x 8½. Two-vol. set. Vol. I: 0-486-65536-9 Vol. II: 0-486-65537-7

Paperbound unless otherwise indicated. Available at your book dealer, online at **www.doverpublications.com**, or by writing to Dept. GI, Dover Publications, Inc., 31 East 2nd Street, Mineola, NY 11501. For current price information or for free catalogues (please indicate field of interest), write to Dover Publications or log on to **www.doverpublications.com** and see every Dover book in print. Dover publishes more than 500 books each year on science, elementary and advanced mathematics, biology, music, art, literary history, social sciences, and other areas.